Land, Women, Youths, and Land Tools or Methods

Land, Women, Youths, and Land Tools or Methods

Editor

Uchendu Eugene Chigbu

MDPI • Basel • Beijing • Wuhan • Barcelona • Belgrade • Manchester • Tokyo • Cluj • Tianjin

Editor
Uchendu Eugene Chigbu
Department of Land and Property Sciences,
Namibia University of Science and Technology
Namibia

Editorial Office
MDPI
St. Alban-Anlage 66
4052 Basel, Switzerland

This is a reprint of articles from the Special Issue published online in the open access journal *Land* (ISSN 2073-445X) (available at: https://www.mdpi.com/journal/land/special_issues/land_women).

For citation purposes, cite each article independently as indicated on the article page online and as indicated below:

LastName, A.A.; LastName, B.B.; LastName, C.C. Article Title. *Journal Name* **Year**, *Volume Number*, Page Range.

ISBN 978-3-03943-953-9 (Hbk)
ISBN 978-3-03943-954-6 (PDF)

Cover image courtesy of Nichola Knox.

Contents

About the Editor

Uchendu Eugene Chigbu is an Associate Professor in Land Administration at the Department of Land and Property Sciences, the Namibia University of Science and Technology (NUST), in Windhoek. His works fall within the interface between Social Sciences and Geodesy (or Social Geodesy). He has more than 10 years of experience in the formulation of land methods for addressing societal challenges. His most impactful work in the field of land administration and management is "Tenure Responsive Land Use Planning: A Guide for Country Level Implementation", which has been adopted by the UN-Habitat for piloting in the Global South. The concepts and practices from this document are currently being piloted in Uganda, the Philippines, Laos, and Zambia. Prior to joining NUST, he was a research scientist in land management at the Technical University of Munich (TUM), Germany. At TUM, he worked for 10 years in a variety of roles, including research, teaching, capacity development, institutional networking, and project sourcing/management. He is a multi-disciplinary consultant with diverse experience across a broad spectrum of land management specializations. His research and practice are focused on land administration and land management themes, including land-use planning, land and natural resource tenure, land policy and governance, and actions for transformations in urban, peri-urban, and rural settlements. He has multiple professional affiliations but is most active with the International Federation of Surveyors (FIG) and the Global Land Tool Network (GLTN). He is the Co-Chair of the International Training and Research Cluster of the GLTN and the Co-Chair of the Urban-Rural Dependency Working Group in Commission-8 of the FIG. He is highly published and sits on the Editorial Boards of the journals *Land Use Policy* (Elsevier) and *Local Development & Society* (Routledge). He is an active reviewer of more than 20 journals.

Preface to "Land, Women, Youths, and Land Tools or Methods"

The importance of land manifests in various components of the everyday lives of people in societies: cultural heritage, livelihood, the environment, economy, and community, among many others. Land is a factor of development. It is the most influential determinant of development because women, youths, and men (and households) depend on it for their livelihoods and for maintaining their living conditions in urban, peri-urban, and rural areas. However, in most cases, women and youth remain excluded from efforts towards securing land rights or the benefits that emanate from the use of land. This challenge persists due to a broad knowledge gap that exists on the land–women–youth–policy nexus of land management study and practice.

The articles in this book explore lessons on the application of land tools (as promising solutions) for improving the living conditions of all, including women and youths. Collectively, the articles build a knowledge base for understanding the challenges that women and youths face (and possible strategies to resolve these challenges) in their quest to access, use, and secure land resources. It also presents various methods (tools and approaches) for tackling land administration and management challenges.

This book reflects a broad research agenda in the field of land management and administration (including land governance and policy). It is a compendium of lessons on issues concerning women, youths, and tools in the context of land as a sustainable factor of development. It documents the concepts being designed, the impacts being experienced, and the progress being made in identifying suitable pathways for tackling the challenges faced by women and youths in relation to land and natural resources. Some of the studies presented in this book are adaptable to improving land-related problems in uncertain environments, such as situations that are encountered during the COVID-19 or coronavirus pandemic.

The content of the book enriches the knowledge base for understanding essential changes that are required for the empowerment of women and youths, as well as in the development of techniques necessary for delivering women-and-youth-responsive outcomes. The wealth of insights into how to facilitate the implementation of methods needed for positive impacts in the land sector is a strength of this book. Contemporary scholars are encouraged to read the collection of articles in the book, as it will afford them free access to land governance information in their ongoing and future research.

Uchendu Eugene Chigbu
Editor

 *land*

Editorial

Land, Women, Youths, and Land Tools or Methods: Emerging Lessons for Governance and Policy

Uchendu Eugene Chigbu

Department of Land and Property Sciences, Faculty of Natural Resources and Spatial Sciences, University of Science and Technology, 13 Jackson Kaujeua Street, Private Bag 13388, Windhoek 9000, Khomas, Namibia; echigbu@nust.na; Tel.: +264-61-207-2470

Received: 3 December 2020; Accepted: 8 December 2020; Published: 10 December 2020

Women and youths encounter problems with access to land, as well as securing tenure in land resources. Several researchers and organizations have dedicated their efforts to conceptualizing land tools for women's and youths' access to land. A land tool or land method is any practical means of solving land-related challenges. To ensure that land tools and methods produce pro-poor and inclusive impacts, they need to be developed to also incorporate youths' and women's experiences (including their needs and participation). This implies acknowledging that land tools or methods may impact differently on women, men, and youths. The focus of this Special Issue (SI) "Land, Women, Youths, and Land Tools or Methods" is crucial for understanding the social aspects of land administration and land management. *Land* (a natural resource), *women and youths* (focusing on people), and *land tools or methods* (problem-solving techniques) share a mutually beneficial relationship (see Figure 1).

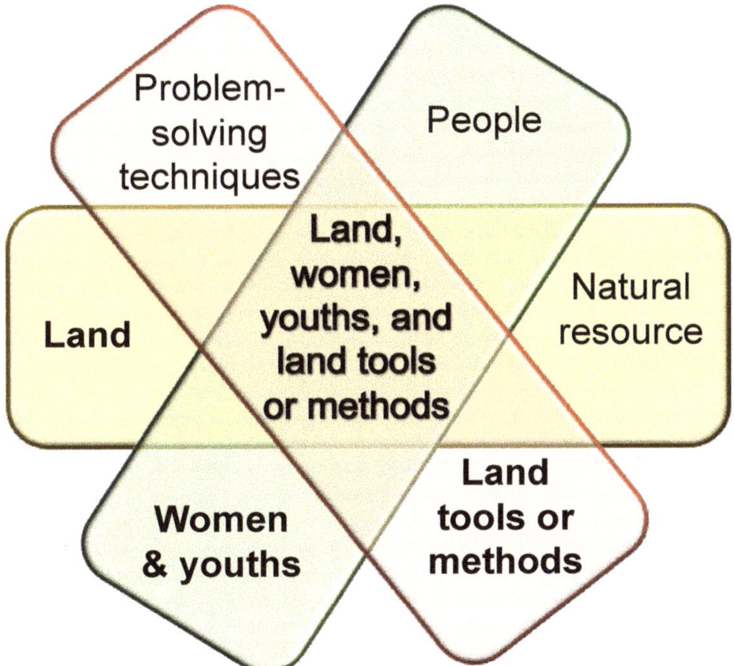

Figure 1. The relationship between land, women, youths, and land tools or methods (author's illustration).

Land, as a natural resource, is essential for ensuring that *women and youths* are part of the community (the critical demographic). This is because access to land provides physical availability of land parcels for women and youths to use or enjoy the rights embedded therein. Furthermore, it provides opportunities for them to improve their livelihoods and their households. Women and youths are more socioeconomically empowered when they have the right to inherit property, transfer their land rights to others, and exercise those rights in relation to water, housing, food, the forest, and environmental and mineral resources, to mention just a few. Without disregarding the problems that men encounter, there is a focus on women and youths because they are the demographic usually left behind in efforts aimed at securing land rights for all. However, efforts to secure land/property rights for women and youths would not be possible unless specifically tailored problem-solving techniques are adopted for solving women- and youth-related land challenges. Furthermore, people (women and youths, in this case) also need to understand their rights in order to use and enjoy land resources, as well as have the capacity to use land sustainably. This is how *land, women and youths, and land tools or methods* share mutually beneficial relationships (again, refer to Figure 1).

With the exception of an Erratum article, the SI contains a total of 13 positively evaluated (peer-reviewed) articles. Therefore, the objective of this editorial is to highlight one key insight from each of these, thus compiling 13 lessons learned from the articles published in this SI. These 13 lessons learned are a selection only. A broader spectrum of insights and findings is available in the publications. A thematic review (and analysis) of the articles required a multidisciplinary perspective to grasp the variety of data types, as well as the collection and analytical methods used. A matrix-type thematic analysis is presented in a tabular format (Table 1).

Table 1. Articles in the Special Issue (SI) *"Land, Women, Youths, and Land Tools or Methods"*, listed in the order they are discussed in this editorial (‡: primary focus area of publication; †: secondary focus; •: unaddressed theme).

No.	Title of Published Articles	Key Subject Area Addressed in the SI Themes [1]				
		L	W	Y	LT	G
1	Willingness to participate in voluntary land consolidation in Gozamin District, Ethiopia [1].	‡	•	•	†	Ethiopia (Africa)
2	Household land allocations and the youth land access nexus: evidence from the Techiman Area of Ghana [2].	‡	•	‡	•	Ghana (Africa)
3	The changing structure and concentration of agricultural landholdings in Estonia and possible threat for rural areas [3].	‡	•	•	†	Estonia (Europe)
4	Rural women's invisible work in census and state rural development plans: The Argentinean Patagonian case [4].	‡	‡	•	•	Argentina (South America)
5	Community development through the empowerment of indigenous women in Cuetzalan Del Progreso, Mexico [5].	‡	‡	•	†	Mexico (North America)
6	The nexus between peri-urban transformation and customary land rights disputes: effects on peri-urban development in Trede, Ghana [6].	‡	•	•	†	Ghana (Africa)
7	Using a gender-responsive land rights framework to assess youth land rights in rural Liberia [7].	‡	•	‡	‡	Liberia (Africa)
8	Benefits and constraints of the agricultural land acquisition for urbanization for household gender equality in affected rural communes: A case study in Huong Thuy Town, Thua Thien Hue Province, Vietnam [8].	‡	†	•	•	Vietnam (Asia)
9	Absent voices: women and youths in communal land governance. Reflections on methods and processes from exploratory research in West and East Africa [9].	‡	‡	‡	‡	Guinea, Kenya, Mali, Tanzania, and Uganda (Africa)

Table 1. *Cont.*

No.	Title of Published Articles	Key Subject Area Addressed in the SI Themes [1]				
		L	W	Y	LT	G
10	Smallholder agricultural investment and productivity under contract farming and customary tenure system: a Malawian perspective [10].	‡	†	•	•	Malawi (Africa)
11	Mapping environmental conflict using spatial text mining: focusing on the regional issues of South Korean environmental NGOs [11].	‡	•	•	‡	South Korea (Asia)
12	Urbanization and increasing flood risk on the northern coast of Central Java – Indonesia: an assessment towards better land use policy and flood management [12].	‡	•	•	‡	Indonesia (Asia)
13	Gender inequality and symbolic violence in women's access to family land in the Southern Highlands of Tanzania [13].	‡	‡	•	†	Tanzania (Africa)

[1] L = Land; W = Women; Y = Youths; LT = Land tools or methods; G = Geography.

Two key inferences can be drawn from the data presented in Table 1. First, all articles in the SI have land as a primary focus, followed by issues about women, land tools or methods, and youths, respectively. Second, while the geographical focus of the articles leaned predominantly towards Africa, the entire volume is representative of countries from all regions of the world. The SI, therefore, addresses land, women, youths, and land tools or methods from a global perspective. The 13 articles in the SI are authored by 37 land and natural resource scholars. They collectively produced 13 notable lessons [1–13]. Below is an outline of the lessons learned.

Lesson one: Landholder farmers are predominantly willing to participate in voluntary land consolidation. The study conducted by Gedefaw et al. [1] evaluated the willingness of farmers to participate in land consolidation in Ethiopia. They revealed that landholder farmers are more willing to participate in voluntary land consolidation than in compulsory land consolidation. The study identified motivations or determinants for their willingness: (1) land exchange in a consolidation process is preferable between neighboring parcels of adjacent farmers; (2) nearness of parcels to the farmstead; and (3) expected productivity improvement.

Lesson two: A social welfare scheme is necessary for aged farmers to provide youths easier access and security to the use of land. Concerning the growing lack of access to land by youths in most developing countries, Kidido and Lengoiboni's [12] study on Ghana underscores the need for a social welfare scheme for aged farmers to encourage earlier transfer of land to the youth to enable easier access to land.

Lesson three: Policy action is urgently needed in Europe to mitigate the impact of land concentration caused by the increasing concentration of agricultural land into the hands of corporate bodies. Focusing away from Africa, Jürgenson and Rasva [3] studied the farmland situation in Europe where the area of agricultural land has remained almost the same, despite a decreasing trend in the number of farms on the continent. They found, with a particular focus on Estonia, that "Agricultural land has been increasingly concentrated into the hands of corporate bodies" [3] (p. 1). Hence, the need for a policy direction to mitigate the impact of land concentration.

Lesson four: A strategy for the economic and political recognition of women's contribution to regional development through their land-based activities is a necessary intervention for gender-inclusive and equitable planning. Concerning the issue of women's land rights in South America, Núñez et al. [4] investigated the historical census data from Argentina and found that women's contributions to the family farming system in the Patagonian region remain unrecognized, and this has caused the invisibilization of women's contributions to development. They found that the system of data usage leads to women being denied their rights to family land use. The consequence is that "Patagonia has become one of the most affected by extractivism" [4] (p. 1). They, therefore, recommended that a strategy for the economic and political recognition of women's contributions to development (through land-based activities) is a necessary intervention for gender-inclusive and equitable planning.

Lesson five: An "inclusive legal and institutional framework" leads to the segregation of Indigenous communities when (or where) policy implementation is weak; however, community development approaches can help. The study by Durán-Díaz et al. [5] is a regionally relevant work that overlaps between cultural Latin America and geographical North America. They investigated "the status of Indigenous rural women, as well as the mechanisms and impacts of their empowerment" (p. 1). It presents a community development approach as a women-focused land tool or method—based on the Masehual Siuamej Mosenyolchicacauani organization in Cuetzalan del Progreso, Puebla—to ensure more effective implementation of programs meant to desegregate Indigenous communities.

Lesson six: A land-use plan meant to convert rural lands into urban plots can become a trigger of tenurial changes and customary land rights disputes. Owusu Ansah and Chigbu [6], in another study on Ghana (Africa), explored the link between peri-urban transformation and emerging customary land rights (CLR) disputes in peri-urban geographies. They found that a land-use plan implemented to develop a functional peri-urban land market contributed to CLR disputes. Going forward, they [6] proposed measures for peri-urban land management and CLR dispute prevention.

Lesson seven: For a Land Rights Act of a country to become responsive to the needs of the youth, a precondition is to embrace community-level interventions. Louis et al.'s [7] study provides insights into the land rights situation of youths in Liberia. They argued for the implementation of the country's Land Rights Act to embrace community-level interventions to improve youth land tenure security in the country. This study provides a land tool (or methods) perspective by using a gender-responsive land rights framework that examines youth land rights through a gender lens.

Lesson eight: Where (and when) agricultural land is acquired for urbanization, it can lead to an increase in the economic status of women in households whose land was acquired. The study by Pham Thi et al. [8] explored agricultural land acquisition for urbanization (ALAFU) in Vietnam. Their key finding indicates that ALAFU has increased the economic status of women in households whose agricultural land was acquired for urbanization. Taking structural gender inequalities into account in agricultural-to-urban land acquisition can be a veritable strategy for achieving gender-responsive sustainable development goals.

Lesson nine: Responsible governance of communal natural resources is only possible when the voices of the disempowered segment of communities, especially those of the women and youth, are heard and respected. This study by Lemke and Claeys [9] explored natural resource governance from a communal perspective, and with a specific focus on the empowerment of women and youths through the promotion of their right to be heard in communal land matters. A point of methodological interest is that they adapted participatory action research (PAR) to the COVID-19 situation. Hence, they present how PAR can be applied under uncertain situations.

Lesson ten: Matrilocal practices positively influence agricultural investment, but even in such a women-friendly customary system, women remain insecure in their landholding. In this study, Benjamin [10] investigates the impact of the customary residence system on agricultural investment, with a particular focus on tea shrub and agroforestry, and productivity among contracted smallholder tea out-growers in southern Malawi. The study concluded that despite the dominance of matrilineal–matrilocal systems in Southern Malawi, there is a need for policy to address gender gaps in the region because women are still insecure in their access and use of land.

Spatial text mining can provide evidence of variability in the degrees of environmental conflict sensitivity, geographically or regionally, leading to policy-relevant information for land or environmental management. Lee and Kim's study [11] found that air quality-related conflicts in South Korea are concentrated in the western municipalities, development-related conflicts are concentrated in the southern region of Jeju Island, and intensive safety-related conflicts occur in the metropolitan areas (especially the city of Ulsan). Their data, when presented in the form of an environmental map, is capable of being used as a tool for country-level land and environmental management decisions.

Land policies are necessary for guiding sustainable flood management in coastal cities and riverine areas. Handayani et al.'s study [12] explored the relationship between urbanization and flood events on the northern coast of Central Java using the river basin as the basis of unit analysis. Their findings show that the growth rate of the built-up area is significantly related to the occurrence of flood events. The study concluded that river basins have a dual spatial identity in the urban system (policy and land-use related). They recommended "problematizing urbanizing river basins" as "an opportunity for an eco-based approach to tackling the urban flood crises".

Lesson thirteen: Gender inequality and symbolic violence in women's access to family land can be improved through the use of dialectical communication between women and men. Lusasi and Mwaseba [13] investigated land-related gender scenarios in selected villages in the Southern Highlands of Tanzania and found that cases of tree-planting surges in the region cause gender inequality and symbolic violence in women's access to family land. The study advanced a communication tool to reduce gender disparity in land issues. They recommended the use of dialectical communication between women and men to reveal and heal practices of symbolic violence in land accessibility, control, and ownership.

Funding: This research received no external funding.

Acknowledgments: I would like to acknowledge all the reviewers who participated in evaluating the articles published in the Special Issue (SI) *"Land: Women, Youths, and Land Tools or Methods."* Without these reviewers, it would not have been possible to complete this SI project. I am personally indebted to some specific individuals who helped throughout the period of the SI. These include Walter Dachaga, Barikisa Owusu Ansah, Cheonjae Lee, Gaynor Paradza, Prince Donkor Ameyaw, and Ernest Uwayezu. I want to thank the editorial team of *Land* journal. Special thanks go to Cosette Yuan, Janie Liu, Aguero Gui, Costin Hao, Jenney Li, Milica Kovačević, and Zoey Gao. I would also like to thank all the authors of the accepted and rejected manuscripts for responding to the call leading to this SI volume. Together, these authors (for submitting their articles) have enriched the knowledge base of what is known about land, women and youths, and land tools or methods. I would also like to thank the Global Land Tool Network (GLTN) and the International Federation of Surveyors (FIG). My respective roles as a Co-Chair of the GLTN Cluster and FIG Commission Working Group in these organizations motivated the conceptualization of this SI project.

Conflicts of Interest: The author declares no conflict of interest.

References

1. Gedefaw, A.A.; Atzberger, C.; Seher, W.; Mansberger, R. Farmers willingness to participate in voluntary land consolidation in Gozamin District, Ethiopia. *Land* **2019**, *8*, 148. [CrossRef]
2. Kidido, J.K.; Lengoiboni, M. Household land allocations and the youth land access nexus: Evidence from the Techiman Area of Ghana. *Land* **2019**, *8*, 185. [CrossRef]
3. Jürgenson, E.; Rasva, M. The Changing structure and concentration of agricultural land holdings in Estonia and possible threat for rural areas. *Land* **2020**, *9*, 41. [CrossRef]
4. Núñez, P.G.; Michel, C.L.; Leal Tejeda, P.A.; Núñez, M.A. Rural women's invisible work in census and state rural development plans: The Argentinean Patagonian Case. *Land* **2020**, *9*, 92. [CrossRef]
5. Durán-Díaz, P.; Armenta-Ramírez, A.; Kurjenoja, A.K.; Schumacher, M. Community development through the empowerment of indigenous women in Cuetzalan Del Progreso, Mexico. *Land* **2020**, *9*, 163. [CrossRef]
6. Owusu Ansah, B.; Chigbu, U.E. The nexus between peri-urban transformation and customary land rights disputes: Effects on peri-urban development in Trede, Ghana. *Land* **2020**, *9*, 187. [CrossRef]
7. Louis, E.; Mauto, T.; Dodd, M.L.; Heidenrich, T.; Dolo, P.; Urey, E. Using a gender-responsive land rights framework to assess youth land rights in rural Liberia. *Land* **2020**, *9*, 247. [CrossRef]
8. Pham Thi, N.; Kappas, M.; Wyss, D. Benefits and constraints of the agricultural land acquisition for urbanization for household gender equality in affected rural communes: A case study in Huong Thuy Town, Thua Thien Hue Province, Vietnam. *Land* **2020**, *9*, 249. [CrossRef]
9. Lemke, S.; Claeys, P. Absent Voices: Women and Youth in Communal Land Governance. Reflections on methods and process from exploratory research in West and East Africa. *Land* **2020**, *9*, 266. [CrossRef]
10. Benjamin, E.O. Smallholder Agricultural Investment and Productivity under Contract Farming and Customary Tenure System: A Malawian Perspective. *Land* **2020**, *9*, 277. [CrossRef]
11. Lee, J.; Kim, D. Mapping environmental conflict using spatial text mining: Focusing on the regional issues of South Korean environmental NGOs. *Land* **2020**, *9*, 287. [CrossRef]

12. Handayani, W.; Chigbu, U.E.; Rudiarto, I.; Intan, H.S.P. Urbanization and Increasing Flood Risk in the Northern Coast of Central Java—Indonesia: An Assessment towards Better Land Use Policy and Flood Management. *Land* **2020**, *9*, 343. [CrossRef]
13. Lusasi, J.; Mwaseba, D. Gender inequality and symbolic violence in women's access to family land in the Southern Highlands of Tanzania. *Land* **2020**, *9*, 468. [CrossRef]

Publisher's Note: MDPI stays neutral with regard to jurisdictional claims in published maps and institutional affiliations.

 *land*

Article

Farmers Willingness to Participate In Voluntary Land Consolidation in Gozamin District, Ethiopia

Abebaw Andarge Gedefaw [1,2,*], Clement Atzberger [1], Walter Seher [3] and Reinfried Mansberger [1]

[1] Institute of Surveying, Remote Sensing and Land Information, University of Natural Resources and Life Sciences Vienna, Peter-Jordan-Strasse 82, 1190 Vienna, Austria; clement.atzberger@boku.ac.at (C.A.); mansberger@boku.ac.at (R.M.)

[2] Institute of Land Administration, Debre Markos University, 269 Debre Markos, Ethiopia

[3] Institute of Spatial Planning, Environmental Planning and Land Rearrangement, University of Natural Resources and Life Sciences Vienna, Peter-Jordan-Strasse 82, 1190 Vienna, Austria; walter.seher@boku.ac.at

* Correspondence: abebaw.gedefaw@students.boku.ac.at

Received: 10 September 2019; Accepted: 10 October 2019; Published: 12 October 2019

Abstract: In many African countries and especially in the highlands of Ethiopia—the investigation site of this paper—agricultural land is highly fragmented. Small and scattered parcels impede a necessary increase in agricultural efficiency. Land consolidation is a proper tool to solve inefficiencies in agricultural production, as it enables consolidating plots based on the consent of landholders. Its major benefits are that individual farms get larger, more compact, contiguous parcels, resulting in lower cultivation efforts. This paper investigates the determinants influencing the willingness of landholder farmers to participate in voluntary land consolidation processes. The study was conducted in Gozamin District, Amhara Region, Ethiopia. The study was mainly based on survey data collected from 343 randomly selected landholder farmers. In addition, structured interviews and focus group discussions with farmers were held. The collected data were analyzed quantitatively mainly by using a logistic regression model and qualitatively by using focus group discussions and expert panels. According to the results, landholder farmers are predominantly willing to participate in voluntary land consolidation (66.8%), while a substantive fraction of farmers express unease with voluntary land consolidation. The study highlighted the following four determinants to be significant in influencing the willingness of farmers for voluntary land consolidation: (1) the exchange should preferably happen with parcels of neighbors, (2) land consolidation should lead to better arranged parcels, (3) nearness of plots to the farmstead, and (4) an expected improvement in productivity. Interestingly, the majority of farmers believes that land consolidation could reduce land use conflicts. The study provides evidence that policymakers should consider these socio-economic, legal, cultural, infrastructural, and land-related factors when designing and implementing voluntary land consolidation policies and programs.

Keywords: voluntary land consolidation; land fragmentation; maximum likelihood estimation; logistic regression model; sustainable land management; land exchange; rural development

---

1. Introduction

Land is a scarce resource in Ethiopia, a country whose population relies on farming as the primary source of livelihood [1]. As farming is an essential factor in the Ethiopian economy, land utilization and allocation is an important undertaking in the country [2]. Agriculture is dominating the economy in Ethiopia; it accounts for 37% of the gross domestic product (GDP), which is one of the highest shares in Sub-Saharan Africa. However, landholdings are often fragmented into small parcels; the average total farm land area per smallholder household is 0.78 hectare, and it is likely to decline further.

The average number of plots constituting a household farm is four [3]. Thus, Ethiopian smallholder agriculture is characterized by extremely small farms fragmented into several plots and cultivated in a labor-intensive manner while supporting relatively large families. Many of these farms are too small to meet subsistence needs, particularly when using traditional technology and currently available resources [4,5]. Fragmentation has led to farmers neglecting strips of land far from their houses, leading to reduced agricultural output. The impact of land fragmentation is that farming becomes more and more difficult, expensive, and labor intensive, especially against the background of an expected mechanization of the farming sector in Ethiopia [6]. Adverse effects of land fragmentation have been observed in many countries where spatially separated parcels of land hinder mechanization and increase the likelihood of disputes [6,7].

Land fragmentation is defined as a situation where farmers are cultivating two or more geographically separated plots of land by taking into account the distances between those parcels [8]. Bentley explains land fragmentation as a type of land ownership, where a single farm consists of numerous discrete parcels [9]. Problems often associated with land fragmentation are small sizes, irregular shapes, and dispersed parcels, resulting in higher efforts for cultivation [10,11]. It can be summarized that land fragmentation is not beneficial in terms of agricultural development [12,13].

The studies mentioned above tackle different aspects of land fragmentation. Land fragmentation can be considered from a cultivation perspective, taking into account agricultural production such as variety of crops, quality of soil, and water conditions. In this respect, land fragmentation can also be beneficial by providing a distribution of plots according to the variety of agricultural site qualities. It also can be seen from a land administration perspective considering the geometry (e.g., shape, area, slopes) as well as the land rights (e.g., land ownership, land tenure) [14]. Another aspect would be an environmental one, where cultivation of small parcels is more likely to provide higher biodiversity. The perspective on land fragmentation is also dependent on different stakeholders. These can be farmers, planners, land administrators, environmentalists, agro-economists, etc. Furthermore, perceptions of land fragmentation vary between countries.

In this study, the investigations were focused on Ethiopia and on the viewpoints of farmers. Thus, cultivation and land administration perspectives on land fragmentation were in the foreground.

Land fragmentation is both an indicator and the result of a (frequently problematic) land tenure structure. In some regions, land fragmentation becomes a major problem because it restricts agricultural development and reduces the opportunities for sustainable rural development. Policies to counter land fragmentation are needed for social, economic, and environmental reasons [11] (see also Table 1).

Table 1. Major reasons why policies are needed to counter land fragmentation.

Reason Type	Reason
Social	to decrease disputes amongst neighbors to develop team work
Economic	to increase production to enable self-sufficiency in food production
Environmental	to enhance soil quality to protect water availability to balance climatic condition

Source: Demetriou, D., et al. 2013. A new methodology for measuring land fragmentation.

Even if farmland fragmentation is widespread and may affect farmers' decisions, it can influence farm performance either negatively or positively. Usually, the term land fragmentation is associated with small parcel sizes, improper shapes of individual parcels, long distances of parcels from homestead, and long distances between parcels [15]. Experiences with quantifying the impact of land fragmentation on agricultural production efficiency reveals the negative association. Studies done in Nigeria show that farmers' landholdings are fragmented, small in size, non-contiguous, and interspersed. Fragmentation of holdings had negative implications for agricultural development [16]. Also, studies in South-East

China reveal that land fragmentation can be an important determinant of technical efficiency in rice production. An increase in average plot size increases rice farmers' cultivation efficiency and vice versa [17]. Another study in Nigeria reveals that land fragmentation affects production efficiency by the finding that there is a negative correlation between amount of fragmented land and yield [18]

Even though policy makers often point out the draw backs of land fragmentation, there is no consensus that fragmentation is strictly a negative phenomenon. Bentley argues that the negatives caused by fragmented land holdings are overrated and that the farmers' own views often are neglected by policy makers [9]. Bentley also documents positive aspects of land fragmentation, such as variety of soil and growing conditions reducing the risk of total crop failure. Plots spread over an area sometimes implies micro-climatic variations and multiple ecological zones. Fragmentation also facilitates crop rotation [9]. Additionally, farmers can take advantage of minor differences in local agroecology [19], as they can hedge risk through spatial dispersion [20] and improve agricultural biodiversity [9,19]. In Africa, specifically in Ghana and Rwanda, Blarel et al. found that fragmentation facilitates crop diversification [20]. Studies in Turkey show more fragmentation is positively correlated with increased yields [21].

In land fragmentation research, land consolidation is regarded as a proper measure to facilitate agricultural cultivation, rural development, and land administration [14]. Land consolidation is also seen as an important tool for improving environmental management [22,23]. Land consolidation can be defined as a land use policy tool designed to overcome the difficulties of land fragmentation [24–26]. Land consolidation means a planned rearrangement of land parcels. If done properly, land consolidation supports farmers to amalgamate their fragmented parcels. It facilitates the creation of competitive agricultural production arrangements by enabling farmers to have farms with fewer parcels that are larger and better shaped. In addition, new infrastructure can be established in the consolidated area, for example, to improve accessibility and water management. In turn, this allows farmers to introduce better farming techniques [14]. Making farming more efficient and ultimately more economically viable creates incentives to attract young people into farming and agribusiness. Land consolidation is therefore considered a worthwhile complementary investment, as it improves the efficiency of rural land use and helps to address the challenges of sustainable rural development [27,28]. In Eastern, Western, and Central European countries, high amounts of farmland have been consolidated over the past decades within different governmental frameworks of land consolidation projects. As a result, farmland fragmentation was solved to a high extent [29–32].

According to a Food and Agriculture Organization (FAO) report, a land consolidation program has to accommodate national and sub-national priorities as well as local ones. A land consolidation strategy should address issues such as [14]:

- Institutional issues: what tasks should be done at what level by which institution, and how participatory "bottom-up" involvement should be implemented;
- Financial issues: how money to support land consolidation will be sourced, and how the process can be made cost-effective;
- Legal issues: what the legal basis for implementing land consolidation will be, and how to ensure that the results are not jeopardized (e.g., by heritage);
- Capacity building: how participants can, at all levels and in all sectors, acquire knowledge and skills they need to carry out their responsibilities;
- International cooperation: how countries can gain access to the technical and the financial resources of donors.

Many studies prove the positive association between land consolidation and agricultural productivity. For instance, Asia's Green Revolution is evidence that investments to improve agricultural productivity by land consolidation and by crop intensification have been important for rural poverty reduction [33,34]. Various studies from Asia, South America, and Western, Central, and Eastern European countries [22,29,35] document that land consolidation policies have contributed to increased agricultural productivity. However, this applies to favorable farming conditions and a certain degree of mechanization. Using household-level data, Nilsson's study found that there is a positive association between land consolidation and crop yields in Rwanda, but only among farm households with land holdings greater than one hectare, which is well above the average farm size in Ethiopia [36].

To improve the Ethiopian agricultural sector, Beyene [2] proposes land reforms in the form of land consolidation as effective and efficient mechanisms to allow the population to invest in farmland. The reforms come in the form of government directing policies that should enable the consolidation of previously fragmented parcels of land [2]. In general, there are four types of land consolidation approaches [14]. These are comprehensive, simplified, voluntary, and individual land consolidation. This article focuses on the third type, as the Ethiopian federal government encourages voluntary land consolidation, and the law states that, "In order to make small farm plots convenient for development, farmers are encouraged to voluntary exchange farmlands" [37]. In line with this law, the land regulation law in the Amhara National Regional State encourages voluntary land consolidation by exchange of land between landholder farmers [38,39]. To encourage voluntary land consolidation, the regulation, enacted in 2006, further states that the government must provide technical services and has to renew landholding certificates free of charge to land holder farmers who exchange plots [39]. However, in practice, the above-mentioned technical support is largely restricted to legal support. The Ethiopian Rural Land Administration and Use determination proclamation and regulations covering land fragmentation and voluntary land consolidation currently lack well defined and detailed procedures, e.g., how to launch and implement voluntary land consolidation schemes, controlling principles to be applied during implementation, inheritance regulations for voluntary land consolidation, and legal measures to avoid future fragmentation.

Even with the already identified positive effects, land consolidation has historically faced challenges. In Central and Eastern Europe, because of post-socialist transformation, cooperative farms now consist of numerous, small, and economically barely viable private plots with a multitude of landowners. To implement land consolidation was difficult because of land ownership and, in particular, values, legitimacy, personal identity, and emotional bonds [40]. In Taiwan, farmers objected land consolidation even when they fully recognized their benefits. Corruption and maladministration of government officials, timing of operation, cost of consolidation, and the fear of receiving low quality land in the exchange process were quoted as reasons against land consolidation [41].

Ethiopia has seen extensive land grabs sponsored by the government as part of its agricultural transformation strategy. The land grabs, involving land consolidation, have been supported by recent Ethiopian policies such as the Growth and Transformation Plan I and II, where the government aimed to transfer a total of 2.3 million hectares to large-scale commercial farming [42]. In many countries, land consolidation did not consider ecological aspects for a long time. Thus, land consolidation processes decreased biodiversity in rural areas and diminished long established habitats of animals around the villages [21,43]. Not surprisingly, in Ethiopia, the above-mentioned failures in land consolidation processes affected the willingness of farmers to participate in such procedures.

Nevertheless, Ethiopia has attempted to improve the economic and the social outcomes of farming in the country. The government, following a recognition that small-scale farmers are perennially underperforming in regard to farm output, invested in land consolidation as a mechanism to improve the fate of the country's agricultural sector. One of the challenges related to agricultural output is fragmentation of land. Thus, these challenges have promoted the willingness of the population to support voluntary land consolidation. Therefore, this study aimed to estimate the willingness of farmers for land consolidation in general and, in particular, to address the factors influencing the willingness of landholder farmers to participate in voluntary land consolidation processes. The investigations were based on interviews with a total of 343 landholder farmers in the Gozamin District, Amhara National Regional State, Ethiopia. In addition, information was gathered in focus group discussion and community consultations.

2. Study Area and Methods

2.1. Description of Study Area

For the purpose of this study, the Gozamin District was selected (Figure 1). The Gozamin District was purposively selected for this study due to:

- The variety of agro-climatic zones leading to diverse types of crop farming practices;
- The considerable degree of land fragmentation;
- The existence of sustainable natural resource management plans;
- The existence of second level land certification documents for landholders (in Gozamin district, all landholder farmers received the documents of completion of second level land certification);
- The authors' local knowledge.

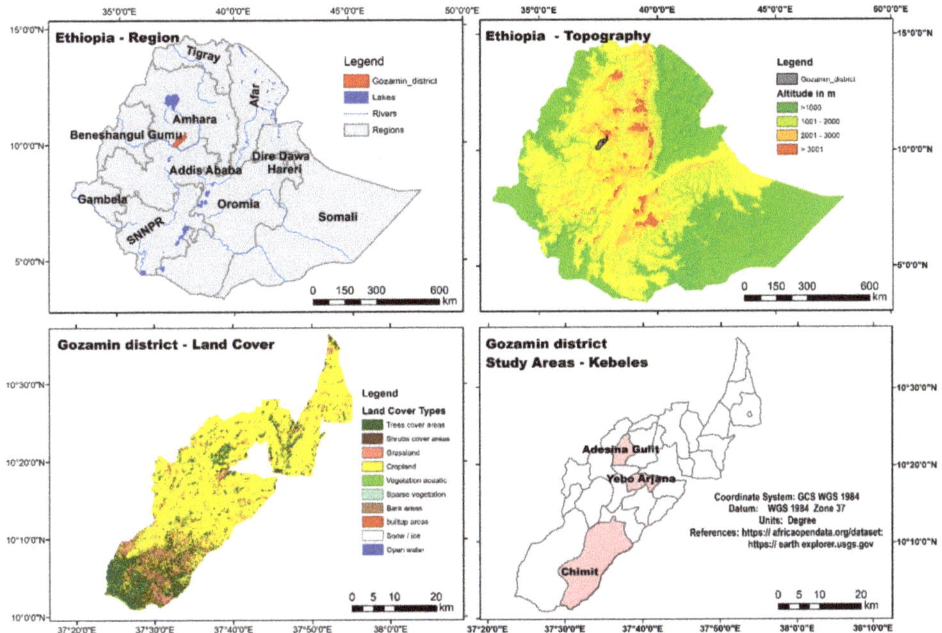

Figure 1. The study area—Gozamin District.

Gozamin district is roughly located 270 km east of the regional capital Bahir Dar and 300 km northwest from Addis Ababa. Debre Markos is the capital of the district and hosts the administrative seat of East Gojjam zone. The district contains 25 rural kebeles (municipalities) in total. Gozamin district has a total area of 1812 km^2. Its elevation ranges from ~ 1000 m to ~ 3200 m above sea level (m.a.s.l), i.e., the Choke Mountain range. The district is otherwise characterized by a relatively flat landscape, flood plains, and wetlands. Chemoga, Dijil, and Kulech are the major rivers in the district [44].

Gozamin District has 134,000 inhabitants with equal gender balance. The population density is 109 people per km^2. Sedentary, rainfed, and small scale agriculture constitutes the primary income generating activity within the Gozamin District. People primarily perform mixed cereal agriculture with farmers mainly growing teff, finger millet, sorghum, maize, barley, wheat, pulses, oil crops, vegetables, and fruits.

The ethnicity of the District population is Amhara, and Amharic is people's language [45]. Due to the high anthropogenic influence in the study area, forests have been lost. Remnant plants around holy places, inaccessible areas, left for shade trees, and on grazing lands are observed. Some of the vegetation consists of *Juniperus procera, Hagenia abyssinica, Podocarpus falcatus, Acacia abysinica, Cordia africana, Ficus sycomorus, Erythrina brucei, Eucalyptus camaldulensis, Calpurnia aurea, Prunes africana, Carissa spinarum, Rosa abyssinica, Dombeya torrida,* and *Maytenus arbutifolia* [46].

2.2. Methodology

2.2.1. Sampling Design and Data Collection

Three kebeles of the Gozamin district were selected for the primary data collection by applying the same criteria as for the selection of the district: Chimit, Yebona Erjena, and Addisna Guilit. From 3277 landholder farmers in three selected kebeles (Table 2), 343 landholders were randomly sampled for interview (115 farmers from Chimit, 131 from Yebona Erjena, and 97 farmers from Adisna Guilit). The sample size of landholder farmers was determined based on the sample size determination equation by Cochran [47]:

$$n_0 = \frac{Z^2 \cdot p \cdot q}{d^2} \tag{1}$$

$$n = \frac{no}{1 + \frac{no-1}{N}} \tag{2}$$

$$n_0 = \frac{(0.5) \cdot (0.5) \cdot (1.96)^2}{(0.05)^2} \tag{3}$$

$$n = \frac{384}{1 + \frac{384-1}{3277}} = 343 (truncatedInteger) \tag{4}$$

where

n_0 = the desired sample size Cochran's (1977) when population is greater than 10,000;
n = number of sample size when population is less than 10,000;
Z = 95% confidence limit (1.96);
p = estimated proportion of samples with specific attribute (0.5 as cases cannot be estimated a priori);
$q = 1 - p$ (0.5);
N = total number of population (3277);
d = precision or degree of accuracy desired (0.05).

The number of landholders selected for interview from each kebele was determined proportional to the total number of landholder farmers in the kebele (rounded to integer) to guarantee an equal representation of landholder farmers' households in each kebele. The detailed information about the total landholder farmers and samples is documented in Table 2.

Table 2. Number of total landholder farmers and sample landholders of the study area.

Name of Kebele	Total Landholder Farmers	Sampled Landholder Farmers
Chimit	1100	115
Yebona Erjena	1250	131
Adisna Guilit	927	97
Total	**3277**	*343*

Sample landholder farmers were selected using the systematic random sampling technique. Sampling frames were obtained for each kebele by taking the list of all landholder farmers in alphabetical order from the respective kebele administration offices. The sample households were drawn from each administrative unit from the list of names after a certain sampling interval (K) that was determined by dividing the total number of households by the predetermined sample size of each kebele. Next, a number was selected between one and the sampling interval (K) using the lottery method (called the random start) and was used as the first number included in the sample. Then, every Kth landholder head after that first random start was taken until reaching the desired sample size for each kebele administration. Systematic sampling was applied because sample units are uniformly distributed over the population [48].

In total, 343 landholder farmers' data were encoded using Statistical Package for Social Science (SPSS, version 24), and the analyses were performed with the "R" software by importing the data from SPSS.

Field data were collected from September 2018 to December 2018. Mainly primary sources were used to collect the data. The primary data were collected using household surveys (HHS), focus group discussion (FGD), and direct field observations. As mentioned before, a structured questionnaire was used for the field interviews. The questionnaire was pre-tested by administering it to selected respondents. On the basis of the results obtained from the pre-test, necessary modifications were made on the questionnaire. Six data collectors with a minimum of a college diploma in related fields of land administration were employed for data collection. Before they started enumeration, a brief explanation about the objectives of the survey and the meaning of each question was given. This briefing was also a contribution towards harmonization of the data collection. Face to face interviews were necessary, as many of the respondents were expected to be illiterate. To avoid language difficulties, two experts in the field of land administration translated the questions from English to Amharic (local language).

In order to supplement the survey with qualitative data, focus group discussions were conducted in each kebele to complement the gathered quantitative information. Participants of FGD were selected based on their knowledge and their experience in agricultural practices. These persons have lived in the kebeles for a long time and constituted the Land Administration and Certification Committee in the kebele (LACC). LACC are representatives from different social groups in the community, including elders, female-headed households, youth, and disabled persons as well as development agents and kebele managers. Nine group discussions were carried out (three in each kebele). Each FGD had 10 to 12 participants selected with the support of the "Kebele Land Administration Officer". The discussion focused on a local-level entity dealing with land-related issues as well as willingness and views of the community regarding voluntary land consolidation and other sustainable land management issues.

In addition, panels and discussion forums were organized with a total of six federal, seven regional, ten zonal, and five district experts. These experts work in rural land administration and land management offices and in other related fields. The discussion with the professionals focused on the accomplishments, the bottlenecks, and the recommendations regarding willingness and determinants of voluntary land consolidation as well as sustainable land management issues.

2.2.2. Variables Specification and Working Hypothesis

A dichotomous dependent variable was defined to specify whether a farmer was willing to accept voluntary land consolidation (WTAVLC = 1/yes) or not (WTAVLC = 0/no). The "yes/no" information about the respondents' characteristics enabled the application of binary response models such as the logit model.

The independent variables applied in the current study were hypothesized to have a relationship with the willingness to accept voluntary land consolidation. Based on the findings of previous and current studies on the adoption of land consolidation, the existing theoretical clarifications, and the authors' understandings of the status of land consolidation in the study area, 13 explanatory variables were identified. These variables, listed in Appendix A, were hypothesized to influence farmers' willingness to accept voluntary land consolidation (see Appendix A).

2.2.3. Method of Data Analysis

Farmers' decisions to adopt or reject land consolidation were influenced by socio-economic, demographic, institutional and biophysical factors, as described in Appendix A. Modeling landholder farmers' opinions to land management innovations such as land consolidation was important both theoretically and empirically.

The analysis of the relationship between the willingness and the determinants (variables) involved a mixed set of quantitative and qualitative data. Usually in such types of studies, the response (dependent) variable is dichotomous, taking on two values:

- The value 1 if the event occurs;
- The value 0 if it does not.

Estimation of this type of relationship requires the use of qualitative response models. In this regard, linear probability models (LPM) are possible alternatives. In LPM, the dichotomous dependent variable is expressed as a linear function of the explanatory variables. Although LPM can be estimated by the standard ordinary least squares (OLS) method, the results can cause several problems [49]. However, the OLS regression technique, when the dependent variable is binary (0,1), produces parameter estimates that are inefficient and in a heteroscedastic error structure. Consequently, hypothesis testing and the construction of a confidence interval become inaccurate and misleading. Likewise, a linear probability model may generate predicted values outside the 0–1 interval, which violates the basic tenets of probability.

To get rid of these problems and to produce relevant empirical outcomes, the most widely used qualitative response models are logistic regression and probit models (logit and probit models) [50]. Amemiya [50] points out that the choice between logit and probit models is difficult because of the statistical similarities between the two models. However, Maddala [51] reports that the logistic and the cumulative normal functions are very close in the mid-range, but the logistic function has slightly heavier tails than the cumulative normal function. Gujarati [52] also illustrates that the logistic and the probit formulations are quite comparable, the chief difference being that the former has slightly fatter tails, that is, the normal distribution curve approaches the axes more quickly than the logistic distribution curve.

For this study, the logistic distribution function (*logit* model) was applied, as it represents a close approximation to the cumulative normal distribution. Moreover, it is relatively simple from a mathematical point of view and lends itself to a meaningful interpretation. According to Aldrich and Nelson [49], the binomial logistic regression distribution function can be specified as:

$$p_i = \frac{e^{Z_i}}{1 + e^{Z_i}} \tag{5}$$

where p_i denotes the probability that the ith landholder farmer is willing to accept voluntary land consolidation. Z_i is a linear function of m explanatory variables (X) and is expressed as:

$$Z_i = \beta_0 + \beta_1 X_{1i} + \beta_2 X_{2i} + \beta_3 X_{3i} + \ldots + \beta_m X_{mi} \tag{6}$$

where β_0 is the intercept and β_i, slope parameters to be estimated in the model. The slope gives evidence as to how the log-odds in favor of willingness to accept land consolidation change as the independent variables change.

The stimulus index Z_i also refers to the natural logarithm of the odds ratio in favor of willingness to accept voluntary land consolidation. The odds to be used can be defined as the ratio of the probability that a landholder farmer is willing to accept voluntary land consolidation (pi) to the probability that he/she is not ($1 - p_i$).

$$ln(\frac{Pi}{1 - Pi}) = ln(e^{\beta_0 + \sum_{p=1}^{m} \beta_p X_{pi}}) \tag{7}$$

If the disturbance term U_i is taken into account, the logit model becomes:

$$Z_i = \beta_0 + \sum_{p=1}^{m} \beta_p X_{pi} + U_i \qquad (8)$$

Hence, the above model was used in this study and was treated against potential variables assumed to affect the decision to accept voluntary land consolidation. The parameters of the model were estimated using the iterative maximum likelihood estimation (MLE) procedure. In reality, the significant explanatory variables do not all have the same level of impact on landholder farmers' decisions to accept voluntary land consolidation. The impact of each significant explanatory variable on the probability of landholder farmers' willingness to accept voluntary land consolidation was computed by keeping the dummy variables at their most frequent values (zero or one).

3. Results

3.1. General Descriptive Analysis

Of the 343 sample landholder farmers, 229 (66.8%) were found to be willing to accept voluntary land consolidation, whereas the remaining 114 (33.2%) landholder farmers were not willing to accept voluntary land consolidation. Table 3 shows that the two categories of sample landholder farmers differed in various aspects.

Table 3. Summary of dummy variables used in the logistic regression model.

Variable Name	Variable Type	Farmers Willing (%)
Educational Level	Dummy	45.2
Farm to Home Nearness	Dummy	60.6
Extension Program Participation	Dummy	73.2
Parcel Exchange	Dummy	65.9
Parcel Preference	Dummy	47.8
Knowledge	Dummy	53.6
Attitude	Dummy	57.7
Perception	Dummy	60.1
Conflict Reduction	Dummy	53.6
Tenure Security	Dummy	55.7
Trust	Dummy	53.4

Source: Model output.

The survey results (Figure 2) show that 45.2% of landholder farmers were literate. The respective percentages for willing and non-willing landholder farmers were 51.5% and 32.5%. Furthermore, 72.9% of landholder farmers who were willing to accept land consolidation voluntarily had parcels close to their home, whereas 36% of non-willing landholder farmers had parcels found near to home. Out of landholder farmers who were willing to accept land consolidation voluntarily, 80.3% had participated in extension programs. The respective percentage for the non-willing landholder farmers was 58.8%.

Results also show that 65.9% of landholder farmers required an exchange of parcels with his/her neighbor. The respective percentage was 90.4% who were willing to exchange parcels with his/her neighbor, whereas only 16.7% of non-willing landholders exchanged parcels with his/her neighbor.

Additionally, 67.2% of landholder farmers were willing to cluster their parcels, whereas only 8.8% of non-willing landholders preferred clustering, and 72.5% of willing landholder farmers had knowledge about the process of land consolidation and its regulations, whereas only 15.8% of non-willing landholder farmers had land consolidation and regulations knowledge. Results show that 71.6% of willing landholder farmers had a positive attitude. Only 29.8% of the non-willing landholder farmers had a positive attitude towards land consolidation processes.

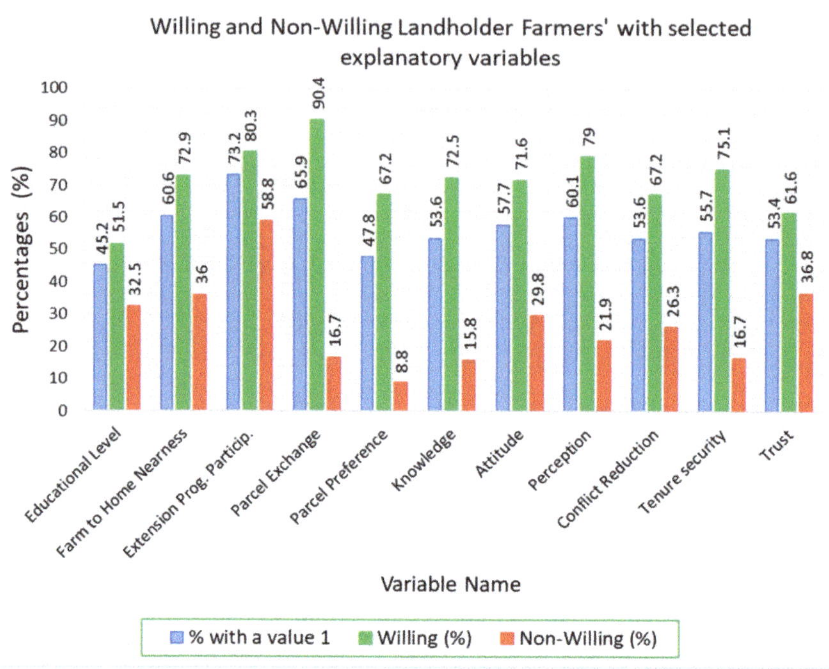

Figure 2. A bar graph showing willing and non-willing landholder farmers with selected explanatory variables.

Landholder farmers had different perceptions that land fragmentation reduces agricultural productivity. The results show that 60.1% of landholder farmers were willing to accept land consolidation voluntarily. From these, 79.0% of willing landholder farmers perceived that land fragmentation reduces agricultural productivity, whereas only 21.9% of non-willing landholder farmers did not believe that land fragmentation reduces the agricultural productivity. Moreover, 67.2% of willing landholder farmers agreed that land consolidation reduces boundary disputes, but only a few (26.3%) non-willing landholder farmers assumed that land consolidation would reduce conflicts.

In total, 75.1% of willing landholder farmers perceived that land consolidation would secure his/her land, but only 16.7% of non-willing landholder farmers perceived that land consolidation would secure his/her land. Regarding trust of landholder farmers to neighbors to exchange land, results imply that 61.6% of willing landholders trusted his/her neighbor to exchange lands, and 36.8% of non-willing landholders trusted his/her neighbor to exchange land.

A more detailed investigation (Table 4) considering the age structure of landholder farmers provided evidence that the average age of the willing landholder farmers was 48 years with 11.3 standard deviation and non-willing landholder farmers was also 48 years but with 9.8 standard deviation. The median age for willing landholder farmers was 48 years, but for non-willing landholder farmers, it was almost 47 years. The average size of farm of landholder farmers was 1.29 ha. Its standard deviation was 0.7. Willing landholder farmers had an average farm size of 1.32 ha with 0.7 standard deviation; non-willing landholder farmers had 1.23 ha, and its standard deviation was 0.8. The median of farm area for willing landholder farmers was 1.25 ha, and for non-willing landholder farmers, it was 1 ha.

Table 4. Summary of continuous variables used in the logistic regression model.

Variable Name	Age	Farm Area
Measurement Unit	**Years**	**Hectares**
Mean Value		
All	48.7	1.29
Willing	48.9	1.32
Non-willing	48.4	1.23
Median Value		
All	48.0	1.00
Willing	48.0	1.25
Non-willing	46.5	1.00
Std. Deviation		
All	10.8	0.70
Willing	11.3	0.70
Non-willing	9.8	0.80

3.2. Results of the Logistic Regression Model

The logit model was used to analyze the influencing factors of landholder farmers' willingness to accept voluntary land consolidation. The landholder farmers were either willing or not willing to accept voluntary land consolidation. Consequently, the variable willingness to accept voluntary land consolidation (WTAVLC) was used as a binary variable. Value 1 indicates the willingness of the landholder farmer in regard to voluntary land consolidation, and value 0 indicates the opposite. The logistic regression model was estimated using the iterative maximum likelihood estimation procedure. Table 5 documents the results of the model.

Table 5. Results of the logistic regression model.

Explanatory Variables Name	Estimated Coefficients	Odds Ratio	Wald Statistics	Significance Level
Age	−0.009	0.991	0.069	0.793
Educational Level	0.827	2.286	1.813	0.178
Farm Area	−0.334	0.716	0.445	0.505
Farm to Home Nearness	1.890	6.621	8.319	0.004 **
Extension Program Participation	0.404	1.498	0.312	0.576
Parcel Exchange	3.656	38.716	28.450	0.000 ***
Parcel Preference	3.233	25.361	18.999	0.000 ***
Knowledge	1.869	6.484	4.379	0.036 *
Attitude	0.455	1.577	0.533	0.466
Perception	2.203	9.054	11.786	0.001 **
Conflict Reduction	1.785	5.962	7.456	0.006 **
Tenure Security	1.333	3.794	3.042	0.081
Trust	1.238	3.448	4.043	0.044 *
Constant	−7.200	0.001	15.175	0.000
Percent correctly predicted	95.6 [a]			
Sensitivity	96.54 [b]			
Specificity	93.80 [c]			
Chi-square value	351.505 ***			
Number of cases	343			

***, **, * Significant at $p < 0.001$, $p < 0.01$, and $p < 0.05$ probability levels, respectively. [a] Based on 5050 probability classification scheme. [b] Correctly predicted willing landholder farmers. [c] Correctly predicted non-willing landholder farmers.

The model results approved the a priori hypothesis that landholder farmers' willingness to accept voluntary land consolidation was influenced by the interaction of several factors. The Chi-square value documents that the parameters included in the model were significantly different from zero at $p < 0.001$ probability level.

Another measure of goodness of fit was based on a scheme that classified the predicted value of the dependent variable (WTAVLC) as 1 if $P(i) \geq 0.5$ and as 0 otherwise. The model correctly predicted 95.6% of the observed values. The sensitivity (the number of willing landholder farmers correctly predicted by the model as willing landholder farmers) was 96.5%, while the specificity (the number of non-willing landholder farmers correctly predicted by the model as non-willing landholder farmers) was 93.8%. Therefore, the model predicted both groups accurately. The signs of all the coefficients turned out to be consistent with the a priori anticipations.

Table 5 shows the signs, the magnitudes, and the statistical significance of the estimated parameters. Of the thirteen variables hypothesized to explain landholder farmers' willingness to accept voluntary land consolidation, two were found to be significant at $p < 0.001$ (0.1%) probability level. These included landholder farmers' need to exchange parcels with the neighbors and landholder farmers' preferences for their parcels to be clustered. Table 5 also gives evidence that three variables were significant at $p < 0.01$ (1%) probability level. These variables were landholder farms' nearness to home, landholder farmers' perception that land fragmentation reduces the agricultural productivity, and landholder farmers' acceptance that land consolidation reduces boundary conflicts. Knowledge of landholder farmers and landholder farmers' trust to exchange lands with neighbors were found to be significant at $p < 0.05$ (5%) probability level.

4. Discussion

Results show that the landholder farmers who were keen to exchange parcels with neighbors had a high degree of willingness to accept voluntary land consolidation. The possible explanation is that landholder farmers exchanged their parcel to cluster the land in one place and to avoid fragmentation. This result is supported by studies in the Ukraine by Malashevskyi et al., which studied parcel exchange aimed at regrouping a significant amount of land use in order to optimize their structure and to consolidate land [53]. This hypothesis also was confirmed in the focus group discussions. Participants said that, in the study area, land exchange has long been tradition, and landholder farmers are motivated to swap the parcel with neighbors by the prospect of better access to irrigable land, facilitation of farm operations, and shorter distance to homestead. Having closer distances to main roads and town infrastructure as well as to plots of a family member in addition to having parcels with higher fertility were further reasons. Currently, the government is giving legal support to secure land use rights for those who swap parcels. During community consultations, a few landholder farmers raised issues and concerns about the risks associated with concentration of farm land in one place, such as infestation by army worm, destruction by floods, and soil fertility differences of parcels. As reasons for parcel exchange, during the expert panel, the experts noted facilitation of agricultural mechanization, better access to irrigable land, and a facilitation of cultivation on consolidated land.

Similarly, the preference to cluster parcels was strongly correlated with landholder farmers' willingness to accept voluntary land consolidation. This result also was confirmed by the discussion with selected farmers. Concentration of farmland in one place facilitates easier and more regular monitoring of the agricultural land. Aggregated plots simplify fertilization and composting. In addition, accumulated land enables the planting of permanent crops in combination with livestock fattening and easy input for transport, saves labor, improves yield due to timely operation of parcels, and facilitates agricultural mechanization. Farmers also expected that aggregated parcels would enable them to put more land under irrigation (including opportunities to intensify use of underground water for supplementary irrigation).

Plot nearness to home had a positive effect on the willingness of landholder farmers to accept land consolidation. This was considered as time taken to move from homestead to parcels as well as from parcel to parcel. This result of the current study is supported by findings by Zeng et al. in Jiangsu province, China, which indicate that, after land consolidation, the average distance from dwellings to the plots decreases, which is also caused by an improvement of the road network [54]. Other studies worked on in Ethiopia by Paul et al. and Teshome et al. indirectly support the plot nearness to home as a positive effect with the finding that fragmentation usually increases distances from the parcel to the homestead [1,42]. In focus group discussions carried out in the current study, the landholder farmers stated their preference for parcels to be located close to their house, as transportation activities and labor forces would be reduced. The farmers also confirmed that they are able to protect and monitor parcels near the house more easily. This aims to produce higher productivity and better output.

The perception that land fragmentation can reduce productivity affected the willingness of voluntary land consolidation positively and significantly. This finding is consistent with other studies done by Nilsson et al. and Alemu et al. in Rwanda and Northern Ethiopia, respectively, which document the correlation between land fragmentation and yield reduction [36,55]. Similarly, a study done in Rwanda shows that the increase in land fragmentation is associated with a negative effect on yields [55]. Studies conducted in Rwanda by Bizoza et al. and Cioffo et al. also assure the association of positive productivity effects with land consolidation [56,57]. A Chinese study confirms that land consolidation processes (opposed to land fragmentation) enable changes of land use types with significant effects on increasing agricultural production [58]. Zeng et al. [54] confirm that land consolidation enhances grain yield capability. Qualitatively, this result was supported in the discussion with the community members. Most of the participants agreed that land fragmentation is a source for low yields, because farmers who cultivate scattered plots run a higher risk of losing their harvest to wild animals, as their plots are too far away for them to regularly monitor. They also mentioned the burden on children who carry food to family members working on distant parcels during harvesting seasons. Also, the participants emphasized the difficulty of using even small agricultural machines, such as three-wheeled multi-purpose tractors, on discontinuous plots as obstacles for agricultural mechanization. Some community members did not directly identify a negative effect of fragmentation on productivity, however, they mentioned that the additional time and effort required to manage distant parcels negatively affects productivity. A few of them considered land fragmentation as a risk spreading strategy, minimizing the risk of harvest loss by planting crops in different locations.

As expected, the determinant "perception of conflict reduction" positively and significantly influenced the willingness of landholder farmers to accept voluntary land consolidation. This finding is approved by former studies done in Turkey by Akkaya et al., which states land consolidation as an effective solution against conflicts raised by land fragmentation [59]. This also was confirmed in the focus group discussion. During community panel, landholder farmers gave evidence that, on the one hand, land fragmentation increases conflicts related to land use and, on the other hand, land consolidation minimizes land use related conflicts. On these issues, experts also confirmed that parcels concentrated in one area have reduced land related conflicts compared to parcels found in a scattered place.

The model results confirm that knowledge of landholder farmers about land consolidation had a positive and significant effect on the dependent variable. This result is confirmed by previous studies conducted by Terano et al. and Gessesse et al. in Iran and China, respectively, indicating that when farmers are aware of land consolidation, their adaptation improves significantly [60,61]. The result was supported by information gained in focus group discussion with landholder farmers. Most of them agreed that knowledge is a key element to adopting a new practice—in this case, land consolidation. In addition, participants emphasized the need for repeated events to raise awareness and clarify issues through combined use of local and scientific knowledge.

The determinant "trust of neighboring landholder farmers" was also positively and significantly related to willingness to accept land consolidation voluntarily. It is conducive to collective action such as voluntary land consolidation. This finding is in line with previous studies done by Bizoza et al., Bouma et al., and Nyangena in Rwanda and rural Kenya, respectively, which indicate that trust constitutes an important feature of social capital assets and is a key element for successfully implementing land consolidation procedures [56,62,63]. Furthermore, capacity strengthening of local leaders, especially at village level, is required so that they are able to sensitize farmers on all aspects regarding land use and management reforms. This is likely to allow farmers to be confident in the process of voluntary land consolidation.

Tenure security of landholder farmers was positively and significantly related to the willingness to accept land consolidation voluntarily. This finding of the current study is supported by investigations done by Demetriou et al. showing that land consolidation is suitable to support land tenure security [11]. Similarly, another study by Zeng et al. in China reports that land consolidation facilitates land tenure transfer and security [54]. The focus group discussion confirmed that a concentration of parcels and consolidated plots increase the land holding capability and the security of land use rights.

Sensitivity Analysis

The above outlined explanatory variables do not equally affect the landholder farmers' willingness to accept voluntary land consolidation. To rank these variables, a "distinctive landholder farmer" can be defined using the most frequent values of the dummy variables. Accordingly, a distinctive landholder farmer:

- Has participated in an extension program (73.2%);
- Needs to exchange parcels with neighbors (65.9%);
- Has one of the parcels nearest to home (60.6%);
- Perceives that land fragmentation reduces productivity (60.1%);
- Has a positive attitude to land consolidation (57.7%);
- Perceives land consolidation as a way to secure land use rights (55.7%);
- Has awareness that land consolidation reduces boundary conflicts (53.6%);
- Trusts neighbors to exchange land (53.4%).

The probability that the distinctive landholder farmer would show interest in willingness to accept voluntary land consolidation was computed to be 0.59. However, the probability declined by 4.1% for those landholder farmers who were distinctive in all respects except that they did not have parcels near home. Similarly, the probability declined by 5.9% for those landholder farmers who were distinctive in all respects except that they did not have need to exchange parcels. Moreover, the probability of willingness to accept land consolidation decreased by 3.8% for those landholder farmers who were distinctive in all respects except that they did not perceive that land fragmentation reduces productivity. The effects of two other significant dummy variables are documented in Table 6.

Table 6. Change in probability as a result of a change in significant explanatory variables.

Description	Probability	Change in Probability	Percentage Change in Probability
A distinctive landholder farmer	0.59		
A distinctive landholder farmer but does not have parcel near home	0.57	−0.02	−4.1
A distinctive landholder farmer but does not have need to exchange parcels	0.56	−0.04	−5.9
A distinctive landholder farmer but does not perceive land fragmentation reduces productivity	0.57	−0.02	−3.8
A distinctive landholder farmer but does not have land consolidation reduce boundary conflicts	0.59	−0.01	−1.4
A distinctive landholder farmer but does not know land consolidation definition and its regulations	0.59	−0.01	−1.4

Source: Model output.

5. Conclusions

A fundamental part of any strategy towards more productive and sustainable agriculture as well as towards rural development enables farmers and food producers to utilize new methods of agricultural cultivation with higher efficiency, transparency, and competitiveness. In Ethiopia, agriculture is the foundation of the country's economy, accounting for 37% of their gross domestic product; however, it is overwhelmingly of a subsistence nature. Farmers lack the means to improve production due to the fragmentation of landholdings, the insecurity of tenure, the absence of modern agricultural technologies, and the lack of proper land management. The government's policy is to promote agriculture. One of the governmental land policies is encouraging voluntary land consolidation as a strategy tool to address the challenges of traditional agricultural practice. The current study gives evidence of a high degree of willingness by landholder farmers to accept voluntary land consolidation. Land consolidation is seen as a proper instrument to increase agricultural productivity and to improve the management of natural resources. Land consolidation is a driver for rural development and contributes essentially to the improvement of land administration systems. Land consolidation reduces land fragmentation and enables an economic cultivation of agricultural land. Finally, land consolidation is an excellent tool to improve road infrastructure and erosion management in rural areas.

In this study, a total of 13 factors hypothesized to influence the willingness of landholder farmers in regard to voluntary land consolidation were evaluated by using a logistic regression model. Findings were presented.

In focus group discussion, landholder farmers identified the following risks of voluntary land consolidation. They fear decreased cropping diversity with increased vulnerability to vermin epidemics as well as an increased risk of crop failures due to local natural disasters such as heavy rain, hail, and floods. Likewise, farmers also mentioned different soil quality, different fertility levels, and different slopes of parcels as obstacles for land consolidation. In addition to these, land scarcity, infrastructure problems, heritage law, accessibility to water resources for irrigation, and financial issues were seen as challenges that cannot be met by land consolidation procedures.

Currently, at the governmental level, there is no authority for supporting the implementation and the monitoring of land consolidation. This shows that the government does not give much attention to land consolidation. There is a lack of technical knowledge and facilities for land consolidation as well as a lack of transparent and clear regulations for voluntary land consolidation.

Despite the problems identified in the study area, many landholder farmers are willing to accept voluntary land consolidation. The conducive local environment is able to push voluntary land consolidation in a systematic manner. There are opportunities to create land consolidation projects in the study area.

Based on the investigations carried out in the study, the authors recommend the following activities for pushing land consolidation processes:

- Land exchange is a key element of land consolidation. The willingness of farmers in regard to voluntary land consolidation became evident by the quantitative and the qualitative surveys outlined in this study. It is the task of the government to facilitate legal land exchange;
- Landholder farmers are interested in aggregated and clustered parcels. Governmental authorities should provide the legal framework to enable consolidated agricultural land fragmentation while preserving environmentally important landscape structures;
- Findings of the study give evidence that landholder farmers prefer to have their parcels situated near the homestead and to have good access to their parcels. The government should enhance accessibility to farms by facilitating road networks;
- Voluntary land consolidation reduces parcel boundary disputes or conflicts. This was proven by the farmers in the survey and in the focus group discussions. Therefore, land administration offices should promote and support voluntary land consolidation to decrease conflicts arising from boundaries;

- Voluntary land consolidation improves the security of land use rights. Access to land and security of land tenure are effective ways to reduce a farmer's vulnerability, to guarantee long-term investments on land, and to conserve natural resources. The government should facilitate components of land registry and surveying of land parcels (cadaster) within land consolidation procedures.

In addition, the authors formulated some recommendations that could not be directly derived from the quantitative analysis of the study but which became obvious during the discussions with respondents:

- Land fragmentation reduces yield. Only land consolidation processes can solve land fragmentation and, as a consequence, enable improved yields. The government has to encourage landholder farmers to participate in voluntary land consolidation. This creates a favorable environment for commercializing mechanized farming and supports agriculture towards higher proficiency and more stable yields;
- In addition, the government should provide loans to landholder farmers to purchase modern technologies to improve the agricultural productivity and to make the work easier for farmers. Loans can be secured by index insurance mechanisms;
- Farmers are not always aware of the benefits of land consolidation. Therefore, land administration offices should continuously inform the rural population about the aims, the benefits, the legal framework, and the implementation of land consolidation procedures.

Land consolidation is a cornerstone for sustainable development, for the alleviation of poverty, for the improvement of rural infrastructure, for mitigated flood and erosion risks, and for an increase of agricultural productivity. Therefore, the Ethiopian government draws attention to voluntary land consolidation in practice. For this, the government has to define policies and regulations taking into consideration the different perspectives of the stakeholders, such as the Ministry of Agriculture, international development partners, policy makers, and especially landholder farmers.

The current study was among the first studies in Ethiopia to investigate the willingness of voluntary land consolidation. However, more research activities and governmental support at local, regional, and national levels are necessary to convince farmers of the benefits of land consolidation processes and to create a higher number of voluntary land consolidation projects in the study area, in the Amhara region, and in all other parts of Ethiopia. A pilot study would be a good way to demonstrate the many potential benefits of land consolidation.

Author Contributions: A.A.G. and R.M.; methodology, A.A.G. and W.S.; validation, A.A.G. and R.M.; formal analysis, A.A.G.; investigation, A.A.G. and W.S.; resources, A.A.G., W.S., and R.M.; data curation, A.A.G.; writing—original draft preparation, A.A.G.; writing—review and editing, R.M., C.A. and W.S.; visualization, A.A.G. and R.M.; supervision, R.M. and C.A.; project administration, R.M.; funding acquisition, R.M.

Funding: This research was funded by the Austrian Development Cooperation within the Austrian Partnership Programme in Higher Education and Research for Development (APPEAR). Project no. 113 "Implementation of Academic Land Administration Education in Ethiopia for Supporting Sustainable Development" (EduLAND2).

Acknowledgments: The authors thank the landholder farmers and the land administration experts in the study area for their collaboration during interviews.

Conflicts of Interest: The authors declare no conflict of interest. The funders had no role in the design of the study; in the collection, analyses, or interpretation of data; in the writing of the manuscript, or in the decision to publish the results.

Appendix A

Table A1. Variables specification and working hypothesis.

Variable Description	Variable	Hypothesis
Willingness to accept voluntary land consolidation	WTAVLC	A dichotomous dependent variable to specify, whether a landholder farmer is willing to accept voluntary land consolidation (WTAVLC = 1) or not (WTAVLC = 0).
Landholder farmers exchange of parcels	PARCELEXCH	These explanatory variable measures landholder farmers need to exchange parcels with other farmers involved in a land consolidation procedure. It is the hypothesis, that landholder farmers, who are interested to exchange parcels, are more likely to be willing to accept voluntary land consolidation. *PARCELEXCH* is expected to be strongly and positively associated with landholder farmers willingness to accept voluntary land consolidation. Therefore, the value is 1, if landholder farmer is willing to exchange parcels, otherwise the value is 0.
Parcel preference of landholder farmers	PARCELPREF	The hypothesis is that landholder farmers with needs to aggregate parcels are more likely to be willing to accept voluntary land consolidation. Therefore, this variable is 1, if the landholder farmer prefers to cluster parcels. If the landholder farmer prefers to have parcels scattered over the area, the value is 0.
Landholder farmers' knowledge about land consolidation	KNOWLEG	As it is hypothesized that landholder farmers being informed about land consolidation processes have a higher willingness to accept voluntary land consolidation, the variable is 1 in case the landholder farmer has some knowledge about land consolidation. Otherwise, the value is 0.
Attitude of landholder farmers	ATTITUDE	It can be assumed that landholder farmers with a generally positive attitude toward land consolidation have a higher degree of willingness to accept voluntary land consolidation. Therefore, this variable is 1 in case of a positive attitude to land consolidation processes.
Landholder farmers' perceptions of land fragmentation reducing productivity	PERCEPTI	This variable describes landholder farmers' perceptions about land fragmentation. It takes a value of 1 if a landholder farmer perceives that land fragmentation reduces agricultural productivity and 0 otherwise. It is expected that the perception variable is strongly and positively associated with landholder farmers' willingness to accept voluntary land consolidation.
Landholder farmers' nearness of farm to home	FARMHOMENEAR	This variable specifies home nearness to the parcels. It takes a value of 1 if landholder farmers prefer parcels near to home and 0 otherwise. It is hypothesized that landholder farmers who want to have their parcels near to home are more likely to be willing to accept voluntary land consolidation.

Table A1. *Cont.*

Variable Description	Variable	Hypothesis
Education level of landholder farmers	EDULEVE	This variable takes the value of 1 if the landholder farmer is literate and 0 otherwise. Education increases land holder farmers' abilities to get information. Thus, it is hypothesized that education has a positive effect for willingness to accept voluntary land consolidation.
Age of the household head	AGE	This variable specifies the age of the landholder farmer in years. The age of landholder farmer is also a measure of farming experience. Age and farming experience are expected to be correlated with tradition. It is hypothesized that higher age has a more negative influence toward accepting voluntary land consolidation. Young landholder farmers are normally less affected by traditions and accept changes more easily.
Farm area in hectares	FARMARE	This variable is the total area of farmland in hectares owned by the landholder farmers at the time of the survey. In the study area, large farms are owned by older landholder farmers, which are more conservative and are not so open for changes. Thus, it is hypothesized that farm size has a negative effect to accept voluntary land consolidation. Also, this variable is a continuous variable.
Landholder farmers' participation in extension program	EXTPROGPART	Ethiopian government carries out an agricultural program where farmers get specific training in how to manage the cultivation of land. It is the hypothesis of this study that landholder farmers who have participated in an extension program will also favor voluntary land consolidation. The variable is 1 if the landholder farmer has participated in an extension program and 0 otherwise.
Conflict reduction	CONFLICTRED	This variable is 1, if the landholder farmer assumes land consolidation as a tool to reduce boundary disputes and 0, if not. The conflict reduction variable is expected to be strongly and positively correlated with landholder farmers willingness to accept voluntary land consolidation.
Security of tenure	TENURESECUR	The variable describes that either the landholder farmer considers that he/she has security for using the parcel during his/her lifetime *(TENURESECURE = 1)* or not *(TENURSECURE = 0)*. The hypothesis is that the existing perception of tenure security has a positive effect on the willingness to accept voluntary land consolidation.
Landholder farmers' trust in neighbor	TRUST	Trust increases landholder farmers' willingness to exchange parcels. Thus, trust is important for willingness to accept voluntary land consolidation. Therefore, the trust variable is expected to be positively associated with landholder farmer willingness to accept voluntary land consolidation. It is 1 if landholder farmer trusts his/her neighbors to exchange plots, otherwise 0.

References

1. Teshome, A.; de Graaff, J.; Ritsema, C.; Kassie, M. Farmers' Perceptions about the Influence of Land Quality, Land Fragmentation and Tenure Systems on Sustainable Land Management in the North Western Ethiopian Highlands. *Land Degrad. Dev.* **2016**, *27*, 884–898. [CrossRef]
2. Beyene, A. *Land Consolidation, Canals and Apps—Reshaping Agriculture in Ethiopia*; The Nordic Africa Inst.: Uppsala, Sweden, 2019; Volume 3, pp. 1–8.
3. FAO. *Ethiopia: Small Family Farms Country Factsheet*; FAO: Rome, Italy, 2018.
4. Bezu, S.; Holden, S. Are rural youth in Ethiopia abandoning agriculture? *World Dev.* **2014**, *64*, 259–272. [CrossRef]
5. Josephson, A.L.; Ricker-Gilbert, J.; Florax, R.J.G.M. How does population density influence agricultural intensification and productivity? Evidence from Ethiopia. *Food Policy* **2014**, *48*, 142–152. [CrossRef]
6. Knippenberg, E.; Jolliffe, D.; Hoddinott, J. *Land Fragmentation and Food Insecurity in Ethiopia*; Policy Research Working Paper; No. 8559; World Bank: Washington, DC, USA, 2017; Volume 47. Available online: https://openknowledge.worldbank.org/handle/10986/30286 (accessed on 22 April 2019).
7. Van Dijk, T. Scenarios of Central European land fragmentation. *Land Use Policy* **2003**, *20*, 149–158. [CrossRef]
8. Bizimana, C.; Nieuwoudt, W.L.; Ferrer, S.R.D. Farm size, land fragmentation and economic efficiency in southern Rwanda. *Agrekon* **2004**, *43*, 244–262. [CrossRef]
9. Jeffery, W. Bentley Economic and Ecological Approaches to Land Fragmantation: In Defense of A Much-Maligned Phenomenon. *Annu. Rev. Anthropol.* **1987**, *16*, 31–67.
10. Gónzalez, X.P.; Marey, M.F.; Álvarez, C.J. Evaluation of productive rural land patterns with joint regard to the size, shape and dispersion of plots. *Agric. Syst.* **2007**, *92*, 52–62. [CrossRef]
11. Demetriou, D.; Stillwell, J.; See, L. A new methodology for measuring land fragmentation. *Comput. Environ. Urban Syst.* **2013**, *39*, 71–80. [CrossRef]
12. Vijulie, I.; Matei, E.; Manea, G.; Cocoş, O.; Cuculici, R. Assessment of Agricultural Land Fragmentation in Romania, A Case Study: Izvoarele Commune, Olt County. *Acta Geogr. Slov.* **2013**, *52*, 403–430. [CrossRef]
13. Hristov, J. *Assessment of the Impact of High Fragmented Land upon the Productivity and Profitability of the Farms—The Case of the Macedonian Vegetable Growers*; Business: Uppsala, Sweden, 2009; pp. 1–77.
14. FAO. *The Design of Land Consolidation Pilot Projects in Central and Eastern Europe*; FAO-Land Tenure Studies: Rome, Italy, 2003.
15. Latruffe, L.; Piet, L. Does land fragmentation affect farm performance? A case study from Brittany, France. *Agric. Syst.* **2014**, *129*, 68–80. [CrossRef]
16. Kakwagh, V.V.; Aderonmu, J.A.; Ikwuba, A. Land Fragmentation and Agricultural Development in Tivland of Benue State, Nigeria. *Curr. Res. J. Soc. Sci.* **2011**, *3*, 54–58.
17. Tan, S.; Heerink, N.; Kuyvenhoven, A.; Qu, F. Impact of land fragmentation on rice producers' technical efficiency in South-East China. *NJAS Wagening. J. Life Sci.* **2010**, *57*, 117–123. [CrossRef]
18. Balogun, O.L.; Akinyemi, B.E. Land fragmentation effects on technical efficiency of cassava farmers in South-West geopolitical zone, Nigeria. *Cogent Soc. Sci.* **2017**, *3*, 1–10. [CrossRef]
19. Di Falco, S.; Penov, I.; Aleksiev, A.; van Rensburg, T.M. Agrobiodiversity, farm profits and land fragmentation: Evidence from Bulgaria. *Land Use Policy* **2010**, *27*, 763–771. [CrossRef]
20. Blarel, B.; Hazell, P.; Place, F.; Quiggin, J. The economics of farm fragmentation: Evidence from Ghana and Rwanda. *World Bank Econ. Rev.* **1992**, *6*, 233–254. [CrossRef]
21. Unal, F.G. Small is Beautiful: Evidence of an Inverse Relationship Between Farm Size and Yield in Turkey. *SSRN Electron. J.* **2011**. [CrossRef]
22. Crecente, R.; Alvarez, C.; Fra, U. Economic, social and environmental impact of land consolidation in Galicia. *Land Use Policy* **2002**, *19*, 135–147. [CrossRef]
23. Gonzalez, X.P.; Alvarez, C.J.; Crecente, R. Evaluation of land distributions with joint regard to plot size and shape. *Agric. Syst.* **2004**, *82*, 31–43. [CrossRef]
24. Burton, S.P. Land consolidation in Cyprus. *Geography* **1988**, *65*, 320–324. [CrossRef]
25. Sabates-Wheeler, R. Consolidation initiatives after land reform: Responses to multiple dimensions of land fragmentation in Eastern European agriculture. *J. Int. Dev.* **2002**, *14*, 1005–1018. [CrossRef]
26. Van Dijk, T. Export of planning knowledge needs comparative analysis: The case of applying western land consolidation experience in Central Europe. *Eur. Plan. Stud.* **2002**, *10*, 911–922. [CrossRef]

27. Dang, H.H.; McPherson, M. *Land Policy for SocioEconomic Development in Vietnam*; Fulbright Economics Teaching Program and Ash Center, Harvard University: Cambridge, MA, USA, 2010. Available online: https://ash.harvard.edu/files/vnm_landpolicypaper.pdf (accessed on 10 May 2019).

28. Huang, Q.; Li, M.; Chen, Z.; Li, F. Land consolidation: An approach for sustainable development in rural China. *Ambio* **2011**, *40*, 93–95. [CrossRef] [PubMed]

29. Pašakarnis, G.; Maliene, V. Towards sustainable rural development in Central and Eastern Europe: Applying land consolidation. *Land Use Policy* **2010**, *27*, 545–549. [CrossRef]

30. FAO. *Opportunities to Mainstream Land Consolidation in Rural Development Programmes of the European Union*; FAO-Land Tenure Policy Series: Rome, Italy, 2008.

31. Vitikainen, A. An Overview of Land Consolidation in Europe. *Nord. J. Surv. Real Estate Res.* **2004**, *1*, 25–44.

32. Van Dijk, T. *Dealing with Central European Land Fragmentation*; Eburon: Delft, The Netherlands, 2003; pp. 1–11.

33. Dethier, J.J.; Effenberger, A. Agriculture and development: A brief review of the literature. *Econ. Syst.* **2012**, *36*, 175–205. [CrossRef]

34. Timmer, C.P. Agriculture and economic development. In *Handbook of Agricultural Economics*; Gardner, B.L., Rausser, G.C., Eds.; University of California: California, CA, USA, 2002; Volume 2, pp. 1487–1546.

35. Niroula, G.S.; Thapa, G.B. Impacts and causes of land fragmentation, and lessons learned from land consolidation in South Asia. *Land Use Policy* **2005**, *22*, 358–372. [CrossRef]

36. Nilsson, P. The Role of Land Use Consolidation in Improving Crop Yields among Farm Households in Rwanda. *J. Dev. Stud.* **2018**, *55*, 1726–1740. [CrossRef]

37. *Federal Democratic Republic of Ethiopia, Federal Rural Land Administration and Land Use Proclamation*; Proclamation No. 456/2005; Federal Negarit Gazeta: Addis Ababa, Ethiopia, 2005.

38. Council of the Amhara National Regional State in the Federal Democratic Republic of Ethiopia. *The Revised Rural Land Administration and Use Determination Proclamation of the Amhara National Regional State*; Proclamation No.252/2017; Regional Zikre Hig Gazeta: Addis Ababa, Ethiopia, 2017.

39. Amhara National Regional State (ANRS). *The Amhara National Regional State Rural Land Administration and Use System Implementation, Council of Regional Government Regulation*; Regulation No. 133/2006; Regional Zikre Hig Gazeta: Bahir Dar, Ethiopia, 2006.

40. Van Dijk, T. Complications for traditional land consolidation in Central Europe. *Geoforum* **2007**, *38*, 505–511. [CrossRef]

41. Williams, J.F. The problems of land consolidation: A case study of Taiwan. *J. Geogr.* **1976**, *75*, 419–426. [CrossRef]

42. Paul, M.; wa Gĩthĩnji, M. Small farms, smaller plots: Land size, fragmentation, and productivity in Ethiopia. *J. Peasant Stud.* **2018**, *45*, 757–775. [CrossRef]

43. Lisec, A.; Pintar, M. Conservation of natural ecosystems by land consolidation in the rural landscape. *Acta Agric. Slov.* **2005**, *85*, 73–82.

44. Gozamin District Agricultural and Rural Development Office. *Summary and Statistical Report of the District*, Debre Markos, Ethiopia, 2018; Unpublished.

45. Central Statistical Agency of Ethiopia (CSA). *Summary and Statistical Report of the 2007 Population and Housing Census*; Addis Ababa, Ethiopia, 2007. Available online: http://www.ethiopianreview.com/pdf/001/Cen2007.pdf (accessed on 12 March 2019).

46. Friis, I.; Demissew, S.; van Bruegel, P. *Atlas of the Potential Vegetation of Ethiopia*; The Royal Danish Academy of Sciences and Letters; Biologiske Skrifter: Copenhagen, Denmark, 2010; Volume 58.

47. Cochran, W.G. *Sampling Techniques*, 3rd ed.; John Wiley & Sons: New York, NY, USA, 1977; Volume 3, pp. 89–97.

48. Feige, S.; Marr, M.A. *Sampling Manual: A Guide to Sampling under the CDM with Special Focus to PoAs*, 1st ed.; GmbH: Hamburg, Germany, 2012; pp. 1–97.

49. Aldrich, J.H.; Nelson, F.D.; Adler, E.S. *Linear Probability Logit, and Probit Models*; Michael, S., Lewis, B., Eds.; Saga Publication: London, UK, 1984; Volume 8, pp. 1–95.

50. Amemiya, T. Qualitative Response Models: A Survey. *J. Econ. Lit.* **1981**, *19*, 1483–1536.

51. Maddala, G.S. *Limited-Dependent and Qualitative Variables in Economics*; Cambridge University Press: New York, NY, USA, 1983; pp. 257–291.

52. Gujarati, D.N. *Basic Econometrics*, 4th ed.; McGraw-Hill, Inc.: New York, NY, USA, 2003; pp. 15–297.

53. Malashevskyi, M.; Palamar, A.; Malanchuk, M.; Bugaienko, O.; Tarnopolsky, E. The Opportunities for Use the Peer Land Exchange During Land Management in Ukraine. *Geod. Cartogr.* **2019**, *44*, 129–133. [CrossRef]
54. Zeng, S.; Zhu, F.; Chen, F.; Yu, M.; Zhang, S.; Yang, Y. Assessing the impacts of land consolidation on agricultural technical efficiency of producers: A survey from Jiangsu Province, China. *Sustainability* **2018**, *10*, 2490. [CrossRef]
55. Alemu, G.T.; Berhanie Ayele, Z.; Abelieneh Berhanu, A. Effects of Land Fragmentation on Productivity in Northwestern Ethiopia. *Adv. Agric.* **2017**, *2017*, 4509605. [CrossRef]
56. Bizoza, A.R.; Havugimana, J.M. Land Use Consolidation in Rwanda: A Case Study of Nyanza District, Southern Province. *Int. J. Sustain. Land Use Urban Plan.* **2017**, *1*, 64–75. [CrossRef]
57. Cioffo, G.D.; Ansoms, A.; Murison, J. Modernising agriculture through a 'new' Green Revolution: The limits of the Crop Intensification Programme in Rwanda. *Rev. Afr. Polit. Econ.* **2016**, *43*, 277–293. [CrossRef]
58. Zhang, Z.; Zhao, W.; Gu, X. Changes resulting from a land consolidation project (LCP) and its resource-environment effects: A case study in Tianmen City of Hubei Province, China. *Land Use Policy* **2014**, *40*, 74–82. [CrossRef]
59. Akkaya Aslan, S.T.; Gundogdu, K.S.; Yaslioglu, E.; Kirmikil, M.; Arici, I. Personal, physical and socioeconomic factors affecting farmers' adoption of land consolidation. *Span. J. Agric. Res.* **2007**, *5*, 204–213. [CrossRef]
60. Terano, R.; Mohamed, Z.; Shamsudin, M.N.; Latif, I.A. Factors influencing intention to adopt sustainable agriculture practices among paddy farmers in Kada, Malaysia. *Asian J. Agric. Res.* **2015**, *9*, 268–275. [CrossRef]
61. Gessesse, A.T.; Li, H.; He, G.; Berhe, A.A. Study on farmers land consolidation adaptation intention. *China Agric. Econ. Rev.* **2018**, *10*, 666–682. [CrossRef]
62. Bouma, J.; Bulte, E.; Van Soest, D. Trust and cooperation: Social capital and community resource management. *J. Environ. Econ. Manag.* **2008**, *56*, 155–166. [CrossRef]
63. Nyangena, W. Social determinants of soil and water conservation in rural Kenya. *Environ. Dev. Sustain.* **2008**, *10*, 745–767. [CrossRef]

Article

Household Land Allocations and the Youth Land Access Nexus: Evidence from the Techiman Area of Ghana

Joseph Kwaku Kidido [1],* and Monica Lengoiboni [2]

[1] Department of Land Economy, College of Art and Built Environment,
Kwame Nkrumah University of Science and Technology, Kumasi +233, Ghana
[2] ITC—Faculty of Geoinformation Science and Earth Observation, University of Twente, 7500 AE Enschede,
The Netherlands; m.n.lengoiboni@utwente.nl
* Correspondence: jkidido@yahoo.co.uk; Tel.: +233-(0)-242-523-182

Received: 11 November 2019; Accepted: 3 December 2019; Published: 5 December 2019

Abstract: Building inclusive societies that reflect the needs of all categories of people within the social spectrum is critical to achieving sustainable development. This is reflected in the Sustainable Development Goals (SDGs) which among things seek to 'by 2030, empower and promote the social, economic and political inclusion of all, irrespective of age, sex. This places enormous tasks on all governments especially in developing countries like Ghana to ensure that the youth are not left behind in access and control over land as a building block for economic empowerment. This task is particularly critical in view of the sheer numbers of the youth and yet economically marginalized underpinned by high levels of unemployment and underemployment. This case study investigates the youth land rights within the context of household landholdings and allocations dynamics. The study took place in the Techiman area in Ghana. The study sampled 455 youth and 138 household heads. The study revealed that household lands are important building block for majority of the youth in the Techiman area. It gives them a sense of security in the usage. However, the youth's ability to depend on this source to kick start independence economic life is beset with land scarcity, non-allocation and accumulation by the lineage heads who have prerogative over household lands. The study underscores the need for social welfare scheme for the aged farmers so that they can timely release land to the younger ones without fearing for what to sustain them. There is also the need for government to create land banks to support the willing youth to engage in agriculture.

Keywords: youth; household; land; access; use; Techiman

---

1. Introduction

Leaving no one behind in the development process is critical to building inclusive societies and achieving sustainable development. This is reflected in the Sustainable Development Goals (SDGs) as set by the United Nations (UN) member's countries in 2015. Goal 10 of the SDGs "Reduce inequalities within and among countries" among other targets seeks to ensure that "by 2030, to empower and promote the social, economic and political inclusion of all, irrespective of age, and sex" [1]. Again, Goal 1: "end poverty in all its forms everywhere", targets poverty reduction at least by half the proportion of all persons including youth and children living in poverty. Furthermore, this goal also targets equal rights to economic resources as well as ownership and control over land by all. As noted by International Labour Organization (ILO), the 2030 agenda for sustainable development 'emphasizes the catalytic power of youth employment in poverty alleviation, economic growth, and peace and prosperity for all" [2]. Indeed, one of the targets under the Goal 8, for instance, envisions to by 2030

" ... achieve full and productive employment and decent work for all women and men, including for young people and persons with disabilities, and equal pay for work of equal value" [1].

This places enormous task on all governments especially in the developing countries like Ghana to ensure that the youth are not left behind in access and control over land as a building block to move out of poverty. This task is particularly critical in view of the sheer numbers of the youth and yet economically marginalized underpinned by high levels of unemployment and underemployment. The youth employment rate globally was estimated at 13.1% in 2016 [3]. This unfortunate situation is even more worrying in developing countries where the youth unemployment rate reached 13.7% in 2017 [2]. In the specific case of Ghana, the youth unemployment rate was estimated at 12.1% in 2015 [4]. Generally, the youth are three times as likely as adults to be unemployed [2]. Given this precarious unemployment situation of the youth, their poverty levels are also high. It was estimated in 2016 that there were about 156 million youth in developing countries who lived in extreme poverty of less than US$1.90 per capita per day [3]. Perhaps the more worrying development is the high incidence of working poverty among the youth compared with the working adults. Incidence of working poverty among the youth in 2016 was estimated at 37.7% compared with 26% among the adults [3]. Clearly, the youth lag behind in almost all key economic and social indicators of development. Abink aptly opined that to be youth, especially in Africa, connotes been disadvantaged, vulnerable and marginalized [5].

The youth however, constitute an important human resource base for rapid economic transformation and the realization of the sustainable development goals. Their sheer numbers as well as their exuberance energies are needed to spearhead economic growth. The youth population is estimated at 1.9 billion and close to 90% are in developing countries [6]. The large youth numbers are a demographic dividend of resources [7] which can positively be exploited to drive development. The physical strength of youth is an essential driving force of development [8]. In the wake of an aging global population, the position of youth in the development process especially in the developing countries is immensely important. ILO reports that, in 2015 the youth between the ages of 15–24 constituted 16.2% of the total population, while adults aged 65 or older amounted to 8.3% [3]. In Ghana's agricultural sector for instance, the farming population is already aging. According to the Ministry of Food Agriculture (MoFA), the average age of farmers in Ghana is about 55 years [9]. Youth are therefore indispensably required in the development process and are not to be passive participants. Given adequate resources, the youth have the potential to effectively transform the world and make it a better place [6].

Interest in youth economic welfare and, in particular, their access to productive resources such as land serves a number of purposes. Empowering the youth through creation of employment opportunities and access to resources help to alleviate poverty among the youth themselves as well as create the needed platform for them to deploy their energies and talents to the economic development process. This will ensure inclusive development and sustainable growth into the future. Many policy options exist to create channels for the youth to be empowered and uplifted from extreme poverty. Among these options include; skills development; macroeconomic policies, new technology to increase youth access to credit, creating a global alliance, creating diverse forms of employments, among others [2]. Ownership and control over land as captured by Goal 1 of the SDGs and is critical in creating diverse forms of employment opportunities especially employment pathways in the agricultural sector.

Land is a key productive resource [10] and in agrarian economies such as Ghana, access to land is critical towards building economic livelihoods from the agricultural sector. However, in Ghana land is a highly contested resource at all levels; household, family and even larger community level (see [11]). These contestations are deeply rooted in the tenure arrangement which vests authority and control over land in the older generations. This tends to be abused by the customary authorities who often solely benefits from the proceeds [12,13], allocate to the 'outsiders' to the detriment of the youth [14,15]. Rapid urbanization and large land acquisitions also undermine youth land access effort [14,16]. The youth are increasingly finding it difficult to access land for farming on their community-owned land [17]. There is the need to fully appreciate the nature of youth land problem within this highly contested landholding environment. Some amount of evidence on youth land access and control under the customary land ownership regime in Ghana exist. For instance, early works by Kidido et al. in the Techiman Traditional of Ghana revealed that youth land sizes are generally small occasioned by a combination of factors including cost and socio-cultural issues [14,18]. They also noted that there is limited legislative support to the youth to assert claims to land under the customary tenure regime in Ghana [19]. Again, studies on changing family land relations and agriculture commodification in Ghana by Amanor have partly touched on the youth land rights. According to Amanor, agriculture commercialization and land contestation at the family level in south east and western parts of Ghana have undermined the youth access to land. Consequently, many of the local youth had abandoned farming to seek alternative livelihood opportunities in illegal mining activities, chainsaw operations and migration to the urban centers [15,20]. Boni's work in the Sefwi area of Ghana also focused on authoritarian interpretations by chiefs in land disputes and the local youth and migrants' farmers land rights [21]. She noted that the chiefs and elders' adjudication of land disputes and interpretation of customs worked against the land rights of the local youth and migrants.

The main objective of this paper is to examine households' landholdings and distribution and explore the nexus with youth land access. The study is guided by the following research questions: How are households' land held and allocated among the members? To what extent are the youth land access for agricultural purposes linked to their household land stock? What are the youth's perception on the security of their landholdings? It is important to appreciate the extent to which the youth rely on their households for land for farming purposes and their perception of security in the use of acquired lands. Households are the basic economic unit of communities and are mostly headed by the older generation who wield greater authority over productive assets including land under the customary tenure system. Thus, a clear appreciation of allocation authority, distribution pattern and the nexus with the youth land rights is imperative. As noted by White, research into the relations between the older generation and the young people (the youth) who desire to receive their share of land to set independent farms and households have been neglected [22]. The findings of this paper offer a further understanding of the youth land problem towards mapping out a holistic remedial intervention to make land accessible to the youth who are rightfully placed to take over from older lineage members.

Conceptual Framework of the Study

Understanding the youth land rights and access at the household level needs to be viewed from the perspective of intergenerational relations in land. Generation is a social structure that is seen to distinguish the youth from other social groups and to constitute them as a social category [22]. The concept establishes 'relations of division, difference and inequality between categories' such as between youth and adults [22] (p. 2). Youth as a social category within the larger social structure lie within the households. Inequality or otherwise of their access to land at the household level can be appropriately analyzed from the perspective of social category discourse (i.e., adult and youth/children relations). Thus, relations in land between the older and younger generation provide the context of investigation and analysis of the issues to understand land rights allocations and access dynamics of youth at the household level.

Intergenerational relations encompass the transmission of resources including land across different generations. Whyte et al. described intergenerational relations as a 'reciprocity' (i.e., a sense of mutual dependence expressed in give and take over time) [23]. According to them, intergenerational relation has two main features; transmission of resources (material and immaterial) and the imbued assumption of morality. The morality aspect of the intergenerational relations, according to them, is borne out of 'intergenerational contract' "the implicit expectation that parents will care for their children until they can care for themselves, and that children will support their parents when they can no longer support themselves, is a moral obligation" [23] (p. 7). This 'intergeneration contract', is not a legal contract but rather based on the 'logic of debt' and includes the idea that parents raise their children as their creditors, the children later fulfilling this debt … " [24] (p. 50). According to Boersch–Supan, intergeneration contract functions at both private and public levels as well as micro and macro levels. She again noted that, the exchanges are anchored to varying degrees of cultural and historical distinctive customary norms and moral obligations [8]. Thus, intergeneration relation is also viewed as a moral obligation where the older generation is expected to support the younger generation who will intend to support the former at a later time when the tables of strength turns.

Focusing on the resource transmission aspect of the intergeneration relations, which is significant to the present discourse, it is usually viewed as an individual transaction where one bequeaths or gives something of value to an individual in the next generation [25]. In the context of land, it is the subsisting land relations between the senior generation and the younger generation. Ownership of customary land is intergenerational [26]. Customary relations in land transcend different temporal regimes of life. This was aptly captured in the words of a renowned Ghanaian Chief, Nana Ofori Atta, the paramount chief of Akyem Abuakwa; "I conceive that land belongs to a vast majority of whom many are dead, a few are living and a countless host are still unborn" [27] (p. 4). In this sense, land is held by a generation and transmitted onto the next generation and the process continuous. The flow of movement is from the elderly who are normally in possession of this asset (land) to the younger generation and subsequently to the unborn. Resources always flow from the elderly generation down to its descendants [28] cited in [29] (p. 268). Kohli called this process evolutionary or sociobiological theory.

In this intergenerational land relations arrangement, the older members of the lineage wield superior influence and control over younger members in the allocation of land rights. Access rights are allocated based on social identity and group or community membership [30,31] and seniority [32]. As age signifies effort invested in sustaining the younger members [33], it creates authority in favor of the older generation. The local structure of patriarchy society gives the older generation control of land resources and emphasis on respect for the elderly [34]. Land is often owned and controlled by the elderly especially men in traditional African societies [25,35] and by household heads on the assumption that the rights are held in trust for all in the household [36]. The management and decisional authority regarding land rest entirely in the hands of the elderly [25]. In Ghana, it is observed that elders often control youth access to land and redistribution of land from senior generation to the younger generation [15] and rather alienate land to migrant farmers [20,21]. Intergenerational relation and its associated solidarity are a global phenomenon. However, societies differ in the form and substance of their intergeneration contract [37]. Thus, intergenerational relations concept as employed in this study seeks to understand the power balance in terms of landholding between the older generation and the younger members within the Ghanaian social structure. The intergenerational land relation was considered by analyzing households land acquisitions, land stocks and distribution among members to understand connections with the youth landholdings. It further offered the platform to ascertain the extent to which the youth depend on their households for land and the underlying dynamics.

2. Materials and Methods

The study was based on a case study survey of youth and households in the Techiman area of Ghana. The approach made it possible to engage the youth and their household heads on matters relating to households' land access, holdings and rights allocation dynamics. It is noted by Casley and Lury, that studies requiring in-depth probing into systems governing behavior and interrelations between people and their institutions are best done using case design [38]. To fully appreciate the household landholdings and the youth land access nexus within limited time and budget constraints, it was imperative to proceed based on case study sample survey approach. As case studies focuses specific issues or unit of analysis [39], the primary units of analysis in this study were the households and the youth within the context of land rights allocations. The study was descriptive based on the quantitative data from both the youth and household heads respondents.

2.1. The Study Area Location

The Techiman l area is located within the central part of Ghana (see Figure 1). It covers two administrative districts (i.e., Techiman Municipality and Techiman North district). The strategic location of the area and easy access to the bigger cities of the north and southern Ghana coupled with fertile arable land make the area a significant food producing and marketing center not only in the Brong Ahafo region but also in the entire country. As a contact zone and stopover point for traders from the North, the area exchanges products of two economic regions of northern and southern Ghana [40]. The vast agricultural opportunities in the area and the desire of youth in the areas to be involved in agriculture especially tomato cultivation (see [41]) informed the choice of the area for this study.

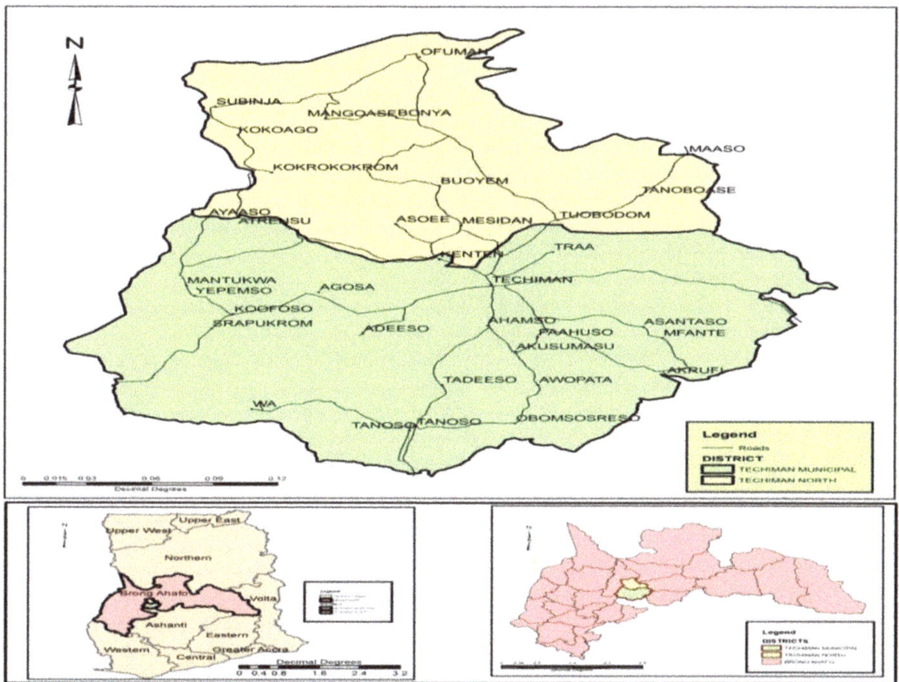

Figure 1. District map of Brong Ahafo Region highlighting the study area districts. Source: Modified from the Techiman North and South districts maps.

2.2. Defining Youth for the Study

It is important to first establish operational definition of the concept of youth. Youth as a concept is socially constructed with its meaning and boundaries vary over time and between societies [34]. It is imperative, therefore, to clearly define the concept for the purpose of this study.

This study defines youth as both male and female within the active workforce of 15 up to 34 years. Under section 89 of the Children's Act 1998 (Act, 560) the minimum age for admission of a child into employment in Ghana is 15 years. This study thus defined youth in line with the statutory prescription of when a young person is permitted to engage in productive economic activity in Ghana. The Ghana National Youth Policy, 2010, pegs the upper limit for youth at 35 years. However, it was impossible to determine the number of youth who were 35 years old from the Ghana Statistical Service (GSS) database. This is because, GSS categorizes people who are 35 years in the age band of 35–39 years. It was thus appropriate to align the upper age limit of youth with the GSS categorization to be able to use their census data to for sampling purposes.

2.3. Sampling of the Respondents

These youth respondents were first sampled based on the criteria of involvement in on-farm agricultural activities. The study targeted the youth involved in agriculture (on-farm) because involvement in on-farm agricultural activities whether as a primary or secondary occupation exposes them to land access issues and were thus considered as appropriate respondents to engage. The youth as a social group lie within the household structure. Thus, the household was the key reference point for the sampling of the youth. Households provide appropriate means of identifying and selecting subsets of the population to whom a research is addressed [38]. As noted by Chauveau et al., 'household remains the basic unit of reference in terms of access to and use of productive resources' [33] (p. 29). In locating the households, houses were used as the reference point and were randomly sampled based on the zones created by moving along the settlement patterns and streets.

A tracer survey was done on a sample of households from which the youth respondents were selected. The number of youth respondents interviewed in this study formed the sample frame from where 138 households were drawn. A simple random sampling technique was used to select the household heads for interview. The house numbers of the youth interviewed were taken during the interview process, so it was easy tracing back to their household heads. In the sampling process where a household head was picked and happened to be a youth already interviewed, that sample was dropped in favor of a non-youth household head. In all 455 youth, 138 household heads were covered in the survey (see Table 1).

Table 1. Number of respondents in the various communities in the Techiman Area.

Community	Youth	Household Heads
Hansua	20	8
Krobo	27	8
Tuobodom	89	26
Nkwaeso	20	7
Bamiri	12	4
Twimia-Koase	21	6
Mesidan	8	4
Sansama	13	4
Kuntunso	12	5
Aworowa	39	12
Nsuta	18	6
Buoyem	14	5
Tanoso	53	14
Adieso	5	2
Tadieso	11	3
Amangoase	9	3
Tanoboase	6	3
Offuman	51	10
Kokroko	5	2
Fiaso	22	6
Total	455	138

Source: Fieldwork, 2015–2017.

Thus, the processes and power relation governing land ownership and access dynamics required engagement of key actors. The household heads and the youth engaged in on-farm agricultural activities household lands were identified as the right actors for this research. The household head plays a significant role in the ownership and transmission of land rights in line with customary practices and norms at the household level. They were thus an important source of households' landholding information which was to be related to that of the youth.

2.4. Data Collection and Analysis Techniques

Personal interviews were carried out with the youth and the household heads using two set of structured questionnaires each for the youth and their household heads. The questionnaires in this survey were closed ended. The major themes of the youth respondents' questionnaire included; demographic information, land access sources, and land holding size, among others. Similarly, the household heads questionnaires sought data on demographic characteristics, land acquisition, land allocation authority, size of holdings among household members. Indeed, the current paper emanates from a bigger study whose field works commenced in 2015 and formally ended in January 2017. The main respondent groups of the survey namely the youth and household heads were each covered at different time lines due to time and resource constraints. The survey covering the youth took place between May 2015 and January 2016. The tracer survey of the household heads started in May 2016 and ended in January, 2017. The data on the youth respondents especially land sizes, access dimensions and legislative issues have already been published elsewhere (see [14,18,19]). However, data on the households are published for the first time in this paper. It was analyzed in relation with the data on youth landholdings.

Statistical Package for Social Sciences (SPSS) software was used to analyze the data from the youth and their household heads. The variables in the questionnaires (both the youth and household heads) were first coded and entered into the SPSS software. Results are presented in form of tables, graphs and charts as captured in the next section.

3. Results

The results of the study are presented in this section in two parts. The first part of the results relates to the demographic characteristics of the households interviewed as well as the youth respondents. The summary of the demographic characteristic is captured in Tables 2 and 3. The second part of the results focuses on household land ownership, allocation dynamics as well as youth land access and household landholding nexus. Details of the results are presented as follows.

3.1. Background of Household Heads Respondents

Since the youth respondents are members of the social structure and are located within the households, it is important to reflect on some key characteristics of the households of which they are members. The majority (58%) of the households were headed by males while female headed households constituted 42%. Majority of the household heads (34.8%) were between 51–60 years and those above 60 years were 31.2%. The data clearly shows that majority of the household heads were above 50 years. Farming constituted the primary occupation for 94.2% of the households. About a third of the households (32.6%) had between 6–8 members. Details of the demographic characteristics of the households are contained in Table 2.

Table 2. Selected demographic characteristics of household heads respondents.

Characteristic	Response	Percentage (%)
Gender		
Male	80	58
Female	58	42
Total	138	100
Age		
30–40 years	10	7.2
41–50 years	37	26.8
51–60 years	48	34.8
Above 60 years	43	31.2
Total	138	100
Community membership status		
Indigene	104	75.4
Migrant	34	24.6
Total	138	100
Household Size		
Below 3	3	2.2
3–5	31	22.5
6–8	45	32.6
9–12	29	21
13–15	12	8.7
Above 15	18	13
Total	138	100
Primary Occupation		
Farming	130	94.2
Running a business	3	2.2
Government work (public service)	3	2.2
Other	2	1.4
Total	138	100

Source: Fieldwork, 2016–2017.

3.2. Background of Youth Respondents

The youth respondents were made up of 299 (65.7%) males and 156 (34.3%) females. The wide disparity in terms of gender representation can be explained by the fact that there were more males involved in farming and thus qualified to participate in the survey than the females. In terms of age, the majority of the respondents—186 (40.9%) fell within the age range of 30–34 years which was the upper limit of the youth respondents. Those in the age category of 15–19 years which constituted the lower limit recorded the least number of respondents 24 (5.3%). People within this age group are normally in school and not in occupation of any agricultural land or actively involved in farming. Consequently, per the survey criteria, many of them were not qualified to participate in the survey, hence their low representation in the sample.

The majority of the respondents 268 (58.9%) were married. Those who were single constituted 176 (38.7%). In terms of occupation, the majority—410 (90.1%) were involved in farming as their primary occupation. The remaining respondents were involved in other professions such as running non-agricultural businesses, schooling, laborer work and the public service, and a few were involved in farming as their secondary occupation. The details are shown in the Table 3 below.

Table 3. Demographic characteristics of the youth respondents.

Characteristics	No. of Respondents	Percentage (%)
Gender		
Male	299	65.7
Female	156	34.3
Total	455	100
Age		
15–19 years	24	5.3
20–24 years	88	19.3
25–29 years	157	34.5
30–34 years	186	40.9
Total	455	100
Community membership status		
Indigene	278	63.1
Migrant	168	36.9
Total	455	100
Marital Status		
Married	268	58.9
Single	176	38.7
Divorced/separated	9	2
Widowed	2	0.4
Total	455	100
Primary Occupation		
Farmer	410	90.1
Agricultural wage laborer	4	0.9
Non-agricultural wage laborer	5	1.1
Self-employed outside farm work	23	5.1
Student	7	1.5
Public/civil servant	4	0.9
Others	2	0.4
Total	455	100

Source: Fieldwork, 2015–2016.

3.3. Households Land Ownership and Allocations Dynamics

Among the households covered by the study, the majority—100 (72.5%) out of the 138 owned land acquired through gift, inheritance and appropriation through community membership (Table 4). Close to a third of the households—38 (27.5%) did not own the lands they held. These were mainly the migrant households who rented, engaged in sharecropping and operated under customary licence. As shown in Table 4, the dominant land access modes through which the households accessed land were gift (31.9%) and inheritance (23.9%). Rentals and customary licence were unpopular among the households.

In terms of size of landholdings, as depicted in Figure 2, the majority of the households—29 (21%) held lands sizes above 15 acres, which, on the face of it does not suggest an extreme land scarcity. However, this may not be enough indicator of real land abundance until it is appreciated within the context of how much of the land available to the households has been appropriated or cropped. For instance, households with land area above 15 acres may well have all their land fully utilized by the household members and even tenants. It is thus essential to further ascertain the level of use and occupation of the lands available to the households in order to appreciate the dynamics of allocation and size of holdings by members especially the youth.

Table 4. Households land ownership and acquisition mechanisms.

Issue	Response	Percent (%)
Own Land?		
No	38	27
Yes	100	73
Total	138	100
Access Mode		
Purchase	4	3
Gift	44	32
Sharecropping	22	16
Customary license	7	5
Inherited	33	24
Usufruct (through community membership)	18	13
Rented	10	7
Total	138	100

Source: Fieldwork, 2016–2017.

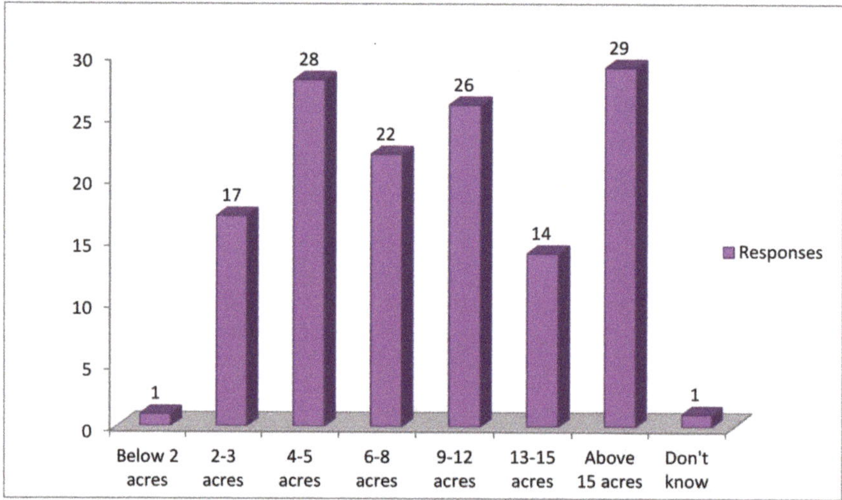

Figure 2. Landholding sizes of the households interviewed (n = 138). Source: Fieldwork, 2016–2017.

Figure 3 depicts the level of utilization of lands the households held. A majority (45%) of the households interviewed described the size of their lands under cropping or occupied by other users as "more than half is occupied or under use". These households have more than half of their land stock under occupation. Again, 24% of the households also indicated that only a "less than half" of their land stock was occupied or in use at the time of the survey. These households had excess land available for either rentals or expansion.

Furthermore, a close to a third of the households (31%) had all their land stock fully utilized (see Figure 3). For these households, any member who requires additional land to either expand or establish a new farm will have to look outside of the household for land. While the land available to these households may just be enough for their usage, it should be appreciated that, they are faced with plausible land shortage as no portions of their land stock were idle. The need to expand their farm sizes or make allocation to new members especially to their youth will be challenging The youth desiring to set up their own independence farms will have to depend on external land market for land supply or wait to be bequeathed a portion when the older members pass away.

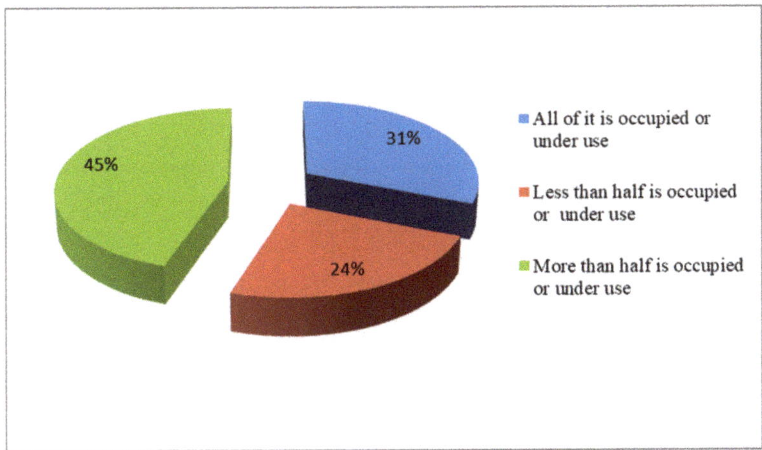

Figure 3. Amount of household lands cropped or utilized (n = 138). Source: Fieldwork, 2016–2017.

It is instructive to indicate that, generally, among the majority of the households, there was some land excess which could be released to those in need especially the young members (the youth). This was possible among the households who had less than half (less than half is occupied or under use) and those with more than half (more than half is occupied or under use) of their lands under cultivation. The data as depicted in Table 5 further revealed that majority of the households (45.5%) with excess land (less than half is occupied or under use) were those whose heads were above 60 years. Land shortage was pronounced among households whose heads were active and between the age ranges. The majority of households whose heads were within the age range of 41–50 years (37.2%) and 51–60 years (39.5%) had all their land stock occupied or under use (see Table 5). These household heads are within the active labor age and perhaps in a position to fully utilize all their available stock compared with those above 60 years. It suffices to indicate that, the observation in the level of utilization of land among the various age categories of household heads was less significant (chi-square value = 11.520, df = 6, p-value = 0.056 at 95 per cent confidence level and a margin of error of 5 per cent). However, the data still points to the fact that older household heads were more likely to preside over relatively bigger land parcels that were not fully utilized compared to the household heads who are more active and below the age of 60 years. It is shows land accumulation by the older members who are privileged to possess relatively more land. As depicted in Table 5, the majority of the households (51.7%) with land sizes exceeding 15 acres were those whose heads were above 60 years. Again, 42.9% of the households with land sizes (13–15 acres), (i.e., the next biggest land size category) had their heads above 60 years. Thus, households with older heads (above 60 years) held relatively bigger land parcels compared with those below the age of 60 years. This relationship between the age of household heads and the size of landholding was statistically significant with a p-value of 0.001 (see Table 5).

In suffices to state that, given that population has increased over the years which has consequently altered the African farming practices of land rotation and shifting cultivation, it is no more an issue of the households using their excess lands for fallowing purposes but rather to ensure equitable allocation to their members. Soil fertility is now sustained based on the application of fertilizer and other modern farming techniques and not based on land rotation. Therefore, the question of allowing a portion of one's farmland to fallow does not arise. The question to ask is; how is the land held by the households shared or allocated among the members, and in particularly, how much is in the hands of the younger members who possess vitality and energy to effectively till the land? Since a household is composed of many members, this question can be answered by appreciating who within the household controls how much of the household land. As depicted in Table 6, an overwhelming number of the households,

108 (78.3%) had their heads holding the largest portion of their land stock. However, less than a quarter of the households, 15 (10.9%) have their youth holding the larger portions of the household land.

Table 5. Analysis of amount of household lands cropped or utilized, size and age of household heads.

Age of Household Heads	*Land Utilization*		
	All of It Is Occupied or Under Use (%)	More than Half Is Occupied or Under Use (%)	Less than Half Is Occupied or Under Use (%)
30–40 Years	1 (2.3)	7 (11.3)	2(6.1)
41–50 Years	16 (37.2)	17 (27.4)	4 (12.1)
51–60 Years	17 (39.5)	19 (30.6)	12 (36.4)
Above 60 Years	9 (20.9)	19 (30.6)	15 (45.5)
Total	43 (100)	62 (100)	33 (100)

Age of Household Heads	*Land Size*							
	<2 acres	2–3 acres	4–5 acres	6–8 acres	9–12 acres	13–15 acres	>15 acres	Don't know
30–40 Years	1	0	2 (7.1)	1 (4.5)	2 (7.7)	0	4 (13.8)	0
41–50 Years	0	8 (47.1)	10 (35.7)	11 (50.0)	3 (11.5)	3 (21.4)	2 (6.9)	0
51–60 Years	0	5 (29.4)	12 (42.9)	4 (18.2)	13 (50.0)	5 (35.7)	8 (27.6)	1
Above 60 Years	0	4 (23.5)	4 (14.3)	6 (27.3)	8 (30.8)	6 (42.9)	15 (51.7)	0
Total	1	17 (100)	28 (100)	22 (100)	26 (100)	14 (100)	29 (100)	1

Source: Fieldwork, 2016–2017. Age and land size relationship significance: (chi-square value = 45.518, df = 21, p-value = 0.001 at 95 per cent confidence level and a margin of error of 5 per cent.)

Table 6. Landholding pattern at the households.

Issue	Response	Percentage (%)
Who Holds the Largest Portion of the land?		
Household head	108	78.3
Other senior members	8	5.8
Young family members (youth)	15	10.9
Tenants (outsiders)	5	3.6
None	2	1.4
Total	138	100
Land Size of the Person Holding the Largest		
< 1 acre	3	2.2
1–3 acres	25	18.1
4–6 acres	42	30.4
Above 6 acres	65	47.1
Don't know	3	2.2
Total	138	100

Source: Fieldwork, 2016–2017.

Again, the majority of the household members holding the largest portion of the household land stock, alone held above 6 acres (see Table 6). These members were mainly the household heads. It thus suggests that land concentration is in the hands of the household heads. There is some element of uneven allocation of land among all household members as the household heads alone control about half in most of the households. As indicated in Table 2, the majority (32.6%) of the households had between 6–8 members including the youth.

Land concentration in the hands of the household heads, as shown above, has to do with authority. This authority is derived from the fact the heads expended resources to acquire household land. As presented in Table 7, the household heads were mainly the ones providing consideration in the acquisition of the household lands. For instance, out of the 72 households which provided some form of consideration (either monetary or in-kind) in their land acquisition process, 57 (79.2%) had their consideration provided solely by their heads. Only in one household it was reported that the consideration was a contribution from the younger members. Clearly, investment in the acquisition of household land largely rests on the shoulders of the heads and this also gives them authority in terms of control and allocation. As shown in Table 7, the authority to allocate household land is largely exercised by the household heads. This gives them the opportunity to appropriate and hold a greater portion.

Table 7. Consideration provision and land allocation authority at the household.

Issue	Response	Percentage (%)
Nature of consideration provided in acquiring land		
Monetary	23	16.7
In-kind/material	33	23.9
Monetary and in-kind	16	11.6
None	66	47.8
Total	138	100
Who provided the consideration?		
Household head	61	84.8
Contributions of other senior household members	6	8.3
Contributions by young household members	1	1.4
Others	4	5.6
Total	72	100
Who allocates household land?		
Household head	103	74.7
Other senior members	14	10.1
Collectively by all members	1	0.7
Self-allocation	20	14.5
Total	138	100

Source: Fieldwork, 2016–2017.

3.4. Youth Land Access and Household Land Holding Nexus

This section focuses on youth land access linkages with their households' land stock and perception of tenure security. It is instructive to indicate that data on the youth landholdings, ownership and access modes in the Techiman area are already published (see [14,18]). It is noted that majority of the youth in the Techiman area hold small land sizes less than 3 acres [14] and on a temporary basis through rental and customary license [18].

It must be noted that household lands remain important to the youth and primary source of access for majority of them. As depicted in Figure 4, the majority—246 (54%) of the youth respondents had all the lands they occupied from their households' land stock. A few respondents, 6% and 1% partially occupied portions of their households' lands described as 'more than half' and 'less than half' respectively. This still suggests that household land stock is crucial for youth agricultural activities.

However, land accumulation by the household heads coupled with scarcity compelled a good number of their youth to access land from 'outside' through the land market. A considerable proportion of the youth, 177 (39%) did not occupy any portion of their households' lands (see Figure 4). This group of respondents acquired their lands outside their household land stock.

The dominant reason accounting for these respondents not accessing or using any part of their household lands was lack of land by their households. As presented in Table 8, 120 (68%) out of the 177 respondents who did not occupy any portion of their household/family land, indicated their household/family had no land from which they could acquire a portion. This was particularly so among the migrant youth whose households did not own any land. Out of the 120 who mentioned that their households had no land, 111 (93%) were migrants.

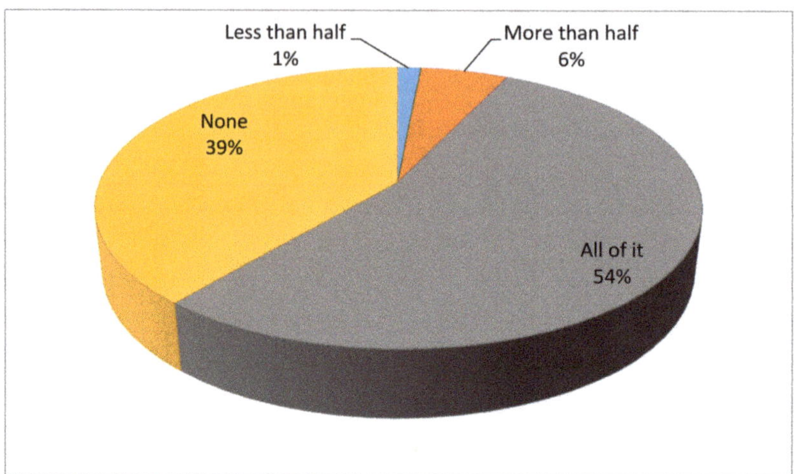

Figure 4. Proportion of youth landholding from their households' land stock (n = 455). Source: Fieldwork, 2015–2016.

Table 8. Reasons for acquiring land outside the household or family land stock among some youth respondents.

Reasons for Acquiring Land Outside Household/Family	Responses	Percentages (%)
Household/family had no land	120	68
Not allocated any portion of household	6	3
Household land fully occupied by senior members	30	17
Household land not suitable/fertile	13	7
Others	8	5
Total	177	100

Source: Fieldwork, 2015–2016.

A considerable number of the youth respondents 30 (17%) also mentioned that 'their household lands were already occupied by their senior lineage members' which have deprived them the chance of accessing a portion. Other reasons included; 'not allocated any portion of their household/family lands'; 'household lands not suitable for the intended agricultural activities' (see Table 8). It can be concluded that the key reasons accounting for some youth respondents depending on external suppliers were landlessness and scarcity (Table 8).

It is important to note that the youth generally felt more secure on lands accessed from their households than from lands accessed from outside their households. For instance, 43.1% of the respondents who had all their land from their households felt 'very secure' and 48% of those who had "more than half" of their lands from their households also felt 'very secure' compared to 18.6% among those who accessed their lands entirely outside their households (Table 9).

Moreover, as depicted in Table 9, there is more insecurity among the youth who depend on land outside their household/family land stock compared to those who wholly or partially depend on their households. For instance, 18.6% of those not holding any portion of their household lands felt insecure compared to 11.4% among those who had their lands from their households. This difference is statistically significant with a p-value of 0.000 (see Table 9). It can be concluded that, households provide more secure landholding rights for the youth than lands they access outside of their households' land stock or from the land market.

Table 9. Youth tenure security and source of landholdings.

Perception of Security	How Much of Your Land Is from Your Household?				
	All of It (%)	More than Half (%)	Less than Half (%)	None (%)	Total
Very secure	106 (43.1)	12 (48)	1 (14.3)	33 (18.6)	152
Secure	93 (37.8)	11 (44)	5 (71.4)	75 (42.4)	184
Less secure	19 (7.7)	1 (4)	0	36 (20.3)	56
Insecure	28 (11.4)	1 (4)	1 (14.3)	33 (18.6)	63
Total	246 (100)	25 (100)	7 (100)	177 (100)	455

Source: Fieldwork, 2015–2016. Notes: chi-square value = 44.650, df = 9, *p*-value = 0.000 at 95 per cent confidence level and a margin of error of 5 per cent.

4. Discussion

The key reasons accounting for the youth whose accessed land outside their household were lack of land by the households and appropriation by the senior lineage members. Those who accessed land outside their households due to the fact their households do not have lands were mainly migrant households. For this youth, their supply pathways through the land market mechanisms through rentals [18]. In the case of youth from households which owned land, appropriation by senior lineage members forced a good of number (17%) to depend on the outside suppliers or the land market. The results further revealed that, among the majority of the households the heads held more land relative to other members with their land sizes exceeding half of their households' land stock. Thus, there is some element of land concentration in the hands of the household heads and close to a third of these heads were already old above 60 years. This result reinforces the dominance of household heads in the land holding arrangement under the customary tenure regime in Africa. For instance, Hill earlier reported of a similar land concentration in the hands of the older household members in Ghana [42]. Similarly, a study by André and Platteau in the Gisenyi area of Northwest Rwanda showed land concentration in the hands of the older household heads while the younger members held small land areas [43]. Indeed, the results also reflect the superior influence and control the older lineage wield over the younger generation in the allocation of land rights. The elders have a myriad of prerogatives over land [25]. Control over land at the household level under the customary tenure system is about seniority [32] and age [21] as shown in this study where older household heads had relatively bigger parcels than the more active and younger household heads. Among the majority of the households surveyed, the heads exercised the power over land including land allocation among household members as well as to non-members. This authority vested in the household heads is derived from the fact that they principally lead in the acquisition and disposition of land. As shown in the results, the household heads were mainly the ones providing the consideration in the acquisition of the household lands. This clearly gives them power and leverage in the use of land at the household level. The youth by virtue of their weak financial standing who are unable to contribute much towards household's land acquisition wield very little influence and control over land.

Again, as 'early comers' the elders (household heads) had the privilege of appropriating portions of unoccupied community land for farming purposes as much as their resources and effort would permit. For instance, a considerable number of households acquired their lands through appropriation of community land in the previous decades when more vacant lands were available. This also gives the heads who were actually the ones acquiring and expending the necessary resources prerogatives over those land. In the Sefwi area of Ghana, Boni also found that the elders through forest clearing in the previous decades accumulated prerogatives over large tracks of land while the youth faced land shortages [21]. As 'late comers', the youth have limited say over their households' lands. This is because they missed the privilege of also directly accessing land through appropriation of vacant community lands as there is no more vacant community lands. Virtually all the community arable lands have been reduced to private ownership by the various families and households. Thus, the youth principally access land in the secondary context through the medium of customary access modes such as gift, inheritance and customary licence from their households or through the market mode of rental, sharecropping and purchase (see [14,18]). The youth who depend on their households tend to get small parcels whiles those who participate in the land market as young people lack financial wherewithal to access viable land parcel's for agricultural purposes [14]. Thus, the combine effects of land concentration in the hands of older generation, general land scarcity and lack of finance undermine youth land access effort.

Customary land is regarded as intergenerational property [26,27]. The period of transition from the current older generation to the younger generation should be driven on the wheels of fairness and moral obligation. As noted by Whyte et al, land relation between older and younger generations is borne out of an 'intergenerational contract' [23]. This 'intergeneration contract' is based on the 'logic of debt' where older generation help raise up the young ones who will later assist them in their old age. Thus, the elders who are currently in possession of land have the intergenerational contract with their youth to support them with land and perhaps resources to become fully-fledged community members. Where land is overly concentrated in their hands and the youth are rather offered small parcels or faced to look for their own land elsewhere, then the intergenerational contract is breached. As such, if the youth succeed on their own without the express support from their elders, the imbued expectation that the youth will support the elders when they become weak is unlikely to occur. This often creates intergenerational tension. As noted by Kidido et al., youth are normally tagged as 'irresponsible' people who do not care about the welfare of their older family members or the welfare of the family [14]. Of course, if the 'old man' refuses to give his son land to farm but rather grants the land to tenants for money, why would the young man respond to his needs? This tension in land transmission and sharing between generations is a common feature in agrarian societies [22,44].

Household lands remain an important building block for majority of the youth in the Techiman area. It gives them a sense of security in the usage. However, the youth's ability to depend on this source to kick start independence economic life is beset with land scarcity, accumulation by the lineage heads who have prerogative of the land. In the past, it was the case that "each headsman saw to it that all members of his lineage [had] portions to farm" [45] (p.102). Land was largely efficiently distributed among the people [46]. As opined by Kasanga, "no man was 'big' or 'small' in his own village or town" with regard to land in Ghana [47] (p. 14). This situation no longer holds because of land scarcity occasioned by the increase in population. It is also seen in this study by way of land accumulation by the elders. The moral economy of the family under the customary system which in past was egalitarian is crumbling.

5. Conclusion and Recommendations

The study has revealed that about a third of households surveyed had all their available lands fully occupied, an indication of potential land shortage. These households do not have sufficient land for further expansions or make allocations to their youth. However, there was excess land in some households with some even lying idle. The study also revealed an element of power in-balance heavily tilted towards the older folks in the matter of land rights within the households. For instance, among the majority of the households, heads held greater proportions of the lands while other members including the youth held relatively small parcels. The authority of household heads to hold much land is derived from the fact that they expended resources in the acquisition of the household lands. The results showed that among the majority of the households, the heads provided the needed consideration during the acquisition of land.

There is the need for fair and equitable access to land to ensure inclusive economic prosperity. States are also required to take measures aim at preventing land concentration and abuse of customary forms of tenure which impacts negatively on the vulnerable groups such as the youth [48]. Ending poverty and reducing inequalities at the micro level of households and families very much call for fair land rights distribution and deconcentration of holdings. In the rural and agrarian economics, land is central to economic empowerment of people. Consequently, where access regime and customary arrangement do not operate to ensure equitable distribution of rights and holdings it creates conditions for the promotion of inequalities and entrenchment of economic impoverishment of certain groups.

There is the need for support system for the aged farmers in Ghana. Farmers do not retire from their lands and even at an old age of sixty years and above they still till land to earn a livelihood as revealed by the evidence from the Techiman area. This phenomenon is not unique to the Techiman area. A report by Ministry of Agriculture also showed that the farming population in Ghana is aging with an average age of 55 years [9]. This situation creates a long period of waiting for the youth and thus amount to a sheer waste of their energies. The period of transition where control and use of land pass onto the younger generation should not occur at death of the older folks. In view of this, the study underscores the need for social welfare scheme for the aged farmers so that they can timely release land to the younger ones without fearing for what to sustain them. In the public sector arena in Ghana, people retire from active service at age sixty. Why do farmers whose work demands physical strength and energy retire at death? A social welfare scheme arrangement was successfully implemented in Mexico which encouraged older landowners to transfer their lands to the youth [49] (p. 27). Additionally, given that a considerable proportion of households do not own any land from which their youth can depend on, the Government should consider creating land banks to support the willing youth to engage in agriculture in the Techiman area which is regarded as the food production corridor of the country. Finally, further research on the youth land problem across the different customary areas in Ghana is needed towards a future review of the current national land policy to incorporate youth land matters. The current national land policy has no specific provision on youth (see [50]). It is important to make land accessible to the youth through a multifaceted approach. As opined by the [6], given adequate resources, the youth have the potential to effectively transform the world and make it a better place.

Author Contributions: J.K.K. conceptualized the paper, collected data, analyzed and prepared original draft. M.L. assisted in writing—review, editing, organization and framing the paper.

Funding: This research received no external funding.

Acknowledgments: Special thanks to staff of Faculty of Geoinformation Science and Earth Observation (ITC), University of Twente, The Netherlands. Specifically, Prof. Jaap Zevenbergen, Liza Groenindijk, Petra Weber and the entire Land Administration Staff of the Department of Urban and Regional Planning and Geo-Information Management (PGM) for hosting one of the authors from Ghana on an exchange program where much of the write-up of the paper was made. Support from the Advancing Collaborative Research in Responsible and Smart Land Management in and for Africa (ADLAND) in organizing and financing the staff exchange program during which time this paper was drafted is duly acknowledged.

Conflicts of Interest: The authors declare no conflicts of interest.

References

1. United Nations. *Transforming our World: The 2030 Agenda for Sustainable Development*; United Nations, General Assembly, 70th session, Resolution adopted by the General Assembly on 25 September 2015, A/RES/70/1; United Nations: New York, NY, USA, 2015.
2. ILO. *Global Employment Trends for Youth 2017 Paths to a Better Working Future Global Employment Trends for Youth 2017 Paths to a Better Working Future*; International Labour Office: Geneva, Switzerland, 2017.
3. ILO. *World Employment and Social Outlook: Trends for Youth*; International Labour Office: Geneva, Switzerland, 2016; Available online: http://www.ilo.org/wcmsp5/groups/public/---dgreports/---dcomm/---publ/documents/publication/wcms_513739.pdf (accessed on 28 November 2018).
4. GSS. *Labour Force Report*; Ghana Statistical Service: Accra, Ghana, 2015.
5. Abbink, J. *Being Young in Africa: The Politics of Despair and Renewal, In Vanguard or Vandals: Youth, Politics and Conflict in Africa*; Abbink, J., Van Kessel, J., Eds.; Brill: Leiden, The Netherlands; Boston, MA, USA, 2005.
6. United Nations. Youth and Sustainable Development Goals (SDGS). 2018. Available online: https://www.un.org/sustainabledevelopment/youth/ (accessed on 28 November 2018).
7. Munthali, T. *Effective Job Creation Strategies for Developing Economies with Bulging Youth Populations*; Discussion Paper for the World Bank/IMF Annual Meetings; World Bank: Washington, DC, USA, 2010.

8. Boersch-Supan, J. The generational contract influx: Intergenerational tensions in post-conflict Sierra Leone. *J. Mod. Afr. Stud.* **2012**, *50*, 25–51. [CrossRef]

9. MOFA. *Youth in Agriculture Programme: Policy, Strategy and Sustainability*; Ministry of Food and Agriculture: Accra, Ghana, 2011.

10. Bennell, P. *Promoting Livelihood Opportunities for Rural Youth*; Paper prepared for IFAD Governing Council Roundtable: Generating Remunerative Livelihood Opportunities for Rural Youth; Knowledge and Skills for Development Governing Council—Thirtieth Session: Rome, Italiy, 2007; Available online: https://pdfs.semanticscholar.org/2842/3e9e1ef81c9f3cc3076b8ab2f6c6f3454061.pdf (accessed on 5 December 2019).

11. Ubink, J.; Amanor, K.S. (Eds.) *Contesting Land and Custom in Ghana, State, Chief and the Citizen*; Leiden University Press: Amsterdam, The Netherlands, 2008.

12. Asiama, S.O. Land Administration and Security of Tenure in Ghana-The Legal Framework. *Ghana Surv.* **2008**, *1*, 76–84.

13. Yeboah, E.; Oppong, R.A. Chiefs, Changing Trust Relations and Land Use Planning in Ghana. *J. Sci. Technol.* **2015**, *35*, 60–72. [CrossRef]

14. Kidido, J.K.; Bugri, J.T.; Kasanga, R.K. Dynamics of Youth Access to Agricultural Land under the Customary Tenure Regime in the Techiman Traditional Area of Ghana. *Land Use Policy* **2017**, *60*, 254–266. [CrossRef]

15. Amanor, K.S. Family values, land sales and agricultural commodification in south-eastern Ghana. *Africa* **2010**, *80*, 104–125. [CrossRef]

16. Kidido, J.K.; Bugri, J.T. Gender Dimensions of Youth Access to Agricultural Land Under Customary Tenure System in The Techiman Traditional Area of Ghana. *J. Plan. Land Manag.* **2019**, *1*, 81–103.

17. Boone, C.; Duku, D.K. Ethnic Land Rights in Western Ghana: Landlord–Stranger Relations in the Democratic Era. *Dev. Chang.* **2012**, *43*, 671–693. [CrossRef]

18. Kidido, J.K.; Bugri, J.T.; Kasanga, R.K. Youth Agricultural Land Access dimensions and Emerging Challenges under the Customary Tenure System in Ghana; Evidence from Techiman Area. *J. Land Rural Stud.* **2017**, *5*, 140–163. [CrossRef]

19. Kidido, J.K.; Bugri, J.T.; Kasanga, R.K. The Effects of the Intestate Succession Law 1985 (PNDCL 111) and the Head of Family Accountability Law 1985 (PNDCL 114) on Youth Access to Agricultural Land in the Techiman Traditional Area of Ghana. *Ghana J. Dev. Stud.* **2018**, *15*, 17–45. [CrossRef]

20. Amanor, K.S. Family values, land sales and agricultural commodification in Ghana. In Proceedings of the Frontier of Land Issues: Social Embeddedness of Rights and Public Policy, Montpelier, QC, Canada, 17–19 May 2006.

21. Boni, S. Traditional ambiguities and authoritarian interpretations in Sefwi land disputes. In *Contesting Land and Custom in Ghana, State, Chief and the Citizen*; Ubink, J., Amanor, K.S., Eds.; Leiden University Press: Amsterdam, The Netherlands, 2008.

22. White, B. Agriculture and the generation problem: Rural youth, employment and the future of farming. In Proceedings of the FAC-ISSER Conference Young People, Farming and Food, Accra, Ghana, 19–21 March 2012.

23. Whyte, S.R.; Alber, E.; Geest, S. (Eds.) Generational connections and conflicts in Africa: An introduction. In *Generations in Africa*; LIT Verlag: Berlin, Germany, 2008.

24. Roth, C. Shameful! The inverted intergenerational contract in Bobo-Dioulasso, Burkina Faso. In *Generations in Africa*; Alber, E., van der Geest, S., Reynolds Whyte, S., Eds.; LIT Verlag: Berlin, Germany, 2008.

25. FAO. *Linkages Between Rural Population Aging, Intergenerational Transfers of Land and Agricultural Production: Are They Important?* Sustainable Development Department (SD), Food and Agriculture Organisation of the United Nations: Rome, Italy, 1999; Available online: http://www.fao.org/sd/WPdirect/WPan0039.htm (accessed on 28 August 2012).

26. Kwapong, O. The Poor and Land: A Situational Analysis of Access to Land by Poor Land Users in Ghana. *J. Rural Community Dev.* **2009**, *4*, 51–66.

27. Ollennu, N.A. *Principles of Customary Land Law in Ghana*; Sweet and Maxwell: London, UK, 1962.

28. Low, B.S. The evolution of human life histories. In *Handbook of Evolutionary Psychology: Ideas, Issues, and Applications*; Crawford, C., Krebs, D.L., Eds.; Erlbaum: Mahwah, NJ, USA, 1998; pp. 131–161.

29. Kohli, M. Intergenerational Transfers and Inheritance: A Comparative View. In *Annual Review of Gerontology and Geriatrics*; Intergenerational Relations across Time and Place; Silverstein, M., Ed.; Springer: New York, NY, USA, 2004; Volume 24, pp. 266–289.

30. Berry, S. *No Condition is Permanent: The Social Dynamics of Agrarian Change in Sub-Saharan Africa*; The University of Wisconsin Press: Madison, WI, USA, 1993.
31. Berry, S. Property, Authority and Citiznship: Land Claims, Politics and the Dynamics of Social Division in West Africa. *Dev. Chang.* **2009**, *40*, 23–45. [CrossRef]
32. Bugri, J.T. Gender Issues in African Land Tenure: Findings from a Study of North-East Ghana. *Ghana Surv.* **2008**, *5*, 44–73.
33. Chauveau, J.-P.; Colin, J.-P.; Jacob, J.-P.; Lavigne Delville, P.; Le Meur, P.-Y. *Changes in Land Access and Governance in West Africa: Markets, Social Mediations and Public Policies: Results of the CLAIMS Research Project*; IIED: London, UK, 2006.
34. White, B. *Who Will Own the Countryside? Dispossession, Rural Youth and the Future of Farming*; Valedictory Lecture on 13 October on the Occasion of the 59th Dies Natalis of International Institute of Social Studies (ISS); International Institute of Social Studies: Hague, The Netherlands, 2011.
35. World Bank. *Youth Employment in Sub-Saharan Africa, Africa Development Forum*; World Bank: Washington, DC, USA, 2014.
36. Kameri-Mbote, P. The Land has Its Owners!; Gender Issues in Law Tenure Under Kenya Customary Law, International Environmental Law Research Centre Working Paper. 2005. Available online: http://www.ielrc.org/content/w0509.pdf (accessed on 28 August 2012).
37. Alber, E.; van der Geest, S.; Reynolds Whyte, S. (Eds.) *Generations in Africa: Connections and Conflicts*; LIT Verlag: Berlin, Germany, 2008.
38. Casley, D.J.; Lury, D.A. *Data Collection in Developing Countries*, 2nd ed.; Clarendon Press: Oxford, UK, 1987.
39. Noor, K.B.M. Case study: A strategic research methodology. *Am. J. Appl. Sci.* **2008**, *5*, 1602–1604. [CrossRef]
40. Kasanga, R.K.; Avis, M.R. Internal Migration and Urbanisation in Developing Countries—Findings from a Study of Ghana. *Environ. Policy* **1988**, 1–108.
41. Okali, C.; Sumberg, J. *Quick Money and Power: Tomatoes and Livelihood Building in Rural Brong Ahafo, Ghana*; Institute of Development Studies (IDS): Brighton, UK, 2012; pp. 44–57.
42. Hill, P. *The Migrant Cocoa-Farmers of Southern Ghana*, 2nd ed.; A study in Rural Capitalism; Cambridge University Press: Cambridge, UK, 1970.
43. André, C.; Platteau, J. Land Relations under Unbearable Stress: Rwanda Caught in the Malthusian Trap. *J. Econ. Behav. Organ.* **1998**, *34*, 1–47. [CrossRef]
44. Quan, J. Changes in Intra-family land relations. In *Changes in Customary Land Tenure Systems in Africa*; Cotula, L., Ed.; Russell Press: London, UK, 2007; pp. 51–63.
45. Gildea, R.Y., Jr. Culture and Land Tenure in Ghana. *Land Econ.* **1964**, *40*, 102–104. [CrossRef]
46. Ensminger, J. Changing Property Rights: Reconciling Formal and Informal Rights to Land in Africa. In *The Frontiers of the New Institutional Economics*; Drobak, J.N., Nye, J.V.C., Eds.; Academic Press: San Diego, CA, USA, 1997; pp. 165–196.
47. Kasanga, R.K. Land Tenure and Regional Investment Prospects: The Case of the Tenurial Systems of Northern Ghana. In *Land Management and Environmental Policy Series*; Institute of Land Management and Development (ILMAD), University of Science and Technology: Kumasi, Ghana, 1999; Volume 1.
48. FAO. *Voluntary Guidelines on the Responsible Governance of Tenure of Land, Fisheries and Forests in the Context of National Food Security*; Food and Agriculture Organisation of the United Nations: Rome, Italy, 2012.
49. Brooks, K.; Zorya, S.; Gautam, A.; Goyal, A. *Agriculture as a Sector of Opportunity for Young People in Africa*; Policy Research Working Paper 6473; Agriculture and Environmental Services Department, The World Bank: Washington, DC, USA, 2013.
50. GoG. *National Land Policy, Government of Ghana (GoG)*; Ministry of Lands & Forestry: Accra, Ghana, 1999.

 *land*

Article

The Changing Structure and Concentration of Agricultural Land Holdings in Estonia and Possible Threat for Rural Areas

Evelin Jürgenson * and Marii Rasva

Chair of Geomatics, Institute of Forestry and Rural Engineering, Estonian University of Life Sciences, Kreutzwaldi 5, 51014 Tartu, Estonia; marii.rasva@emu.ee
* Correspondence: Evelin.Jyrgenson@emu.ee; Tel.: +372-505-7428

Received: 5 January 2020; Accepted: 31 January 2020; Published: 2 February 2020

Abstract: In most European countries, there has been a decrease in the number of farms, while the area of agricultural land has remained almost the same. This ongoing process of land concentration can affect Europe's small farms and rural areas. The EU has acknowledged that the problem is serious and that, to solve it, it must be studied more closely. Accordingly, the aim of this study is to discuss changes in the agricultural sector from the aspect of land use, with emphasis on land concentration in Estonia, further scientific discussion about the effects of changes in land use on rural areas is encouraged. The study is carried out using two kinds of data sources: (1) statistical data from Eurostat, FAOSTAT and Statistics Estonia, (2) data from the Estonian Agricultural Registers and Information Board. The conclusion of the paper is that while the number of farms is going down, the average area of agricultural land use per farm is on the rise in Estonia. Agricultural land has been increasingly concentrated into the hands of corporate bodies. This study shows that there is a status of land concentration in Estonia that needs ongoing studies and a proper policy should be established to mitigate the impact of land concentration.

Keywords: agricultural land use; land concentration; landholding; Estonia

---

1. Introduction

1.1. Motivation

The agricultural sector is directly connected to the issue of food security. Land, an elemental source for the production is needed. It is scarce resource, not a "normal" market good [1] and therefore land issues may require special regulations.

The rush for land in developing countries in the Global South has caught much attention, much less has been given to the process of land concentration in Europe [2–4]. Large agricultural land users in Europe are expanding their scope widely and quickly. Tens of thousands of small farmers are being forced out of farming every year [4,5]. It is also evident that in many European countries, the degree of land-based inequality is similar to some countries with notoriously inequitable distribution of land ownership and land-based wealth such as Brazil, Colombia, and the Philippines [4,6].

A report in the European Union [7] points out that the ongoing process of farmland concentration in Europe is just as problematic as farmland grabbing. As land becomes concentrated into fewer and larger holdings, the Common Agricultural Policy (CAP) subsidy becomes more concentrated as well [4]. Although the EU considers land concentration to be a serious issue, there have been few current studies. Additional country-specific research is also needed. The present study is mainly focused on the problem of land concentration, referring to a process in which large agricultural producers are increasingly buying up or leasing land from other agricultural producers.

The ongoing process of land concentration has particularly affected Europe's small farms: it is implied that the expansion of large farms in Europe has come at the expense of small farms [7]. Agricultural land is becoming increasingly concentrated into the hands of large businesses, a situation in which small farmers are losing control of their land [7,8]. In the meantime, small farms are important for rural life: they play an active role in the economic fabric of rural areas, conserving the cultural heritage, maintaining rural life, sustaining social life and making sustainable use of natural resources. Small farms produce a sufficient amount of healthy and high-quality food and ensure a broad distribution of land ownership in rural areas [6,8–13]. In short, the process of land concentration has implications for society as a whole, not only for small farms.

There is no universally accepted definition of small or large farms [3,14]. The farms may be divided according to different parameters, such as farm structural size, economic size, herd size, labour force or utilised agricultural area [3,11]. There are many discussions about the relative productivity of large or small agricultural producers in light of the growing world population. Economic efficiency and productivity depend on many criteria. It is frequently stated that purely economic results are better for the larger farms. Even the negative impact caused by land fragmentation can be negated by the larger utilised land area [15]. At the same time, large farms can be inefficient due to the high monitoring cost, anonymity and lack of transparency [16]. However, the larger picture should include not just economic factors, but, for example, the social impacts as well.

1.2. Historical Overview of Changes in the Agricultural Sector in Estonia

Agriculture in Estonia has been through many changes, caused by different policies. Before Estonian independence in 1918 agricultural land in Estonia was owned by Baltic Germans. After gaining independence, this situation changed. There were numerous assumptions for triggering land reform. Before the reform, 58% of agricultural land was used by large agricultural holdings [17–19]. To carry out the land reform the majority of agricultural land owned by estate owners was expropriated by the state. As a result of this reform, ca 140,000 farms with an area over 1 ha were created, the previous number was 50,000 farms [17,20].

In 1940, the Soviet Union occupied Estonia and land was declared as people's property [20,21]. The largest land area that one working peasant could own was 30 ha. The rest of the land was incorporated into the State's land fund or given out to those peasants who had too small an area of land for agricultural use [20,22]. In the summer of 1941, Germany occupied Estonia and another change in agriculture followed: changes made by the Soviet Union were cancelled and the land divided during the land reform was given back to its rightful owners [20]. After three years of German occupation, the Soviet Union occupied Estonia and the declaration of 1940 was restored [21]. The state became the landowner and collectivization took place, simply as a political decision. It was believed that large agricultural holdings like kolkhozes and sovkhozes were more efficient than small farms. Forced collectivization intensification took place in 1949 when kolkhozes and sovkhozes were merged into larger ones.

The average area of kolkhozes and sovkhozes changed over the years, with the kolkhozes tending to be smaller than sovkhozes. By 1976, their average area was equalised. The total area of kolkhozes was 8086 ha, containing 4333 ha agricultural land. The total area of sovkhozes was 8015 ha, with 4542 ha agricultural land [20,23]. The optimal size of kolkhozes and sovkhozes was a research issue in the Soviet time. According to the instructions from the 1960s, the optimal area for kolkhozes and sovkhozes in Estonia would be 5500-9000 ha, that includes agricultural land 4500–6,500 ha and again arable land 1700-1900 ha [23,24]. By the year 1980 kolkhozes and sovkhozes had grown into giants and had to be divided into smaller ones [17,23].

Before independence was restored in Estonia, the Estonian SSR Farm Act was instituted in 1989, with benefit to small agricultural businesses. It didn't restore private property but it made private agricultural farming legal: 10,152 farms received the right to use 252,255 ha land [18,20,25].

As Estonian independence was restored in 1991, new winds started to blow in agriculture. Firstly, there was a goal to restore ownership of former farms. It was important to support agricultural land use that was based on small farms. To redesign ownerships based on historical, economic policy and socio-psychological subjects, it was also decided to enforce land reform [18,20,25]. The land reform law and then the agriculture reform law both favoured agriculture based on small farms [18,23]. Many small agricultural users arose in Estonia but as the years went by this number decreased.

Land reform in Estonia has had a combination of aims: to raise economic efficiency, the need to move to a market economy, and to repair historical injustice to owners whose real properties had been expropriated [25]. The same has been true in other Central and Eastern European countries that implemented land reform [26–28]. The multiple aims had a side effect—land fragmentation—in Estonia and other Central and Eastern European countries that implemented land reform [29–32]. As a result, some of the countries (for instance, Lithuania, Poland) that implemented land reform recently have developed the strategy of land consolidation [29,30], a strategy which has not taken hold in Estonia [25,29]. However, agricultural land holdings have enlarged over the years. The agricultural producers acquire available land plots that are situated at a distance from the farming centre and previous land plots. Maasikamäe et al. [33] presented that issue based on the polygons drawn over each agricultural land producer's land plots. Usually, several polygons (several agricultural producers' land holdings) overlap each other. It means workers must move from plot to plot, and sometimes these land plots are at a distance. The Estonian agricultural production situation is different from the eastern European Member States (Czech Republic, Slovakia) which implemented the land reform but went over to corporate farms [9,30]. Although administrations have changed, land use has remained more or less the same.

However, since the restoration of the Republic of Estonia in 1991, there has been considerable development. The centrally-planned economy has moved to a free-market basis. Currently, property and land reform has been almost entirely implemented. The land reform has changed the previous land relations: issues between state and private ownership have been resolved [25]. The property and land reform have led to changes in the agricultural sector as well: the kolkhozes and sovkhozes have been replaced with private agricultural producers. At the beginning of the reforms it was expected that a private small farm system would take over the system of kolkhozes and sovkhozes [20,23], but the trend did not continue: the number of small farms has decreased and continues to do so.

1.3. Aim and Scope

In 2014, the problem of land grabbing and land concentration was brought up by the European Economic and Social Committee. The European Economic and Social Committee decided to draw up its own-initiative opinion on "Land grabbing—a warning for Europe and a threat to family farming (own-initiative opinion)". This document was adopted in January 2015 [8]. It makes clear that since land forms the basis of food production, there is an obligation for countries to recognise the right of each individual living in their own territory to adequate and safe food, that is directly linked to access to land. As land is no ordinary commodity and its supply is finite, it will be necessary to regulate the market for agricultural land and to prevent land concentration. Accordingly, there must be greater regulations on land use and ownership. It is also important to adjust the CAP so that the first few hectares can be given a stronger weighting in direct payments. This could strengthen small farms and increase their ability to compete, thus the usual market rules should not apply.

Following that European and Economic and Social Committee document [8], the European Parliament's Committee on Agriculture and Rural Development requested the study "Extent of farmland grabbing in the EU", which was published in May 2015 [7]. In 2017, the European Parliament reported on "The state of play of farmland concentration in the EU: how to facilitate the access to land for farmers?" [6]. It points out that land, its management, and urban development rules are matters for the Member States, better account should be taken of farmland conservation and management. The report also called on the Member States to focus their land-use policies on using available tools (taxation, aid schemes and CAP funding) to maintain a family-farm-based agricultural model. The land policy must help to ensure broad, fair and equitable distribution of land tenure and access to land.

There are different rules for the independent governing of EU Member States' land-use policies and those for the agricultural land market. The European Association for Rural Development Institutions (AEIAR) reported in 2015 on the status of agricultural land market regulation in seven European countries (Germany, Belgium, France, Hungary, Italy, Lithuania, and Poland) [34]. The report states that all addressed countries use the tools for regulations of the agricultural land market and some use the approval process. For example, in Germany, sales of agricultural land over a certain size must be approved by the administrative authority. In France, there is special private, non-profit organization SAFER that is responsible for observing land transactions, setting up and restructuring agricultural and forestry structures, supporting local development and contributing to the protection of the environment. In Hungary, the notary sends the relative documents to the agricultural administrative authority for approval of a sale. All countries in the report are allowed to use the pre-emption right if agricultural land is sold. Some countries have limits for land ownership: 500 hectares in Lithuania and 300 hectares in Hungary. Additionally, Hungary limits land possession to 1200 hectares (consisting of owned and leased land).

In addition to the concentration of land, there are several other drivers—technological, institutional, and economical—which when combined, can influence changes in the pattern of agricultural land use [35]. For example, land needed for dwellings, infrastructure and commerce has decreased the total area of agricultural land. Such changes can alter rural societies in ways that can be a threat in sparsely populated areas. All these disruptions, though not the focus of this paper, are still relevant when considering the current and future status of land use.

This paper presents the statistical data about the changes in agricultural land use and a number of producers in the European Union, giving primary attention to the changes in Estonian agricultural producers' land holdings up till 2016. The bases for analyses are the landholdings (owned or leased) area per agricultural producers. The aim of the paper is to discuss changes in the agricultural sector from the aspect of land use, with the emphasis on land concentration in Estonia, and to encourage scientific discussion about the effects of changes in agricultural land use on rural areas in Europe, using the example of changes in Estonian agricultural producers' land holdings following the property and land reform that started in 1991 in Estonia. As Estonia is a small country, the changes take place more quickly and within a shorter time frame than in larger countries. Therefore, the Estonian case can be helpful to other countries as it helps to understand the changes in other European countries as well. This paper presents the first in-depth study of the agricultural land use and holdings changes following the property and land reform that started in 1991 in Estonia.

2. Materials and Methods

Two kinds of data sources were needed for the study. The first was statistical data from Eurostat, FAOSTAT and Statistics Estonia, the second considered the data for the landholdings of agricultural producers. This information source is the Estonian Agricultural Registers and Information Board (ARIB).

Data from Eurostat (https://ec.europa.eu/eurostat/) is used to compose the overview of agricultural land use in European countries, including data about utilised agricultural land use, the number of farms and average utilised agricultural land area per farm. The Eurostat data is from the years 2005, 2007, 2013 and 2016. However, the figures mainly present the changes between the years 2005 and 2016. In two cases (Croatia, Italy) the data was incomplete and data from 2007 or 2013 had to be used. The year 2016 or 2013 was used as the base year for calculating the changes that took place in the area of utilised agricultural area, number of farms and average utilised agricultural land area per farm. Eurostat's mission is to provide high-quality statistics for Europe.

Eurostat defines utilised agricultural land as follows:"Utilised agricultural area, abbreviated as UAA, is the total area taken up by arable land, permanent grassland, permanent crops and kitchen gardens used by the holding, regardless of the type of tenure or of whether it is used as a part of common land". Farm is defined as a single unit, both technically and economically, operating under single management and which undertakes agricultural activities within the economic territory of the European Union, either as its primary or secondary activity. Other supplementary (non-agricultural) products and services may also be provided by the holding.

Data from FAOSTAT (http://www.fao.org/faostat/en/#data) is also used in this study to provide an overview of agricultural land use in European countries. The FAOSTAT data is from years 2005 and 2016. The base year for calculating the changes in the area of agricultural land was 2016. It was necessary to add FAOSTAT data to this study because it differs a bit from Eurostat data about the utilised agricultural land area. FAOSTAT defines the agricultural land area as land used for the cultivation of crops and animal husbandry. The total of these areas falls under "Cropland" and "Permanent meadows and pastures".

Statistics Estonia (https://www.stat.ee/about) defines an agricultural household as a unit with uniform technical and economical management and at least one hectare of agricultural land, or where agricultural products are produced primarily for sale (irrespective of land area). From 2007, agricultural households are also units where agricultural products are not produced but the land is being conserved in good agricultural and environmental conditions. Agricultural land area in use is land that is used for agricultural production or being conserved in good agricultural and environmental conditions by agricultural households in the reference year.

ARIB data (ARIB is responsible for delivery of national and the EU subsidies for agricultural activities) from 2011 and 2016 is used for the case study to present a more detailed overview of the recent changes in the pattern of agricultural landholdings in Estonia. Figure 1 illustrates the study area and its location in Europe.

Figure 1. Location of Estonia (study area) in Europe.

The data from the ARIB Field Register is used for the study. The Field Register is one of three registers in the charge of ARIB and area support is one of the subsidies that ARIB delivers. The digitalised database of agricultural plots is required for payment of area support from the budget of the EU. In the process of delivering national and EU subsidies, ARIB collects information about the applicant (every applicant gets an ID number) and land that is filed for area support.

ARIB data about the agricultural land area and the number of producers were analysed in order to get an overview of changes in Estonian agricultural land users' landholdings. Agricultural land users and land area per producer was summarized using GIS software ArcGIS (version 10.4). Producers were divided into six groups according to the size of their landholdings: 0–<2 ha, 2–<40 ha, 40–<100 ha, 100–<400 ha, 400–<1000 ha and >1000 ha, data was taken on the basis of these size groups. The basis for this division comes from Farm Accountancy Data Network (FADN) (https://maainfo.ee/index.php?page=9&) where agricultural land area is divided into four size groups

(0–<40 ha, 40–<100 ha, 100–<400 ha, >400 ha). In order to get a closer look at the smallest agricultural land users, FADN size group 0–<40 ha was divided into size groups 0–<2 ha and 2–<40 ha. FADN size group >400 ha was divided into size groups 400–<1000 and >1000 ha in order to characterise the largest agricultural land users.

This study concentrates on agricultural land users' land holdings that cover all plots which are used for agricultural production in Estonia. No distinction is made between land held in ownership and leasehold land. Also, no differentiation was made between different production groups.

3. Results

3.1. Agricultural Land Use Pattern and Its Changes in Europe

Europe has 12 million agricultural land users, with 25 million people involved in agricultural production and 69% of agricultural land users having less than five hectares: the average size is 14.2 ha [36]. The total number of agricultural land users in the Baltic and Nordic countries is 607,500, which is 4.2% of the total number in the EU. In 2009, the agricultural land user had an average 59.7 ha of the agricultural area in Denmark, 42.9 ha in Sweden, 38.9 in Estonia, 33.6 ha in Finland, 16.5 ha in Latvia and 11.5 ha in Lithuania [37].

According to Eurostat, between 2005 and 2016 (Figure 2), the utilised agricultural land area has grown in Bulgaria, Estonia, Ireland, Greece, Croatia, Latvia, Lithuania, Luxembourg, Hungary, Malta, Slovenia, Slovakia and the United Kingdom. The agricultural land area has decreased in other European countries, in greater quantity in Germany, Spain, Romania and Switzerland.

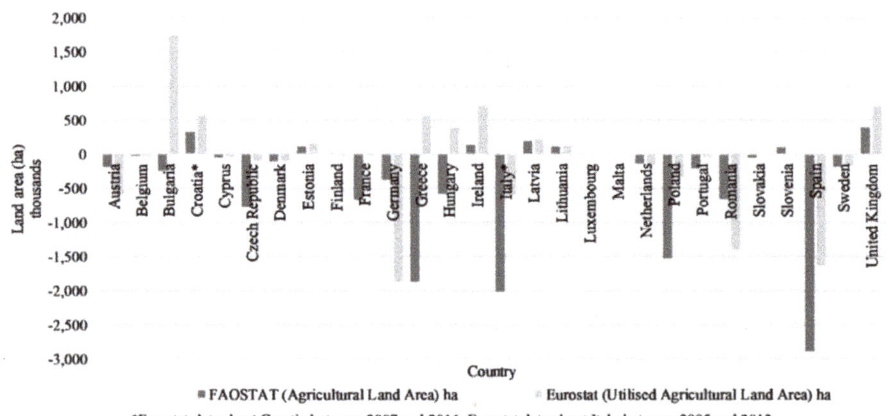

Figure 2. Utilised agricultural land (Eurostat) and agricultural land (FAOSTAT) use change (ha) in Europe between 2005 and 2016.

FAOSTAT data shows that, unlike Eurostat data, utilised agricultural land area has decreased in Bulgaria, Greece and Hungary. These indicators are more similar in other countries. Utilised agricultural land and agricultural land are not precisely the same concepts and their collecting methodology differs, therefore further study is needed regarding the differences seen in Figure 2.

The total number of farms in Europe between 2005 and 2016 (Figure 3) has decreased by four million, affecting all countries except for Ireland, where there was an increase of 4860 farms. The largest (−1,065,770) decrease in the number of farms has occurred in Poland, but the decrease is remarkable in Romania (−834,120) and Italy (−718,200) as well.

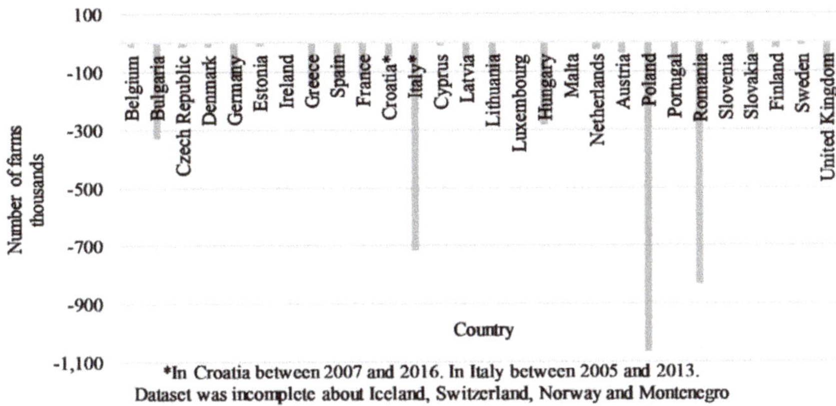

Figure 3. Change in the number of farms in Europe between 2005 and 2016 (Eurostat).

The average growth of agricultural land area per farm between 2005 and 2016 in Europe is 4.8 ha. The biggest growth in agricultural land use per farm has occurred in Slovakia (46.2 ha) and in the Czech Republic (46.0 ha) (Figure 4). In Estonia, the growth of agricultural land use per farm has been also relatively large (29.7 ha) compared to other countries in Europe. Cyprus is the only country where the average utilised agricultural area per farm has decreased slightly (−0.2 ha).

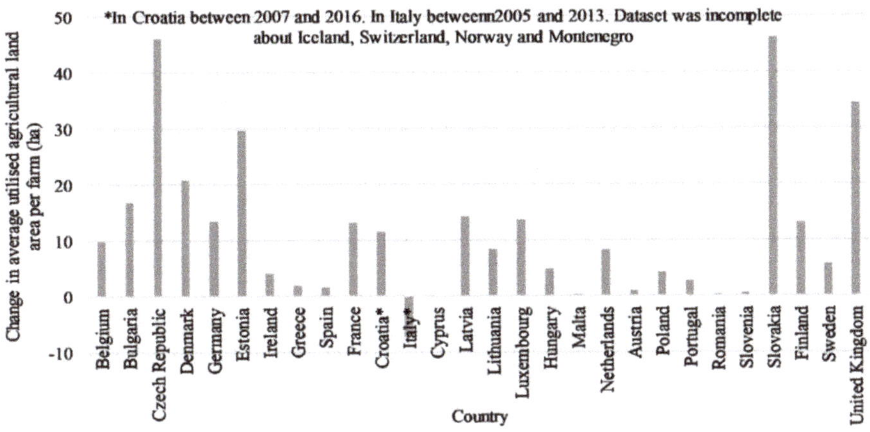

Figure 4. Change in average utilised agricultural land area (ha) per farm between 2005 and 2016 (Eurostat).

While agricultural land use has changed in Europe, the utilised agricultural areas have changed less, having decreased only 1% as compared to the years 2005 and 2016. At the same time, the shrinking number of farms is remarkable: 30% fewer farms as compared with 2005 and 2016. The decrease has been larger in some countries while the average utilised agricultural area per farm has increased in almost all countries.

3.2. Agricultural Land Use Pattern and Its Changes in Estonia

Statistics show that the number of agricultural households in Estonia has decreased yearly (Figure 5). In 2001 there were 55,748 agricultural households in Estonia but by 2016 this number had decreased to 16,696, concurrently, the area of utilised agricultural land has remained almost the same. Estonian utilised agricultural land area in 2001 was 871,213 ha and in 2016 it was 995,130 ha (Figure 5).

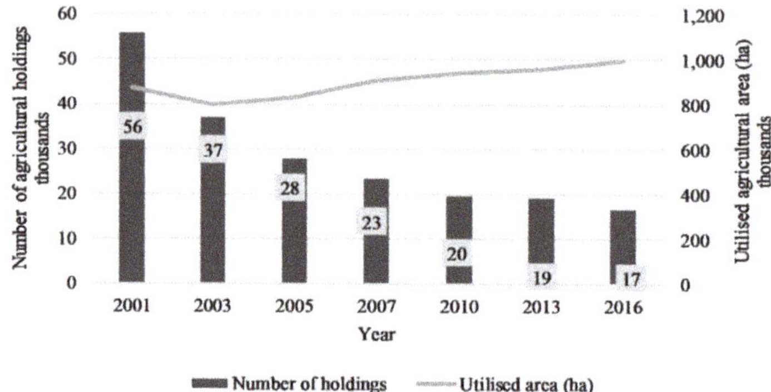

Figure 5. The number of agricultural households and agricultural land area in Estonia between 2001 and 2016 (Statistics Estonia).

The decrease in the number of households and almost constant agricultural land area shows that average land use per agricultural household in Estonia has increased (Figure 6). In 2001 the area of agricultural land use per household was 16 ha but by 2016 it had grown to 60 ha. The average agricultural land-use area per user has grown yearly 2 to 10 ha per year.

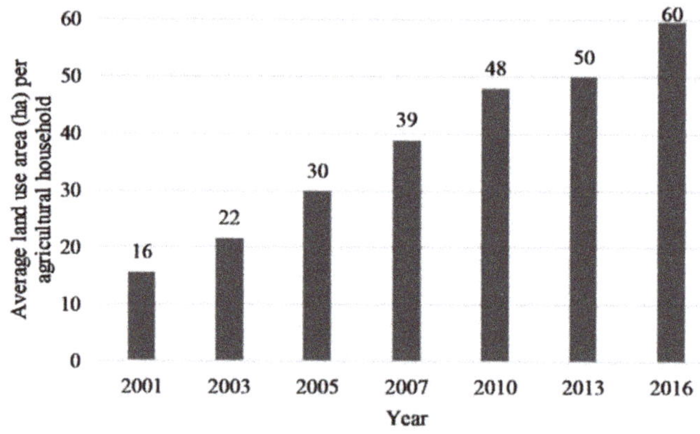

Figure 6. Average land use (ha) per agricultural household in Estonia between 2001 and 2016 (Statistics Estonia).

In 2001 the corporate bodies used 327,788 ha which was 38% of all agricultural land (Figure 7). While self-employed workers used 543,426 ha of all agricultural land, corporate bodies used 215,638 ha less. By 2016, the situation has changed a lot. Corporate bodies used 645,598 ha −65% of agricultural land (ha). At the same time, self-employed workers used 349,505 ha, which is 193,921 ha less than 2001 and 296,093 ha less than corporate bodies in 2016.

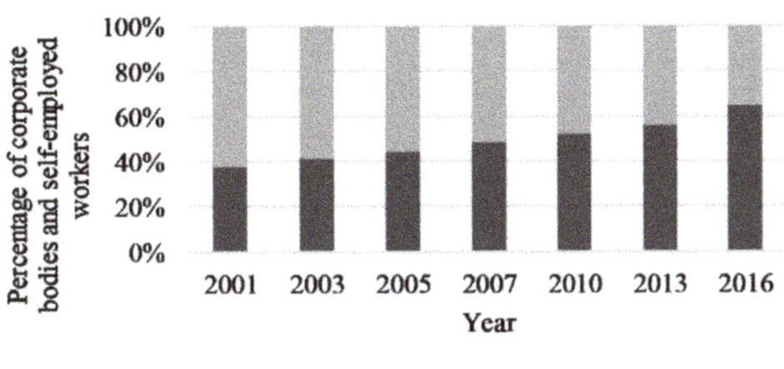

Figure 7. Percentage of corporate bodies and self-employed workers in Estonian agriculture between 2001 and 2016 according to the agricultural area (ha) used by agricultural producers (Statistics Estonia).

For a deeper understanding of the changes in the agricultural land use and users sector, the case study based on the ARIB data was undertaken. It is a more comprehensive study that covered all agricultural producers land holdings registered in the ARIB which applied for support from the EU. The data used was from 2011 and 2016. According to ARIB data agricultural land use area has grown 11% and the number of land users has dropped 5% in Estonia between 2011 and 2016. Table 1 presents the data for the land users, which were divided into groups according to the size (area) of their landholdings.

Table 1. Data for land users groups that form according to the area of land users land holdings for the years 2011 and 2016 (ARIB).

Groups ha	2011		2016	
	Number	Area (ha)	Number	Area (ha)
<2	1475	2140	1355	2026
2–<40	11,654	132,888	10,767	121,960
40–<100	1460	91,563	1481	93,093
100–<400	1174	225,708	1317	260,956
400–<1000	337	207,844	390	237,670
>1000	126	216,893	146	252,110
Total	16,226	877,036	15,456	967,816

Comparing the years 2011 and 2016, the number and the area of these land users decreased in two smaller land users' (0–2 ha, 2–<40 ha) groups and increased in the four following (40–<100 ha, 100–<400 ha, 400–<1000 ha and >1000 ha) groups (see Table 1). Analysis of land users according to the area of their landholdings in size groups 0–<2 ha, 2–<40 ha, 40–<100 ha, 100–<400 ha, 400–<1000 ha and >1000 ha shows that between 2011 and 2016, agricultural land area used by land users in size groups 100–<400 and that >1000 has grown the most. Agricultural land area used by land users in size groups 0–<2 and 2–<40 has decreased and the agricultural land area used by size group 40–<100 has remained almost the same.

There are 536 agricultural land users in Estonia with land holdings over 400 ha. They are using 489,780 ha or 51% of the agricultural land utilised in 2016 in Estonia. In 2011, there were 463 agricultural land users with land holdings over 400 ha. They used 424,736 ha or 48% of the totally-used agricultural area in 2011.

The agricultural land area used by larger land users has grown while the smaller ones have decreased (Figure 8a). The number of households in size groups 0–<2 ha and 2–<40 ha has decreased

(Figure 8b). In 2011, there were 1,475 agricultural users in size group 0–<2 ha using 2,139.72 ha of agricultural land. In size group 2–<40 ha there were 11,654 agricultural land users using 132,888.41 ha. In 2016, there were 120 fewer land users in size group 0–<2 ha using 813,24 ha less land. In size group 2–<40 ha there were 887 fewer users, they were using 10,928.15 ha less land than in 2011.

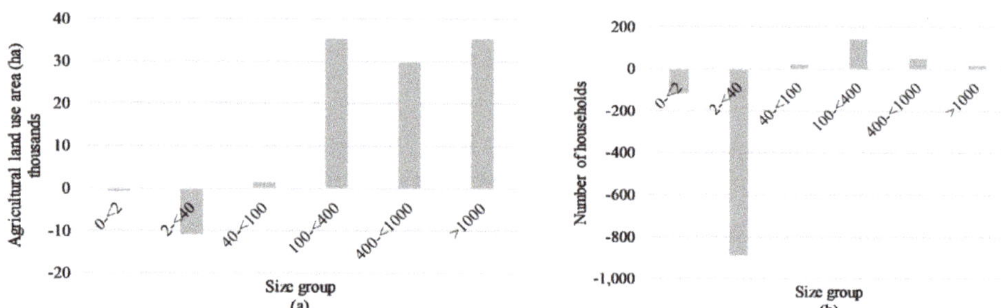

Figure 8. (**a**) Difference in the area (ha) of agricultural land use, (**b**) Difference in the number of agricultural households in size groups between 2011 and 2016 (ARIB).

Households in size groups 40–<100 ha, 100–<400 ha, 400–<1000 ha and >1000 have grown in number. In 2011, there were 337 land users in size group 400–<1000 ha and they were using 207,843.80 ha of agricultural land. In size group >1000 ha, 126 land users were using 216,892.61 ha. By the year 2016, there were 53 more users in size group 400–<1000 ha and 20 more in size group >1000 ha. Agricultural land-use area had grown 29,826.53 ha in size group 400–<1000 ha and 35,217.66 ha in size group >1000 ha.

In 2016, there were 257 corporate bodies and 1098 self-employed workers in size group 0–<2 ha (Figure 9). In size group 2–<40 ha there were 4,319 corporate bodies and 6,448 self-employed workers. In these two size groups, self-employed workers form the majority. In size groups 400–<1000 ha and >1000 ha there are no self-employed workers. In size group 400–<1000 ha there are 390 corporate bodies in size group >1000 ha there are 146 corporate bodies.

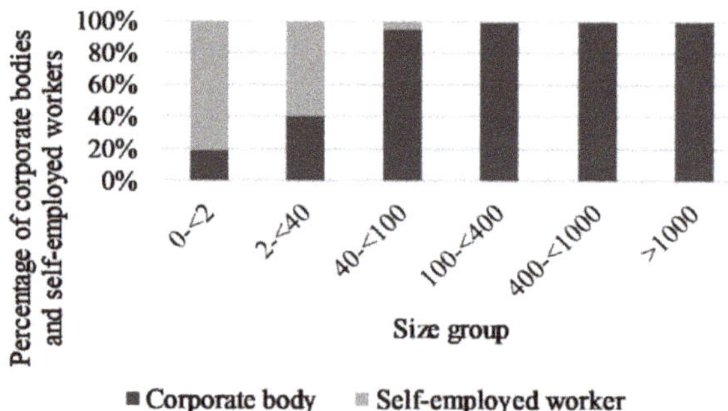

Figure 9. Percentage of corporate bodies and self-employed workers in size groups in 2016 (ARIB).

In 2016 the number of users in size group 0–<2 ha forms 8.8% of the total number of agricultural land users in Estonia (Figure 10a) using 0.2% of the total land use (Figure 10b). The number of land users in size group 2–<40 ha amounts to 69.7% of the total number of land users in Estonia using 12.6% of total land use. Concurrently, the number of agricultural land users in size group 400–<1000 ha accounts for 2.5% of the total number, using 24.6% of total land use in Estonia. The number of agricultural land users in size group >1000 ha accounts for 0.9% of the total households, using land 26% of total land use in Estonia.

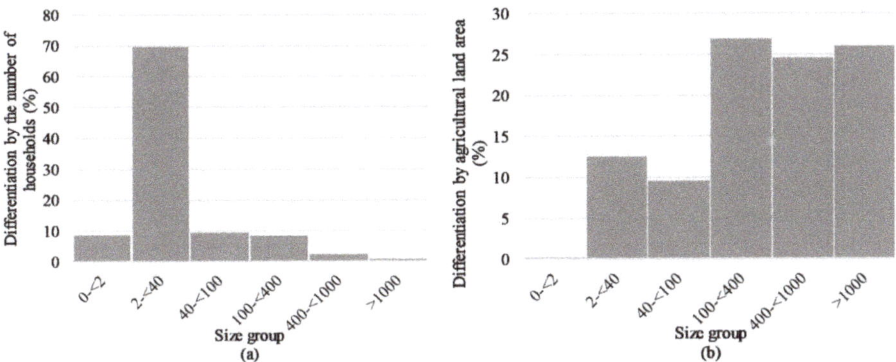

Figure 10. (**a**) Differentiation of size groups by the number of households, (**b**) Differentiation of size groups by agricultural land area in 2016 (ARIB).

4. Discussion

The resulting changes in the utilized agricultural area in Europe (Figure 2) are diverse, increasing as well as decreasing in many European countries. A possible direction for this indicator needs future studies. However, it is to be expected that in some countries, agricultural land use has been transformed into other uses. For example, the Czech Republic and Poland have decreased the utilized agricultural land area and the studies [38,39] confirm that changes of agricultural land use to another use, for example for urban needs (dwellings, infrastructure objects, businesses, commercial and retail areas) is problematical. The number of farms in Europe has decreased in all countries (Figure 3) except for Ireland. The average utilized agricultural land area (ha) per farm increased in almost all countries except Italy and Cyprus (Figure 4). That means that despite a decrease in a utilized agricultural area in some countries (Czech Republic, Germany, Poland) there was an even greater decrease in the number of farms, so that average utilized agricultural land area (ha) per farm has actually increased.

The Estonian case presented here that the area of utilized agricultural land has remained almost the same or has increased slightly (Figure 5). The number of landholdings decreased nearly 3 times within 15 years (Figure 5) while the area of agricultural land use per household increased almost 4 times (Figure 6). While average land use per agricultural household in Estonia has increased, the agricultural land in Estonia has become increasingly concentrated into the hands of corporate bodies (Figure 7). Corporate owners held approximately two times more hectares of land in 2016 compared with the year 2001. At the same time, agricultural land used by self-employed workers is decreasing. It suggests that corporate bodies are growing at the expense of agricultural land used by self-employed workers, the reasons for that need further studies.

The deeper case study with the Estonian AIRB data affirms the previous data that the total number of land users has decreased while the utilized agricultural area increased. The total number of agricultural households in Estonia has dropped (Figure 5) and households that have closed their businesses are mostly in size groups 0–<2 ha and 2–<40 ha (Figure 8b). The biggest growth in the number of households between 2011 and 2016 appears in size group 100–<400 ha. The number of self-employed workers in Estonian agriculture has decreased while there has been a growth in corporate bodies (Figure 7). At the same time, there has also been a growth in agricultural household number in size groups over 40 ha (Figure 8). In 2016, most of the self-employed workers are using land in size groups 0–<2 ha and 2–<40 ha while corporate bodies form the majority in size groups over 40 ha (Figure 9). There are also some self-employed workers that use agricultural land in size group 40–<100 ha and a few in 100–<400 ha. Land users in size groups over 400 ha are mostly corporate bodies.

The data presents the changes that happened in the agricultural land use and land users sector in Estonia. These changes have taken place within a short time period, only fifteen years (2011–2016). The agricultural land has steadily been concentrated and, at the same time, more land users are corporate bodies. Additionally, it can be firmly stated, according to this study, that while the number of small land users has decreased, the number of larger land users has increased.

Scientists and official documents [6,8–13] have presented the case for smaller vs. larger agricultural producers: smaller farms perform essential tasks in rural society. However, it is also shown that smaller producers are under greater economic pressure. They often need support from the state [3,8]. If this issue is developed only under free-market rules, then the small agricultural producers will shut down their activity. The people who were engaged in small production remain without income and the state must pay the subsistence allowance. The other alternative is for farm labourers to move to find work—normally to the city or towns.

It is important to think beyond the land-use issue to the ownership issue as well. Will the changes in land use pattern bring together changes in the land ownership pattern? This paper addresses the changes in agricultural land use looking at the size of landholdings. It did not distinguish between land that is owned or leased by agricultural producers. It could be that bigger landholdings bring together a concentration of land ownership as well. Agricultural land users get subsidies from the EU. Land users which have bigger holdings receive larger subsidies that enable them to acquire land plots, as is pointed out in documents compiled by the European Economic and Social Committee [8] and requested by the European Parliament's Committee on Agriculture and Rural Development [7].

The question is, is there a need for regulations about land use and/or ownership? Some countries limit land ownership. For example, it is possible to own 500 ha of land in Lithuania and 300 hectares in Hungary [34,40]. Limits for land ownership or use are absent in Estonia. It is theoretically possible for a person with enough money to acquire as much land as is available on the market. The largest land user in Estonia consists of more than 5000 ha of agricultural land, while 146 owners use more than 1000 ha in 2016. The number of such large land users is increasing—it was 126 in 2011, their number in five to ten years is not predictable.

The instructions issued during the Soviet period assigned the optimal agricultural land area for kolkhozes and sovkhozes: 4500–6500 ha agricultural land for kolkhozes or sovkhozes [23,24]. The larger land user in Estonia used the optimal amount of agricultural land (5523 ha) according to these instructions. However, compared to previous use, the kolkhozes or sovkhozes land use in the Soviet period was more compact compared to the current agricultural producers' landholdings. The land reform implementation resulted in land fragmentation, as the previous kolkhozes and sovkhozes were divided among many private owners [25]. Recent agricultural producers must acquire land plots from the land market, thus land holdings are scattered [33]. The area of landholding is comparable or even higher than in the eastern EU member states that, after land reform, became corporate farms [9,30] but the landholdings are more scattered. It means that current land users need more agricultural land for affordable production to compensate for the costs of plot fragmentation. As a result, landholdings will exceed the area that had been used previously by kolhozes and sovkhozes.

Questions about the scale and equitable arrangement of future agricultural land ownership remain. There were two major land reforms in Estonia (1918 and 1991) the purpose of which was to share land holdings between farm owners (mostly German in 1918) and those Estonians who worked the land. Similar examples can be found elsewhere. Now, however, the advent of much larger-scale production, though economically more efficient, also means the concentration of ownership into fewer hands at the expense of small landholders. The resulting imbalance and related societal disruption to rural life and development raise issues that may need to be addressed.

A recent example is from Scotland, where the Government declares: "We are improving Scotland's system of land ownership, use, rights and responsibilities, so that our land may contribute to a fair and just society while balancing public and private interests." [41]. They are undertaking land reform, as land ownership is in the hands of a very small number of persons, not the best circumstances for society and rural development.

Estonia needs policy direction and regulations for the agricultural land market that help to mitigate the impact of land concentration in rural areas in the long run, similar to several other European countries [34]. The direction of the policy and extent of the area of land use or ownership is a matter for further research and even debate, to determine appropriate regulations and possible limitations to land areas, e.g. to 300, 500, 1000 or even more hectares.

5. Conclusions

The study presents the changes – both increase and decrease - in the utilized agricultural land use in Europe and Estonia. The number of farms decreased while the average utilized agricultural land area (ha) per farm increased in almost all countries in Europe. The decreasing number of agricultural households and almost constant agricultural land area in Estonia shows that average land use per agricultural household has increased. Deeper analyses show that agricultural land in Estonia has become increasingly concentrated into the hands of corporate bodies, that growth has come at the expense of agricultural land used by self-employed workers. These changes have taken place within a short time period and may have been a result of notable change in land relations, after implementation of the post-Soviet land reform. Accordingly, conditions in the agricultural sector should stabilize and such extensive changes should not be the norm in the future.

This paper's aim is to discuss changes in the agricultural sector from the aspect of land use and encourage scientific discussion about the effect of the resulting changes in rural areas. For the discussion to be productive, it needs additional data about the situation in EU countries. As demonstrated, the statistical databases Eurostat and FAOSTAT can provide a range of relevant information in various countries. As the ongoing process of land concentration continues, these changes must be studied more, as there are diverse drivers causing changes in the agricultural sector. Further study must focus on the need for the policy direction and regulations that can mitigate the potential threats that can occur with the land concentration threatening the rural areas. For broad-based statements, input from researchers in different fields is essential. The land holdings should be suitable for necessary and sufficient agricultural production at affordable costs, acceptable to local societies, while also supporting sustainable development. Definitely, the issue is complex. Appropriate solutions cannot arise without further attention.

Author Contributions: Conceptualization, E.J. and M.R.; methodology, E.J. and M.R.; formal analysis, M.R.; investigation, E.J. and M.R.; resources, E.J.; data curation, M.R.; writing—original draft preparation, E.J. and M.R.; writing—review and editing, E.J.; visualization, M.R.; supervision, E.J.; project administration, E.J. All authors have read and agreed to the published version of the manuscript.

Funding: This research received no external funding.

Acknowledgments: We would like to thank four reviewers for their helpful comments.

Conflicts of Interest: The authors declare no conflict of interest.

References

1. Alexander, E. Land-property markets and planning: A special case. *Land Use Policy* **2014**, *41*, 533–540. [CrossRef]
2. Azadi, H.; Houshyar, E.; Zarafshani, K.; Hosseininia, G.; Witlox, F. Agricultural outsourcing: A two-headed coin? *Glob. Planet. Chang.* **2013**, *100*, 20–27. [CrossRef]
3. Eurostat. *Archive: Small and Large Farms in the EU—Statistics from the Farm Structure Survey.* Statistics Explained; Eurostat. 2016. Available online: https://ec.europa.eu/eurostat/statistics-explained/index.php?title=Archive:Small_and_large_farms_in_the_EU_-_statistics_from_the_farm_structure_survey (accessed on 20 December 2019).

4. Franco, J.; Borras, S.M., Jr. *Land Concentration, Land Grabbing and People's Struggles in Europe*; Transnational Institute: Amsterdam, The Netherlands, 2013. Available online: https://www.tni.org/files/download/land_in_europe-jun2013.pdf (accessed on 5 November 2019).

5. Lowder, S.K.; Skoet, J.; Raney, T. The Number, Size, and Distribution of Farms, Smallholder Farms, and Family Farms Worldwide. *World Dev.* **2016**, *87*, 16–29. [CrossRef]

6. European Parliament. *The State of Play of Farmland Concentration in the EU: How to Facilitate the Access to Land for Farmers*. European Parliament. 2017. Available online: https://www.europarl.europa.eu/doceo/document/A-8-2017-0119_EN.pdf?redirect (accessed on 1 December 2019).

7. Kay, S.; Peuch, J.; Franco, J. *Extent of Farmland Grabbing in the EU*. European Parliament. 2015. Available online: https://www.europarl.europa.eu/RegData/etudes/STUD/2015/540369/IPOL_STU(2015)540369_EN.pdf (accessed on 5 January 2020).

8. *Land Grabbing—a Warring for Europe and a Threat to Family Farming*; European Economic and Social Committee: Bruxelles, Belgium, 2015. Available online: https://www.eesc.europa.eu/en/our-work/opinions-information-reports/opinions/land-grabbing-europefamily-farming (accessed on 1 December 2019).

9. Eurostat. *Agriculture Statistics—Family Farming in the EU*. Statistics Explained. 2018. Available online: https://ec.europa.eu/eurostat/statistics-explained/pdfscache/38078.pdf (accessed on 5 November 2019).

10. Grubbström, A.; Sooväli-Sepping, H. Estonian family farms in transition: A study of intangible assets and gender issues in generational succession. *J. Hist. Geogr.* **2012**, *38*, 329–339. [CrossRef]

11. Guiomar, N.; Godinho, S.; Pinto-Correia, T.; Almeida, M.; Bartolini, F.; Bezák, P.; Biró, M.; Bjørkhaug, H.; Bojnec, Š.; Brunori, G.; et al. Typology and distribution of small farms in Europe: Towards a better picture. *Land Use Policy* **2018**, *75*, 784–798. [CrossRef]

12. McDonagh, J.; Farrell, M.; Conway, S. The Role of Small-scale Farms and Food Security. In *Sustainability Challenges in the Agrofood Sector*; Wiley: Hoboken, NJ, USA, 2017; pp. 33–47.

13. Shucksmith, M.; Rønningen, K. The Uplands after neoliberalism?—The role of the small farm in rural sustainability. *J. Rural. Stud.* **2011**, *27*, 275–287.

14. Davidova, S.; Thomson, K. Family Farming in Europe: Challenges and Prospects. 2014. Available online: https://www.europarl.europa.eu/RegData/etudes/note/join/2014/529047/IPOL-AGRI_NT(2014)529047_EN.pdf (accessed on 1 December 2019).

15. Looga, J.; Jürgenson, E.; Sikk, K.; Matveev, E.; Maasikamäe, S. Land fragmentation and other determinants of agricultural farm productivity: The case of Estonia. *Land Use Policy* **2018**, *79*, 285–292. [CrossRef]

16. Csaki, C.; Lerman, Z. *Structural Change in the Farming Sectors in Central and Eastern Europe: Lessons for the EU Accession*; World Bank: Washington, DC, USA, 2000; p. 272. Available online: http://documents.worldbank.org/curated/en/267151468776098706/Structural-change-in-the-farming-sectors-in-Central-and-Eastern-Europe-lessons-for-EU-accession-Second-World-Bank-FAO-Workshop-June-27-29-1999 (accessed on 5 January 2020).

17. Kivistik, J. Agraarstruktuuri muutustest Eesti Vabariigis. In *EPMÜ teadustööde kogumik*; Infotrükk: Tallinn, Estonian, 1997; pp. 44–47.

18. Lillak, R. *Eesti põllumajanduse ajalugu*; Trükikoda Trükipunkt: Tartu, Estonian, 2003; p. 260.

19. Roosenberg, T. *Künnivaod. Uurimusi Eesti 18.-20. sajandi agraarajaloost*; Tartu Ülikooli Kirjastus: Tartu, 2013; ISBN 978-9949-32-226-8. Available online: https://www.tyk.ee/admin/upload/files/raamatud/1366367626.pdf (accessed on 5 January 2020). (In Estonian)

20. Virma, F. *Maasuhted, maakasutus ja maakorraldus Eestis*; OÜ Halo Kirjastus: Tallinn, Estonian, 2004; p. 339, ISBN 9985-9553-3-1. (In Estonian)

21. Sirendi, A. *Eesti Põllumajandus XX sajandil (2. osa)*; Eesti Vabariigi Põllumajandusministeerium: Tallinn, Estonian, 2007; p. 328, ISBN 9789949136483. (In Estonian)

22. Mander, Ü.; Palang, H. Changes of Landscape Structure in Esonia during the Soviet Period. *GeoJournal* **1994**, *33*, 45–54.

23. Kasepalu, A. *Mis peremees jätab, Selle Mets võtab*; Eesti teaduste Akadeemia Majanduse Instituut: Tallinn, Estonian, 1991; p. 129. (In Estonian)

24. Eesti NSV Teaduste Akadeemia Majanduse Instituut. *Põllumajandusökonoomika küsimusi. 4 osa*; Eesti Riiklik Kirjastus: Tallinn, Estonian, 1964; p. 171. (In Estonian)

25. Jürgenson, E. Land reform, land fragmentation and perspectives for future land consolidation in Estonia. *Land Use Policy* **2016**, *57*, 34–43. [CrossRef]

26. Adams, M. *Breaking Ground: Development Aid for Land Reform*; Oversea Development Institute: London, UK, 2000; ISBN 0 85003 500 7. Available online: https://www.odi.org/sites/odi.org.uk/files/odi-assets/publications-opinion-files/8124.pdf (accessed on 5 January 2020).

27. Swinnen, J.F.M. The political economy of land reform choices in Central and Eastern Europe. *Econ. Transit.* **1999**, *7*, 637–664. [CrossRef]

28. Swinnen, J.F.M.; Mathijs, E. Agricultural privatisation, land reform and farm restructuring in Central and Eastern Europe: A comparative analysis. In *Agricultural Privatisation, Land Reform and Farm Restructuring in Central and Eastern Europe*; Swinnen, J.F.M., Buckwell, A., Mathijs, E., Eds.; Routledge: London, UK, 1997; pp. 333–337. ISBN 1-85972-648-8.

29. Hartvigsen, M. *Experiences with Land consolidation and Land banking in Central and Eastern Europe after 1989*; Land Tenure; FAO: Rome, Italy, 2015; pp. 1–128. Available online: http://www.fao.org/3/a-i4352e.pdf (accessed on 5 January 2020).

30. Hartvigsen, M. Land reform and land fragmentation in Central and Eastern Europe. *Land Use Policy* **2014**, *36*, 330–341. [CrossRef]

31. Van Dijk, T. Scenarios of Central European land fragmentation. *Land Use Policy* **2003**, *20*, 149–158. [CrossRef]

32. Van Dijk, T. *Dealing with Central European land fragmentation: A Critical Assessment on the Use of Western European Instruments*; Eubron Delft: Delft, The Netherlands, 2003; ISBN 90-5166-996-8. Available online: https://library.wur.nl/WebQuery/wurpubs/346204 (accessed on 5 January 2020).

33. Maasikamäe, S.; Jürgenson, E.; Sikk, K. *The Rearrangement of Leasehold Agreements as an Alternative to Land Consolidation*; Kadaster International: Apeldoorn, The Netherlands, 2016; pp. 207–215. Available online: https://www.oicrf.org/documents/40950/43224/The+Rearrangement+of+the+Leasehold+Agreements+as+an+Alternative+to+the+Land+Consolidation%281%29.pdf/6a78bea7-e376-3b40-f4dc-b736ab35ebf2 (accessed on 1 December 2019).

34. AEIAR. *Status of Agricultural Land Market Regulation in Europe*. AEIAR. 2016. Available online: http://www.aeiar.eu/wp-content/uploads/2016/04/Land-market-regulation_policies-and-instruments-v-def2.pdf (accessed on 7 November 2019).

35. Van Vliet, J.; De Groot, H.L.; Rietveld, P.; Verburg, P.H. Manifestations and underlying drivers of agricultural land use change in Europe. *Landsc. Urban Plan.* **2015**, *133*, 24–36. [CrossRef]

36. European Coordination Via Campensina ECVC Toolkit on Land Grabbing and Access to Land in Europe 2017. Available online: http://www.eurovia.org/wp-content/uploads/2017/04/EN__ECVC-Land-Kit.pdf (accessed on 1 December 2019).

37. Valdvee, E.; Klaus, A. Agricultural Holdings in the Baltic and Nordic Countries. In *Quaterly Bulletin of Statistics Estonia*; Statistics Estonia; 2009; pp. 110–115, ISBN 1736-7921. Available online: https://www.stat.ee/valjaanne-2009_eesti-statistika-kvartalikiri-1-09 (accessed on 5 November 2019).

38. Václavík, T.; Rogan, J. Identifying Trends in Land Use/Land Cover Changes in the Context of Post-Socialist Transformation in Central Europe: A Case Study of the Greater Olomouc Region, Czech Republic. *GISci. Remote Sens.* **2009**, *46*, 54–76. [CrossRef]

39. Busko, M.; Szafrańska, B. Analysis of Changes in Land Use Patterns Pursuant to the Conversion of Agricultural Land to Non-Agricultural Use in the Context of the Sustainable Development of the Malopolska Region. *Sustainability* **2018**, *10*, 136. [CrossRef]

40. Ciaian, P.; Kancs, D.; Swinnen, J.F.M. *EU Land Marets and the Common Agricultural Policy*; 2010; ISBN 978-92-9079-963-4. Available online: https://papers.ssrn.com/sol3/papers.cfm?abstract_id=1604452 (accessed on 5 January 2020).

41. Scottish Government Land Reform. Available online: https://www.gov.scot/policies/land-reform/ (accessed on 7 December 2019).

 *land*

Article

Rural Women's Invisible Work in Census and State Rural Development Plans: The Argentinean Patagonian Case

Paula Gabriela Núñez [1,2,*], **Carolina Lara Michel** [2], **Paula Alejandra Leal Tejeda** [3] and **Martín Andrés Núñez** [4]

1   Universidad de Los Lagos, Cochrane 1046, Osorno 5290000, Chile
2   Universidad Nacional de Río Negro- CONICET, Bariloche 8400, Argentina; cmichel@unrn.edu.ar
3   Departamento de Trabajo Social, Facultad de Ciencias Sociales y Humanidades, Universidad Autónoma de Chile sede Temuco, Av Alemania 1090, Temuco 4780000, Chile; paula.leal@uautonoma.cl
4   Grupo Ecología de las Invasiones, INIBIOMA-CONICET, Universidad Nacional del Comahue, Pionero 2350, Bariloche 8400, Argentina; mnunez@utk.edu
*   Correspondence: paula.nunez@ulagos.cl; Tel.: +54-2944705118

Received: 2 February 2020; Accepted: 20 March 2020; Published: 22 March 2020

Abstract: This article reviews the invisibility and the recognition of rural female work in the Patagonian region of Argentina over time. The analysis is carried out based on (a) the systematisation of research articles (b) a historical study of censuses, and (c) the systematisation of rural development plans related to the subject. The article adopts an ecofeminist perspective. The results have been organised into four sections. (1) An overview of the later Patagonian integration; (2) the work of Patagonian women in history; (3) the recognition of rural production in censuses; (4) Patagonian family farming. We found out that the metaphors that relate women with the land are used to deny both rural female work and the family land use. One of its consequences is that Patagonia has become one of the most affected by extractivism. We conclude reviewing the forms of economic and political recognition, which could intervene in future planning.

Keywords: Patagonian rural female work; family farming; land feminisation; territory; censuses; Argentina; women

---

1. Introduction

Feminist theories recognise that women's work is invisible in different ways. Mellor [1] argues that there exists an imposed altruism, that implies women themselves often threaten their own work recognition [2]. Mellor identifies the circumscription of feminine work to limited space and unbounded time, which also is unpaid or an altruistic work. In contrast, male work is understood in the context of unlimited space, with defined time and acknowledged by a reward. In Patagonia 'altruism', as a non-valuation argument, takes on a dimension which allows us to think about the aspects of the denial of Patagonian rural female work. For example, Leon [3] shows that with regard to women's work in Latin America, the vast bulk of rural female work was unpaid and women who had the most regular economic activity profile are those who were highly educated, not married or have no children.

The most significant number of women involved in certain occupations are in some degree an extension of women's domestic role such as teaching or nursing. León observed the largest percentage of women workers in urban areas. Thus, her paper perpetuates the omission of rural effort as work. What kind of female work is rural work and what particularities does it show in Argentine Patagonia?

This article considers that the limits to the recognition of rural female work are not only due to social discipline but also respond to structural conditions, which lead us to pay attention not just to the hierarchical organisation of society but also to review contextual conditions of the region.

Patagonia, understood as a whole settlement and environment, is a frontier region [4]. It was described in Argentinean State documents by intersecting the borders between human, animals, and plants [5]. This is because a significant part of the population, the native people, were defined as animals or even as 'natural hazards'. The forest was taken as a moral reference of the society that is trying to change. The trees became the symbol of social behaviour because botanists described plants incorporating human characteristics, that is, an attitude that is at the same time desirable for the population; 'Cypress, the winner' or 'larch, the majestic' are examples [6]. In the configuration of the Argentine State, this use resulted in a public policy called 'positive environmental fatalism' [7], p. 403.

The role of women, and the devaluation of their work, do not escape the weight of this environmental determinism. From this point of view, we address topics related to agricultural and livestock spaces in the socio-ecological systems of arid lands of the steppes. We seek to show that the recognition of female work and family farming (named in Argentina as family agriculture) face us with the necessity to incorporate ethnic and colonial elements typical of the region of Patagonia, since the analysis of feminine work involves hiding variables linked to racist or social-class discrimination [8]. This situation demands the recognition of particularities that we analyse considering that land and landscapes involve the female experience of work.

Núñez [9] appeals to the idea of land feminisation to show how the dependence of the Patagonian population was linked to a specific understanding of the landscape. To see the impact of the historical lack of assessment of female work in family farming, we must go through the meaning given to the territory to rethink the current implications of these processes in one of the most vulnerable sectors of the productive national framework, that of the family farming.

Within feminist literature, there are many studies related to the invisibility of rural women´s work [10]. In gender studies, Patagonia add the question around the particular situation of women and work located in border territory [11,12]. Patagonian women, as the colonisers of the American West or Australia, have been distinguished as especially strong and autonomous [13], whose strength, superlatively recognised in literature and romance, is diluted in terms of political or economic recognition [14].

The article proposes that, in case of rural feminine work in Patagonia, it was necessary to anchor women devaluation in the devaluation of the environment. From an ecofeminist perspective, the concept of woman and the concept of nature should be seen articulated [15], where the sense of one and the other are fed back into the same logic domain. Then the rural and urban conceptions of women allow us to explore the particularities of Patagonian space.

From this perspective, we describe from the lack of recognition they have in censuses' history, scientific research, local development studies and analysis of productions at the family level, crossing these observations with current official perceptions about rural producers. As a result, we found an articulation between the permanence of female subjects as subaltern and the limited valuation of family farming, which can be a starting point to rethink development policies.

This work seeks to explore how worker's recognition and planning of territory not only establishes specific logics of tenure but also relies on an unequal consideration of the population. This work adds the need to consider the work of women structural, in the process of constitution of Patagonian territory. As a summary of works that explore the Patagonian problem from development policies, we study the association of women's work and the devaluation of the space itself. Its negation as an autonomous space drags as a consequence the strong invisibility of scales of production, which are inserted into the reproductive and are dismissed.

The scale that we will analyse is that of family farming, which is an ambiguous scale of no precise definition in Argentina. However, broadly speaking, 'Family Farming is understood as a type of production where the domestic and the productive units are physically integrated; the family contributes the predominant fraction of the labour force used in the exploitation, and the production is aimed at both self-consumption and the market . . . family farming is a way of life, and a cultural matter, which objective is the social reproduction of the family in decent conditions. The management of the productive unit and the investments made in it are made by individuals who maintain family ties with each other. Most of the work is contributed by family members; ownership of the means of production (although not always land) belongs to the family, and it is within it that the transmission of values, practices and experiences takes place' [16], p. 5.

Argentina is a country with a great diversity of ecological-productive regions, so the characteristics needed to be a unit of economic organisation differ from region to region. Today, the category of Family Agriculture tries to integrate multiple dimensions (productive, social and cultural) of the subjects, as well as their economic, social and regional heterogeneity. In turn, being in a capitalist context, it does not necessarily assume a mode of production of capital accumulation.

According to Law No. 27,118, which is entitled as 'Historical Repair of Family Farming for the Construction of a New Rurality in Argentina' Sanctioned in 2014, a family farmer is defined as one who carries out productive agricultural, livestock, forestry, fishing and aquaculture activities in rural areas and meets the following requirements:

(a) The management of the productive enterprise is carried out directly by the producer and/or a member of his family;

(b) He is the owner of all or part of the means of production;

(c) The job requirements are mainly covered by family labour and/or with complementary contributions from employees;

(d) The family of the Farmer and Farmer resides in the field or the town closest to him.

(e) To have as primary economic income of his family the agricultural activity of his establishment.

(f) The Small Producers, Small-holders, Peasants, Chacareros, Settlers, Medieros, Artisanal Fishermen, Family Producer and also the landless rural farmers and producers, the peri-urban producers and the communities of native peoples included in sections (a), (b), (c), (d) and (e).

Producers of Family, Peasant and Indigenous Agriculture shall be characterised by the enforcement authority for their priority inclusion in the actions and policies derived from this law, taking into account the following factors:

(a) Self-consumption, Marginal and Subsistence Producers.

(b) Production Levels and Destination of Production.

(c) Place of residence.

(d) Net and Extra-property Income.

(e) Level of Capitalisation.

(f) Family Labour. Complementary Workforce.

(g) Other elements of interest.

In terms of family farming, Patagonia is interesting because it has very large area (1.043 million km^2), mostly arid, and contains the largest populations in the world dedicated to grazing. Besides, it has vast areas of fruit production, which in many regions clash with hydrocarbon extractions, hydroelectric power generation, vast national parks and areas of tourist interest. The female effort, from female farmers to prostitutes, is present in the region's literature. However, all this is omitted in the processes of measurement and design of development policies.

Throughout this writing, we will study the case of rural women, while the effort of both is especially hidden in this marginal context. The results have been organised into four sections. The first section presents an overview of the later Patagonian integration and the idea of sterile land, elaborated from the first scientific studies that analyse the space, to contextualise the territorial particularities where women developed their tasks. The second section shows the work of Patagonian women in history, taking censuses as sources. The third one inquires the recognition of rural production in censuses. The fourth explores family farming in Patagonian land, showing how the official measures of production sustain the invisibility and devaluation of rural Patagonian women workers.

2. Material and Methods

We seek to understand how the structural elements that have configured the historical expression of female work are impacting the recognition of family farming as a significant element of Patagonian development. The problems of recognition of female work are transferred to the territory itself and lead to the current considerations of family farming as areas located in Patagonia and organised around domestic efforts. To reflect on these problems, we start by asking how Patagonia was integrated as territory, and also how to measure what exists in terms of population, landscapes and resources. We propose to start the analysis from a historical perspective. After that, we ask about the present, both in censuses and in work recognition. Therefore, we will refer to the extensive literature in Spanish, since most of the results of the social sciences related to Patagonia have not been published in English. At this point, the sources are presented as an original contribution, practically unknown in publications outside of Latin America, despite the relevance of the territory regarding people and spaces affected by the problem. We clarify that we have maximised the references in English to facilitate the dialogue.

We have organised these sections in Data Sources; Theory; Outcomes; Implications for the recognition of women rural work and implications for the recognition of family agriculture.

2.1. Data Sources

We initiate our study from a historical perspective. First, we present the process of territorial integration of Patagonia [4,17]. In this section we will review military and scientific sources associated with the Patagonian conquest to start from the terms with which the territory was undervalued [18–20], both in relation to the environments and in relation to the populations.

For this goal we will start with the first national census, in 1869, until the present [21–26], analysing the way in which census recognition logics have varied. We understand the need to review the censuses as one of the most significant academic results, in which the State and policies consider what exists in dialogue with local producers' different senses of land. The quantitative modifications are reviewed in already published works [27].

We include the main State sources from this period by including the analysis of the National Congress debates during the process of incorporation of the Patagonia territory into the Argentinian state, to identify the terms in which this process was developed [28]. We also consider scientific, technical and military sources that characterised the region, while presenting a particular recognition of female work and rural activities at the domestic level, both from primary sources [18–20,29], and from secondary ones of authors who dedicated themselves to systematising the military sources of the army campaigns to Patagonia [30].

The analysis of production in Patagonia is done following three cases. The sources are related to human trashumance in the north of Neuquen province [29,31,32], the production of cherries in the parallel 42° Andean county [33–39], and the national agricultural censuses [24–26].

The specific question on family agriculture has been recently studied in Argentina [40–46] and also in Patagonia [33,47–50].

As we pointed out, they are almost unknown works in Spanish, even within Latin American studies. The effort of this article to bring these references closer to the debates in English publications can help to review global subaltern dynamisms.

2.2. Theory

Twenty years ago, McDowell [51] criticised the ethnocentrism of many of the key categories employed by feminist geographers and challenged feminist researchers to develop a theory that allows thinking in terms of differences and diversity. Much closer in time, Nzegwu, Bockover, Femenias and Chaudhuri [52] argue that the gender category effectively allows us to recognize space practices that are not so western. Already within the question of the invisibility of female work, the long tradition of existing studies finds its own line in the face of the question of rurality [2,10,53–55] that puts the senses of femininity in tension, and with it the gender relations established in these scenarios.

This article is still responding to those challenges for the concealment of the work of Patagonian women and their productions in censuses and development plans of their particular territorialisation in national reviews of the case in the existing bibliography. The work of Patagonian women seems limited to a practice that is not completed in a dialogue with the work of women in general. We have connected co-temporal invisibility and devaluation seeing diversity transcending the strictly human. We achieved this connection by inquiring how landscapes and people are seen or hidden from similar logics [15,56–59]. From a place so little considered in gender studies, such as Patagonia, diversity emerges giving clues about how gender, work and territory make sense in practices that transcend the limits that state recognition seeks to impose.

Hence, the question about the type of theory that allows identifying the particularity of a case so little studied cannot be taken as insignificant. Lewis and Simpson [60] analysis of the theory of performative effect observed that theory constitutes the 'conditions of possibility' [61]. Theories shape the world [62]. Within the theories, the construction of ontology is central, and many times the debates ignore that not only the discourses but the way in which materiality is conceived is crossed by biases of inequality. Femenías [63] points out the need to explore the construction of the history of submission from the deconstruction of theories that sustained it and give value to the struggles to overcome that inferiorisation.

Social and environmental issues are culturally appropriated and signified together. The possibility of an imposed altruism, suggested by Mellor, is supported on arguments associated with people, but also, with territories and landscape and economical orders. In our study case, Patagonian lands are territories of extractive capitalism [64], and the selected sources allow us to recognise how the description of the land is transferred to people.

Balibar [65] indicates that frontier regions contain institution-limits because frontiers are seen as fixed. Institutions in the frontier, and the frontier in itself, are needed by States as stable frames; therefore, the reference for change resides somewhere else. This subalternity, which implies that change itself is linked to the management of an external change conceived as more important, leads us to think about the similarity with the feminine. Patagonian women were presented as a framework that enabled the possibility of development of another—the male—at the same time that this meant their own stability [14].

2.3. Outcomes

2.3.1. Implications for the Recognition of Women Rural Work

Research on women's work in Argentina has reviewed mostly urban scenarios, linking access to work with the construction of citizenship [66]. Among other results, it is worth remarking that until 1926 women lacked rights to the point that the work of married women was illegal, and from there it is analysed how the laws incorporated labour as a right [67]. Consequently, our article moves toward the analysis of rural female work, not so much from the laws but public policy tools such as censuses and development plans. This is to show that differences in aspects of citizenship, such as the right to access work, deepened in rural areas.

The first two sections of this work are focused on the question about rural female work. It is observed that it acquires certain particularities in Patagonian cases since Patagonia is mainly a rural region. The difficult conditions of women's work in rural contexts is one of the main lines in feminist decolonial studies [68], because of the relevance of families' rural practices in those geographies. From a broader framework, Agarwal [54,55] argues that difficulties of rural Indian women are linked with the conception of female obligations. No matter what, they were conceived a natural resource; consequently, self-exposure is a social requirement. In Latin America, rural studies reach similar conclusions [41,42,69]. But as we indicated, it is a region where women colonists have been presented as highly empowered. The possibility of denying the work of women recognised in memories and poetics introduces a particularity that makes the case interesting.

Although León [3] recognises that rural female work suffers a major devaluation than urban female work and labour in Latin America, her effort, like that many others, did not mention the structural inequalities in land distribution in the continent and its impact on gender relationships [70]; or in the unequal economic evaluation of the effort [43], because these intersections are limited in rural studies. This denial happens even though in Latin America, feminism started to be contemplated by presenting a philosophical thesis about this topic as early as 1901 [71] indicating that the issue of inequality was not only given by material conditions but because of how inequality became normal [72].

2.3.2. Work and Implications for the Recognition of Family Farming

To develop the understanding the recognition of family farming, we appeal to the idea of land feminisation [9], because it allows us to realise the stereotyped male recognition of the landscape as a structure of social ordering. Baydar [56] observes that hetero-patriarchal understandings of space based on masculinist premises have largely ignored women and queer subjects who may subvert or alter normative spatial practices. We add the need to recognise the historical depth of these processes.

The described conclusions on historical rural female work intersects the difficulties in recognising actual female work in family farming. One of the reasons was that prejudices such as subalternity and women's dependence when measuring the country are transferred from the conceptualisation of women to the territory. The Patagonian case allows us to reveal how, even in scenarios where rights were extended, territorial recognition did not change, which still impacts on invisibilities of female work, especially at the level of current family farming.

This is associated to another debate. Researches of economic recognition in Patagonia show that it is affected by conditions of subaltern integration that facilitate an extractive land use [64,73], but that detracts from efforts in general. Political geography considered that extractivism has been possible from the recognition of territories of sacrifice [64], and where family agriculture is ignored.

We will show that Patagonian women bodies have similarities with the idea of the territory of sacrifice; they were needed and, because of that, were doomed to disappear. Struggling Patagonian women of the early 20th century have been recognised as necessary but problematic. The activities they developed were of such relevance that it was impossible not to recognise them. Women and their work were considered indispensable to survive; but, at the same time, they were proof of the failures of the State. If the State had been present correctly, it would have placed women different from the frontier ones, similar to a fragile bourgeois denounced by liberal feminism [14,48,74]. Hooks [75] denunciates white feminism, indicating that in the search for highlighting, its weakness and oppression deny the exercise of inequality in a racist manner or, in the case that concerns us, in a geographic one.

Patagonia, inserted into the country characterised as an area of extensive livestock and exploitation of energy resources, was particularly affected by the devaluation of its population and by an unusually marked demand towards reproduction for the very possibility of subsistence. Because in addition to being rural, Patagonia is cold, and this generates a higher load of resources aimed at heating. It is also arid, with which water and obtaining food is a central issue [76].

Thus, this issue long travelled from the feminist analysis, and it is discovered as a specific form of devaluation, which projects in the shadows not only impacts on women, like sexed bodies that strive, but on the activity 'smallholder' developed by them, and by many men. We can think that this inequality is projected in the social set, whose effort is invisible—and then feminised—while the possibility of development needs to be resolved in other spaces, locating the local population in a tutelary dynamism. In studies about the abandonment of rural practices, the main problems of rural practices has been seen in the economic concentration and the crisis of the capitalised familiar production; the great property, the large estate and the concentrated rent; the situation of the peasants and indigenous people; the preservation of the environment and the common natural goods; the living and work conditions of the workers and rural labourers; and the set of the specifically agrarian expressions of the dependent condition of the country [77,78]. In general, gender aspects are not mentioned.

The motivational aim of this article is to provide arguments to reflect on the existing public policy. The research objective is to show how Patagonian female rural work has been hidden over time, and the way aspects of the territory have been denied, which results in the legitimisation of land use that continues to despise family farming. The proposition guiding this work is that an approach that relates the recognition of space to the recognition of women is an original way to challenge development policy to dismantle structural paternalistic order.

The mentioned historical censuses are in the National Institute of Statistics and Censuses: INDEC's repository, in Buenos Aires, Argentina. The secondary sources are presented in the reference section.

3. Results

3.1. The Patagonia, the Later Integration and the Sterility

The Patagonian case has particularities. Argentina is a country with complex territorial integration. Girbal-Blancha [17] indicates that Argentina could be divided into three regions linked with their logics of territorial integrations (Figure 1). (1)The central zone that includes Buenos Aires and the Humid Pampa, base of country organisation since the second half of the 19th century, in grey; (2)The provinces with a political organisation inherited from the Spanish colonial period and confronted against Buenos Aires for territorial control until the second half of the 19th century in white, and (3) The historical National Territories, which were areas outside the Spanish domain, under control of native people, without political rights in the Argentinian state, in black. Figure 1 shows these different aspects.

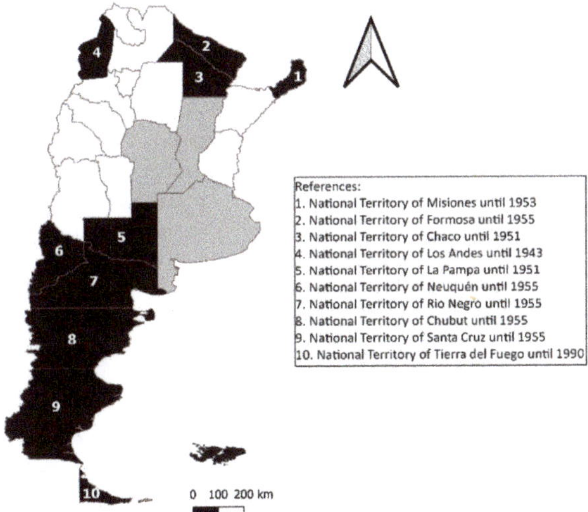

Figure 1. Argentinean map. Historical Central Region in grey, historical provinces in white and National Territories in black. In the references, it is mentioned the year of recognition as province.

Until the mid-19th century, regions 5–10 were mentioned as Independent Indigenous America [20]. Argentina, as a country, only influenced some ports on the Atlantic Ocean. In those years, several scientific expeditions toured the area, but one of who impacted the most was Darwin aboard the Beagle. The English naturalist in 1839 presented Patagonia as a 'cursed land' condemned to aridity. This idea was taken for the desert imaginary that affected Argentina. As we said, the idea of desert applied more to a government program than to an environmental condition. Basically, 70% of the Argentine lands are arid, but that is not synonymous with a desert. It was the influential thinker Domingo Faustino

Sarmiento [79] who developed this idea in 1845, when he wrote: 'the problem of Argentina is the extension, the desert surrounds it everywhere'. By desert, he referred to the space occupied by his political opponents which he located on the site of barbarism.

The same metaphor was transferred to Patagonia to explain the territory to be conquered. The idea of a cursed infertile land was added, not because of its conditions but its inhabitants. Thus, at the end of the 19th century, the Argentinean State conquered the regions in black by a military campaign called 'The Desert conquest'. From the State´s perspective, the savage nature of Patagonia could only be controlled and organised by the State. The metaphor with which the Argentinian state is presented, refers to a 'father' who protects the region that, released to its nature, only hurts itself [80]. In official debates, the Patagonian land appeared as a rich supply of resources for national development and the making of in-depth studies of the region was seen as crucial for the future of the country [28]. This 'virginal space's fertility' depended on the masculine hand of the producers and the male perspective of the State regarding public policies among other things because as we mentioned, in those years, women in general, had substantial legal obstacles to access recognised work.

In Patagonia, issues around the native people valuation are also crucial to analyse. Many studies recognised strong racist elements in the constitution of Argentinian identity [81]. But, a more attentive reading from a gender perspective shows that the idea of 'Indian race' did not involve a human group [82]. In Patagonia, the native people race was presented with a masculine connotation, because the males were the ones who acted and held ethnic belonging. Francisco Moreno [18], one of the leading explorers of Patagonia during the 19th century, describes the lack of native men's capacity to produce, which contrasts with women´s effort. 'Indian' women lost their ethnic character when presented as mothers. The idea of differentiated ethnic recognition, according to race, is not new. Concerning black women in Latin America, Gilliam [83] points out that the difference between male and female slaves is sexualised so that violence over them becomes romance; in consequence, they are denied in the independence processes. In the story about Patagonian women, regardless of ethnicity, the overvaluation of care, which brings us the idea of 'altruism' cited by Mellor [1], poetically conceals the difference between male and female. Above all, it hides the effort of women as workers or warriors, denying the very possibility of recognising them as active agents of development or conflict.

We can argue that in the 19th century, women's work capacity was not recognised because it was not commercial. In 19th century's sources all trade and production were conceived through male management [18,79]. In official arguments, race was taken to make evident the inadequacy of that native settlement [84]. Settlements and activities are presented as masculine in official descriptions and public plans. This situation has two consequences, one on the territory which is explained as a border of the correct development because there is a centre that contrasts as a role model [17]. On the other hand, the feminine work of native peoples is not seen as incorrect, but neither is it human, allowing to interpret that feminine work operates as another resource of the territory [30,74]. Thus, the initial legal issue was transferred to other orders that increased inequality.

If we review the introduction of scientific studies during the Desert Campaign, we have a clue to understanding the constitution of work recognition in Patagonia. Alberto Ebelot, at the beginning of the zoological study, mentioned:

'It was necessary to conquer these 15,000 leagues so real and effectively, to cleanse them of Indians so completely, so unquestionably, that the scariest of the scary things of the world, the capital destined to revive the livestock and agriculture enterprises, had itself to pay homage to the evidence, which did not feel suspicion in springing on the footprints of the expeditionary army and to sealing the taking of possession by the civilised man of such dilated regions' [19], p. XI.

The actor of change was the capital because the purpose of every activity was commerce. After the conquest, the principal activity was sheep breeding, managed by international trading of British founds [20,38]. From this, a territory that was presumed to be capitalist required specific actors who not only devoted themselves to production but also became actors of the trade. Those actors could not be natives nor women.

Since the conquest itself, the female work was hidden, even when they were paid for their duties, the recognition on land that was given to the soldiers was not granted to the four thousand women that accompanied the military campaign in 1879. At the end of the conquest it was denied that they had done work for the State [30], but even that the effort was evaluated as masculine [14]. In the State stories, women are described more as things than people. This has an impact on their efforts not being valued as work, even when there was a salary payment. This becomes so extreme that at the end of the 19th century, when the military counted the stolen things by natives in frontiers farms, they mention horses, work tools and also women and children in the same category [30]. In Argentina until the middle 20th century women did not have political rights, not even full access to their civil rights. Married women had serious limitations for the practice of their individual liberties [85]. It has impact in Patagonia, where the State paid for work which was illegal. In consequence, to hide women's work meant to hide the State contradictions.

Now, the reason why women could not be seen as economic actors was supported by a national construction that contained Patagonia but was not limited to it. The boundaries between 'economy' and 'culture' share paths with the differences between the productive and the reproductive, and that is a global issue [86]. Besides, the landowner scale, promoted by the Argentinian State at the end of the 19th century, was consolidated as the base of Patagonian land organisation. In that context, only extensive farms and capitalised producers were recognised [78]. It was not only a strategy of devaluation of women but also of a way of thinking about work and property. Picone [87] analysed crimes in the territory, observing that women were particularly vulnerable to theft and expulsion from their lands. In our study, the information provided by censuses introduces the idea of a vulnerable land. This is based on generating descriptions that economic agencies deny and management capacities in family units and especially in women.

3.2. The Measure of Work of Patagonian Women in Censuses' History

The measure and the representation of women are feminist issues because the objectivity, in a strong sense, has been a tool of androcentrism [88]. From there, we discuss national censuses, because many spatial senses installed social links that enhance sexism and racism into landscapes using numerical arguments elaborated from censuses data.

3.2.1. Historical Censuses

The censuses have been the most important tools to register what the State accounted as existing in the territory. The first Argentinean censuses were signed by the principle responsible authority. Diego de la Fuente was the author of the first census of 1869 and the second one in 1895. The 1869 census ignored information of Patagonia, a situation that was solved with the 1895 census, made after the desert campaign. Alberto Martínez was in charge of the next census in 1914.

The 1869 census indicates that measuring the materiality of the country is also measuring morality. This data was presented as 'the first inventory of living elements that integrate nations'. The document also indicates that 'statistic measures revels organic, physical and moral, social and political conditions' [21]. This census differentiated male and female professions. In male professions, it mentioned 'male healers and female healers'. It also indicates that Buenos Aires had the main professionals in scientific and liberal professions. It also compared the results with Chile, highlighting that it had thirteen thousand percentage more agriculture, and twenty-three thousand percentage more miners [21], p. XLIV.

The female work was mentioned as 'seamstresses, washerwomen, weavers, ironers, cigar-makers, tampers, etc.' [21], p. XLV. The women workers were seen as a population which 'waits with uncertainty the sustenance of the wage'. The female work, in opposition to the male one, was a mark of weakness. In this case, prostitution was measured, taking as evidence of the extreme female fragility. This data was not taken in subsequent censuses.

As mentioned before, the first census did not take Patagonian data. Cerdá [89] indicates that this census makes explicit a social way to consider working woman, seen as fragile and weak. Cerdá identified the way that these censuses hid primarily rural female work, actually in emblematic activities for women, as was the work in the grape harvest.

The second census in 1895, entitled 'Census of Wealth and Population', registered population data as well as biophysical, agricultural, industrial and commercial data and presented, for the first time, results from Patagonia. It repeated the devaluation of female labour by highlighting that the male population is larger than the female population; the author indicated that this was ' . . . undoubtedly advantageous taking into account the work of men, more active and better paid, which contributes to a greater degree than that of women to the aggrandisement and prosperity of the nation' [22], p. XXXVIII.

The 1895 census remarked some measures that started to value women. Thereby, the basic scholar instruction was mentioned as a significant characteristic to look at, divided by sexes, 'since it is known that in many nations, by giving a great preference to the male sex, the most beautiful part of humanity is deprived of the benefits of instruction, leaving the mothers of future generations in ignorance' [22], p. LXXXV. So, the recognition of the feminine was not linked to development, but for being the mothers of future generations.

The conceptualisation of women was built around the idea of mother and wife with a nationalistic emphasis. Foreign women were considered more hardworking than Argentinian women: 'the foreign woman always helps her husband in daily work and thus becomes an element of production and not a heavy burden' [22], p. CXIII. In opposition, the opinion on local women was defined by their relationship with their husbands: 'The Argentines, on the other hand, are not so hard-working . . . they find it less easy to form a family . . . in which, in general, the woman is only conservative of the goods produced by her husband's work' [22], p. CXIII. We can think that, unlike the native women who lost their ethnicity when they were seen as women, the European and North American foreigners did not do it in the same way since their ethnic origin was linked to their different work capacity, even when this work was under-registered.

The census hides the intention behind the number in the pursuit of seeing the existence of everything. Looking at how the data was released, we can recognise that there was a differentiated hierarchy in the assembly of the questions that intersected the racial, sexual and national hierarchy in the survey design.

The occidental over the native, the foreign over the national, men over women, and the urban over the rural were assumptions of the censuses, which displays national land planning policies. Cerdá [89] asks why these first censuses did not consider many women working in viticulture when the grape harvest was an almost exclusively feminine activity. In fact, the 1895 census recognised concern for the large number of women without profession: 'This shows that it has not yet been possible to give a useful and directly remunerative direction to the work of women, who, deprived of their own means of subsistence, must be entrusted to the protection of man' [22], p. CXLII. Thus, the paternalistic idea of protection resulted in a non-recognition of female labour, which resulted in survey methodologies. Figure 2 shows how the way of measuring changed the number of people recognized as workers. This graph shows how the percentage of working women decreases from the end of the 19th century until the end of the 20th century, when it begins to grow.

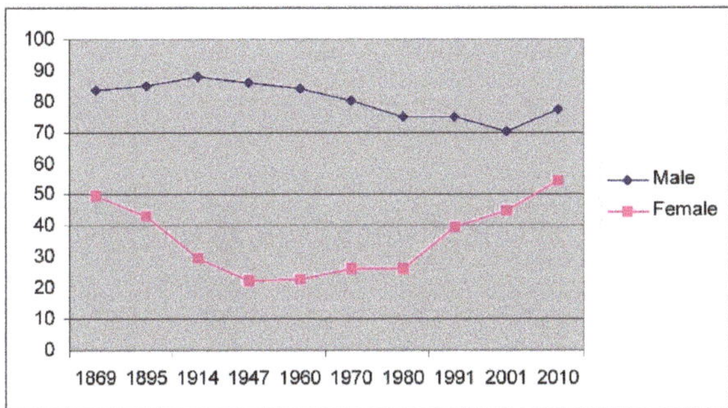

Figure 2. Changes of participation rates of male and female work taking into account INDEX Censuses dates. [27], p. 4.

Manzano [27] analysed female work from 1869 to 2010 censuses. He observed that the participation of women in the working world was more significant in the first two censuses, and then it decreases. We found percentages of working women similar to those of 1869 only in the 21st century.

The first census observed fifty per cent of women up to fourteen years old as workers, whose professions were female healers and prostitutes. The second census excluded those activities and unified male and female professions in the same table. Accordingly, all population was considered in the same group of employments. However, the census introduces a difference related to territorial integration. The distinction between male and female workers, and also natives or foreign ones, was made only in historical provinces. In National Territories like Patagonia, the list of professions was also put in rows; in the double entry tables that present the data, the types of work intersect with the name of the territory, without any other distinction.

3.2.2. The Invisibility of Rural Patagonian Women

The first census that showed specific data of Patagonian women workers was the one carried out in 1914. The female work detected in this census had new characteristics. The professions identified were similar to the previous census, but the number of women without a profession increased. That is interesting because the census author asked about this change and gave an amazing answer. He indicated 'Within the organisation of all civilised people, women have their natural basis in the home ... men must think about the needs of the family. But the part that those women take, in the economic life of modern societies, is very large and tends to increase every day, instead of decreasing' [23], p. 252. The difference with this global tendency was explained through the survey methodology. In this census, female work was not recorded by itself; it was omitted in households with income from male labour. This census maintained the non-legal recognition of the work of the married women, which was illegal. The remarkable point is that even though women's rights and access to work have been improved since 1926, no census impact is seen until 1990.

The low consideration of female work was not only an Argentinean characteristic. Quay Hutchinson [90] gives evidence that in Chile the censuses had limitations around accurate measures of female economic activities. The author observes that the progressive modernisation of the censuses implied substantial changes in the collection and interpretation of data. As a result, women's economic activities became increasingly invisible as the 19th century progressed.

So, in Patagonia, the female work, from Argentinean or Chilean States, started to be measured at the same time that the hidden structure was built. The lack of political and economic recognition is a decision that has an impact on the lack of census and statistical attention. All this had repercussions on structural policies that denied female labour, especially in the field.

Even when 1914 census introduced the question of male and female work for Patagonian territory, it excluded any female initiative linked to any other male activity. However, if we analyse Figure 3, we can observe that the denial of effort focused on rural rather than urban jobs. Patagonia was centrally a rural area, but the feminine work censuses were mainly linked to the industry, the services and the handicraft.

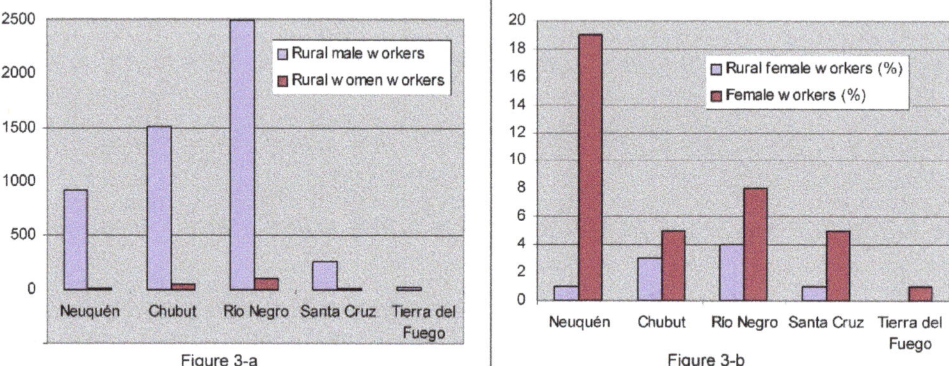

Figure 3. Characterisation of work. Female rural work in Patagonia in 1914. (**a**) Differences between male and female rural workers. (**b**) The percent of rural women workers in the general percent of women workers in each province. Own elaboration on the base of 1914 Census.

These numbers shed light on the gender discrimination behind the survey [27,40,74,89,90]. Of the total number of women in the territory, those who work are a maximum of 19 percent in Neuquén. In a rural area with urbanisations of small towns, whose 'cities' at most reached a few thousand inhabitants, the women recognised in their work are urban, those gathered in the small towns that began to grow. So, the rural women´s invisible work is much higher in Patagonia, given that the national percentage of women at work is just under 30 per cent.

If we consider the results since the 1869 census that recognised 50 per cent of women as workers, the denial of the effort of women who are pioneers in a region without infrastructure is even more striking. In contrast, working men reach almost 90 per cent. This is not just a recognition of the population; it is a construction of the territory. Patagonia in 1914 was still being integrated into the national territory. The differentiated attention accompanies a speech that says that the only way to produce is in large farms, placing small family producers in a place opposite to that of development. In 1935, Sarobe denounced that with this argument, the police force is used to propitiate land concentration, a problem that reaches to this day [20]. By 1958, President Arturo Frondizi himself, inaugurating mining works in the region, speaks of the 'virile effort' that made Patagonia grow.

But as Sarobe already senses, this is not only a lack of recognition of women, but the entire family productive structure is denied in politics, despite being recognised in the poems. To understand the transfer of female work invisibility to family farming, we observe how the problems in recognition of production, and not only producers, adds elements to the comprehension of that phenomena and its linkage.

3.3. The Devaluation of Patagonian Production

In Patagonia, the recognition of local production at a family scale had its own characteristics. We notice an earlier recognition of rural Patagonian production, because, in fact, Patagonia was conceived as a rural production space even without knowing the field. This definition without knowledge brought problems, since many areas were identified as marginal in productive terms, and many others were marginal in the sense of being a national boundary. Moreover, those conceptions of marginality were mixed.

3.3.1. The Thashumance Case

Daus [29], as the main reference of Argentinean geographical studies from the middle of the 20th century, sought to explore the economies of the marginal regions and took the case of Patagonian goat transhumance practices as one of them. In his description, Daus talked about the zones almost uninhabitable for humans, which were actually inhabited. These humans, who inhabit uninhabitable areas, were those who dealt with that economy that Daus presented as not so reasonable, omitting the necessary family structure to make this possible, or referring it only to indicate that it was not economically relevant at the time.

The type of animal and the type of management justified a lack of recognition that still opens paradoxes. Silla [31], for example, recognises in the transhumance of northern Patagonia a porous identity that tends to be something that does not end up being described as Argentine, Chilean, friend, enemy, thief or merchant. It is presented as an enemy of moral patterns and rationality. Therefore, the population dedicated to these activities is perceived under attack by both the Argentine and the Chilean national states. Michel and Núñez [32] observe that the response to many tensions in this type of production seeks alternatives in the international market as an exclusionary strategy. Michel [91] even argues how the answer to the problem of land tenure and lack of water was answered with the management of the 'Denominación de Origen' (DO) for the Creole goat, as an argument for sale and export. In other words, it continues to be argued that the exit to a foreign market is the solution to the paradoxes of the region.

The bias of taking the international market as an answer to any productive question appears throughout the Patagonian territory. The chronology of rural production in the Andean region repairs the market and not the local productive dynamics because those dynamics were built with elements and networks of family production. These networks are presented as constituents of the local identity in the area of the Andean region [4,37]. Family agricultural production is affected by invisibility, similarly to the women's case. The previously cited transhumance practices are usually presented as a male activity, but in pictures, we can see women, kids and goats, presented as objects (and not subjects) of activity [29].

The farming of cherries is another example of a characteristic activity of the southern Andes. The production of jams, liqueurs and pickles are shown as characteristic of the region, and it is one of the most reviewed activities, offered in many touristic guides about Patagonia. Farms with a family organisation, represent in tourist guides the history and the present of their settlers, but if it ignores the operation of any establishment that does not focus on the objective of the sale.

3.3.2. The Cherries Case

The Bolsón valley and the Andean Shire of 42 Parallel are one of the most avowed Patagonian tourist destinations linked to rural activities, around which it is called 'fine fruit' (it involves mainly berries and cherries). It is also seen as a region for alternative ways of living against capitalism from the 1960s. Nevertheless, the local references locate the beginnings of the berries' and cherries' production in the 1960s. In those years, the trade was established as the main activity objective [36], although the beginnings of these plantations find antecedents in the decade of the 1940s. It is remarkable how visibility depends on the possibility of commercialisation in formal logics, even in a place with migrants discussing capitalism.

Danklmaier, Wienke and Riveros [37], p. 34 deepen the analysis of cherry production. They observed that it is one of the most emblematic of the area, and the quality of local production is superlative regarding the rest of the country. Furthermore, its commercialisation is given from a valuation of the product that has to do with the local identity. However, it is observed that although home productions move a significant volume in the informal market, there are currently no statistical records of this sector. The production for 'self-consumption' is also no knowledge, despite the extensive network of exchange between domestic products, which is the foundation of the identity and the sustainability of the production in the area [34–37,92].

The 'self-consumption' production, as the familiar farm, needs to be complexed in all Patagonia. It is a small scale of production, which involve a production site and exchange networks. It is not necessarily subsistence [93], but it is measured as subsistence by the State [74]. Several studies reiterate the relevance of horticultural production as to fruit production in the household order. However, rural household production, which is the basis of family farming, belongs to the constitutive order of productive identity, not to official or mercantile recognition. This situation is demonstrated by the repeated omission of data on this matter. There are no specific statistics that allow knowing the actual dimension of the harvested and processed production. Valtriani [35] recognises something similar in forestry. Trpin, Rodriguez and Brouchoud [48] have the same observations in pear and apple Patagonian productions.

3.3.3. The actual Livestock Censuses Case

The national agricultural and livestock censuses are the central tool of Argentine State to explore what is understood as the agricultural and livestock capacity of the country. Through this global survey, the environment is presented as a natural resource. However, this is only a process at the national level.

We can think that this census reflects a global look at the country, meanwhile the censuses since 1950 are designed in accordance with FAO guidelines. It promotes the use of international standards for concepts, definitions and methodologies from that program. In addition, since 1952 the censuses in Argentina began to be carried out in a decentralised manner, and each provincial or territorial government was directly responsible for the tasks in their respective jurisdictions, although normative centralisation was maintained.

The National Agricultural and Livestock Census should be measured according to regulations contained in the census manual. In that document, it is explained how to fill out the specific census forms [24–26]. The farm unit was defined by the census as the production organisation unit, with an area of no less than five hundred square meters, which produces agricultural, livestock or forestry goods that are destined for the market. This point is central to the issue at hand. The censuses only recognised what is traded in markets, especially the export market, a situation that was established in the earliest censuses. It is interesting to note that the current censuses remain without discussion.

The 2008 National Agricultural and Livestock Census introduced a change in the data collection. It explained that production was what goes to market and that self-consumption was not production. The previous censuses indicated that production was just what gone to the market, the 2008 census especially indicates that self-consumption is excluded. The producer units that never allocate surpluses for commercialisation were not counted [25], no matter the size or the social relevance.

This draws attention. In 2008, the debate about family agriculture was extremely relevant in Argentina. Additionally, many countries, such as Colombia, changed their censuses in order to count family agriculture and self-consume as part of the economic structure. The last Argentinean censuses took that debate and resolved not to modify the historical view, even in this context of recognition.

However, this is not the only denial. Even when all regions have productive species, some regions included more survey variables than others. This form of differentiated dating introduces elements that disciplining the recognition of family farming because the most commercialised species are taken as the focus of interest. Nevertheless, those are not precisely produced on family scale. The rhetoric of the value of family agriculture was blurred in the thick production cuts, reflected in Figure 4. We can see the number of variables measured in 2008 and 2019 censuses for each mentioned species in each region.

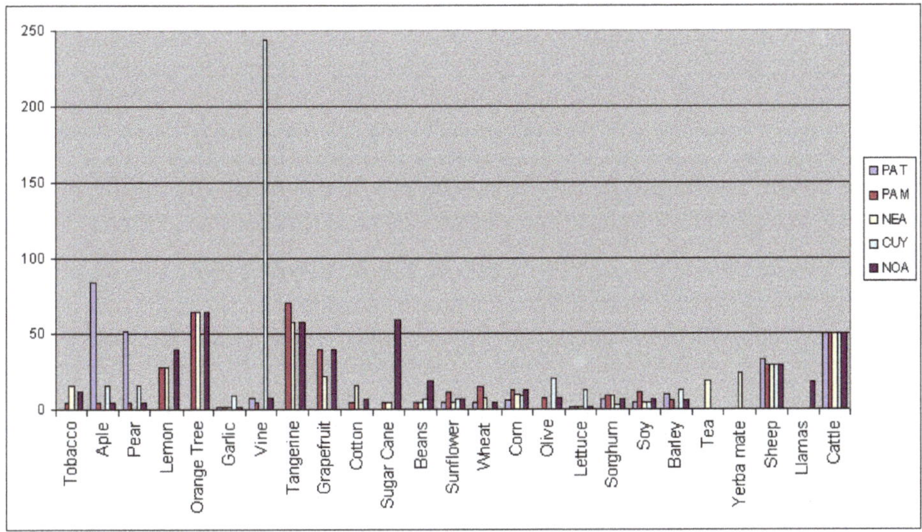

Figure 4. Number of variables for each regional census production. Own elaboration based on the 2018's census. PAT: Patagonia; PAM: Pampa, Argentina central region; NEA: Argentina North East; CUY: Cuyo; NOA: Argentina North West.

It is relevant to keep in mind that in National Agricultural and Livestock Censuses each region is studied in specific variables and number of characteristics. Figure 4 shows the number of variables for each production in each one of the five Argentinean regions.

The Argentinean regions involve specific provinces in censuses. PAT, the Patagonian region, includes Neuquén, Río Negro, Chubut, Santa Cruz and Tierra del Fuego. PAM, the central region, contains Buenos Aires, Córdoba, La Pampa, Entre Ríos, San Luis and Santa Fe. The NEA, Argentina North East, includes Corrientes, Chaco, Formosa and Misiones. The NOA, Argentina North West, contains Catamarca, La Rioja, Salta, Tucumán, Jujuy and Santiago del Estero. The CUY region refers to Cuyo, a region formed by Mendoza, San Luis and San Juan. Figure 4 shows that PAT (Patagonia) is the region that contains less productive species, besides not recognised as a specific one, except apple and pears.

In Patagonia, it is clear that the regional particularity is marked by pears and apples, which are distinguished from other regions. In the livestock area that concerns us, we see that the relevance in detail is comparable to any other part of the country. If we take into account that the relevance of pear and apple production has generated inequalities within the region [94], we can think that in Patagonia, having a higher number of variables allows more adequate policies. By centrally relieving export production, it follows that the policies related to this trade are of greater interest than those related to staying and intern produce. In this way, only because of this differentiated relevance, the primary data needed to build policies is delineating the reason for the continuity of invisibility.

From here it is observed that the regions with more productive species are, historically, the areas of greater connectivity and political recognition. Patagonia is out of this situation, and we can see that the central productions are seed fruits and minor ruminants. Concerning rural women's work and family agriculture, more than census variables, it is interesting what is not surveyed. There are no variables or elements in the censuses that account for native production (plants, animals, cultural practices) outside from the strictly commercialised commodity. For example, when the census asks about breeds in sheep, only those exotic species that belong mostly to large farms such as the Merino and Corriedale breeds are present, but the Criolla or Linca breed of great importance in small producers and rural women are not present [95].

Regarding the goats, no specification of the breed is requested, showing the low productive value as opposed to the sheep or other productions where an exhaustive detail of each variety is requested, in line with the limited state assessment of the Northern Patagonian transhumance linked to this production. This type of activity corresponds to that of the small farmers and family agriculture that for more than half a century has been presented as not belonging to the economy, marking a continuity with 1947 Daus's view. Currently, there are no varieties of Andean potatoes, Andean maize or creoles varieties in Patagonia, which are present in almost the entire national area. This exclusion ends up leaving out the main part of the family structure production, repeating the unmeasured and unrated over the years.

3.3.4. The Invisibility of Rural Female Work

Female work, directly denied until middle-century, with a recognition that has just begun to be statistically noticed in the 1990s, structured the identification of the production of those who were considered the actors of the economy. These were the export capitalists. The censuses were configured according to their interests and practices. Thus, the approach that what does not go on the market is not production is presented as the main bridge linking the denial of female work to the impossibility of recognising family farming.

This is because the primary production that allows subsistence, which since the exchange consolidates family ties and friendships and promotes barter in disaster processes, is not seen as production nor is it considered in public policies. It remains in the unknown practices characterised as 'informal' and 'domestic'. Words that justify not paying attention to them. But they are the very basis of family farming, they are the first references in the interviews cited in Patagonic studies, and yet they are systematically dismissed. The gender problem, in this way, is linked to that of territorial integration by introducing the need to rethink the 'gender' category in this matter.

For years, different 'minorities' have reviewed the Euro-American burden of the gender category [52] that does not allow us to understand pre-colonial cultures, to which we would also add those forged on the margins of capitalism. It is interesting because the Patagonian case should not be regarded as part of the otherness of the Western order, but as a frontier of it. The incompleteness that creeps from the notion of a desert does not generate a 'being outside', but a 'being subordinate'. Femenías [96] reviews this distinction in the Latin American scenario referring to the need to think about gender from dynamic intersections, deepening into these dynamisms. Moving away from an essentialist binarism, which in our opinion is one of the leading gender theorists of Latin America, shows the need to think about the transits and moves that the philosopher travels in the breadth of Latin America. The idea of 'feminisation of the land' elaborated for the Patagonian case, is an appropriation of the view of Femenías. The feminine, in the case at hand, is not only the body of women. The metaphor of the feminine slides towards other bodies, practices and things, in a line that brings us closer to the problems articulated by ecofeminism [15,57,58]. Not seeing or measuring is, according to Plumwood [58], a decision to recognise only the instrumental links. The criticism does not aim to ignore them but to think what tools we need to see what does not enter into that instrumentality, but that constitutes us.

In the case at hand, family farming, like women, is fundamental for social development in general. However, it drags a history of denials that results in its permanently subaltern location that, when investigated, is recognized with remarkable resilience. Because almost as a challenge to statistics, it continues to exist as one of the most massive rural productions in the world.

3.4. Family Agriculture in Patagonian Land. The Permanence of the Denial of Women Work

The recognition of family agriculture goes through the low recognition of rural female work. In Latin America, this scale started to be analysed in the 1990s decade [44]. Conti [33] indicates that family agriculture adopts changing characters in the Patagonian scenario. This scale considers family farmer to any person who lives in rural areas and works in agriculture together with her/his family.

In Argentina, family agriculture is a scale of great recognition in rural studies [45]. In South America it is presented as one of the primary food providers, a guarantee of food sovereignty, and as the basis that supports the lifestyle and cultural practices [40–43,69]. In fact, studies at the scale of family agriculture are considered as overcoming previous analyses perspectives. At the end of the 1980s, the question about local producers appears in many regional rural investigations. This was clear in the North-West area, linked to peasant conflicts [97].

3.4.1. The Problem of the Market as the Only Option

Even after decades of specific promotion of family farming, the most basic of it is still considered today outside of the production cycle. The paradoxical overvaluation of the market as a source of answers feeds the contradictions of the local economy because productions with more than seventy years of history in the region are unknown. The production is made visible only if it goes on the market, despite its history and social meaning [98].

The self-consumption and family farming are not seen as part of the productive logic, although it has been the focus of the development policies for more than twenty years. It is worth mentioning, as an illustration, the Mohair plan that proposed the genetic improvement of goats in Patagonia [99] about which it was indicated that one of its weaknesses was that it traded in local networks. Thus, not only self-consumption but internal marketing is seen as a problem. The idea of an armed territory from an external demand crosses Patagonia. Local organisations and populations, in general, are considered strengthened from a commercial intervention that excludes reviewing ties of solidarity, customs or ways of understanding the problem of land tenure and lack of water, which is often not supplied by trade. The rural population of Patagonia in many cases is Mapuche, that is, a member of native peoples, so the relationship with the environment, animals and commerce does not necessarily follow the modalities of economic reasoning. Besides, the sale is not separated from the local affections. Conti [33] shows the case of a wool cooperative that stopped selling pastures to a buyer who paid very well because he left production to rot. This, even without getting new buyers.

From here we seek to clarify that the problem is not to consider the relevance of the commercial variable for export, but the fact of having seen it as the only one impacted on the devaluation of self-consumption, local networks, emotional ties and cultural senses of production and trade, in the same line that Plumwood proposes for instrumentalization.

3.4.2. Idealization of Family Farming

It should be noted that the memories carried out in South America and Patagonia drag a gender perspective, since the affective senses of production, when they are relieved, are found more in the reflections of women rather than men. Because in them, there is the reproductive burden that gives social meaning to the set of activities that are carried out, including trade [41–43,69,92].

The Patagonian gender studies are still set out to highlight the fluctuating productive-reproductive character of family farming [40,100,101], because the reproductive part of an economic order is not an issue. Paz [102], adds another issue to the discussion; he opposes the category of the peasantry to family farming, indicating that family agriculture, as an organisation, is seen in many academic works as avoiding the history of the struggle for land, which has characterised peasant organisations. In many cases, they are linked to organisational practice with clearly feminine matrices. The author denounces a view that idealised the practices of the central area of the country, denying regional particularities and blocking the development of appropriate interventions. The most critical references, in one way or another, record the problem of subalternity in territorial integration, which results in the devaluation of the scale of agriculture sustained mostly by women.

From this writing, it follows first, the denial of the struggle for land tenure as constitutive of rural practices, but also that the experience of the centre is transferred to marginal spaces. In this transfer, the tools to deny the feminine actions are strengthened while policies are set in place to dismiss the initiatives coming from the domestic order. We add that it also replies to the invisibility of female work, appealing to a romantic perspective still present in studies about family farming. Carrapizo, Speranza and Ganduglia [46] are an example of analysis that assumes that the success of family farming in an individual process without significant conflicts. In their work of some impact on regional policies, they move away from problematizing structural policies, placing female work in the anecdote and success in international trade.

3.4.3. Contributions from the Gender Perspective

The history of recognition and work invisibility challenges us to reflect that feminist economics has sought to review in structures often marked by patriarchy. The story told contains both recognitions and devaluation, which challenges us to follow the path of the feminist economy, which managed to review the structures that the patriarchate has denied [1,2,10,53–55]. This discusses harmony as part of family farming, highlighting essentialist views that idealise this structure, as well as the need to study the complexity of the family-scale in rural production. Here we could return to the reflection of Femenías about the need to review the changes in the male-female orders that are established in societies. The constructions of femininities transit diverse materialities in the spaces of subalternity. Gender tensions also dialogue with alliances and empowerments of women who often rely on the social construction of difference. In Patagonia, the reference is often loneliness, the companionship of the man who co-inhabits a region of very low demography and with connectivity in extremely precarious conditions, with cold climates, a couple seen in many cases with such relevance that justifies the denial of domestic violence. Progressively, artisan organisations begin to consider themselves as an alternative to this situation [92]. The steps that follow are not linear, nor do they respond to gender theories. They are possible, at the crossroads of economic, political and family strategies that demand a greater vision regarding the process they carry out.

The visibility claims of rural women's groups allow us to explore the challenges they face. Among them, disputes over land uses are one of the largest and are not considered mostly concerning the senses of family farming. Raguileo [50] has reviewed some tensions regarding oil holdings and capital investments. But they are not the only tensions. Cobelo [34] indicates that tourism is structural to land use in the Patagonian Andean valleys and that it adversely impacts rural activities. This tension is more interesting because these are initiatives that should not collide with the objectives, but which clash with the structural denial of local work, resolved at the level of family farming.

Tourism structured land organisation in the area [34,87,94,103]. The urban logics are projected throughout tourism in rural areas. Hence, as the tourist brochures say, the farms of the region are attractive because they reflect history. Nevertheless, as we can see, history has veiled both women's work and family agriculture. Thus, the lack of systematic data on family productions brings us back to the question of the link that hides female work and production on a family scale. The census forms and the questions posed to the population are made with categories based on a model of masculine society that especially hides female rural work by omitting tasks. These tasks do not end up being recognised as part of daily work in the strengthening of a stereotype of the urban working woman, even with the few dates surveyed.

The South American female work that supports family farming is recognised in many studies [41–43,69,77,104], but without statistical recognition [27,40,89,90,98]. Now, another census allows us to suspect that the permanence of inequality in recognition of rural women's work not only has social roots, but it is also linked to environmental interpretations. The work of rural women does not end inside their homes, it extends to their environment. Moreover, in the way we recognise the environment, we find clues to understand the support that still maintain the invisibility, because denying women's work implies ignoring the elements with which they interact in their tasks [105,106].

3.4.4. Impact of These Processes on Public Policies and Academic Recognition

Finally, it should be noted that this lack is especially structuring of the non-recognition of rural women's work because the works that reveal family agriculture in Patagonia, such as Muzi [47], take as a basis for their conclusions the censuses that, paradoxically, omit this scale deliberately. In these institutional works, phrases like the following can be read, as part of the introduction, 'In this work, information constructed with emphasis on cartographic aspects is made available to the reader based on the data provided by the National Population Census, Homes and Housing 2001 and the National Agricultural Census of 2002' [47], p. 10.

In the reflections that are the basis of public policies, there are no statements related to what is observed in national censuses, they are just shadows, while there are not surveys tools designed to see what is intended. As a result of this denial, the graphs that seek to represent family farming are progressively cutting what is registered as family farming in relation to a small, precarious, decapitalised and informal work, eclipsing the potential that the bibliography that they take like antecedent, demands to make visible. From this description, the echoes of the characterisation of women's work in 1869 return, as a sign of vulnerability rather than capacity. The gender vision on agriculture, present in numerous manuals with concrete tools to account for these processes [45,46,100,101,104], ends absent in the final presentation of the world of rural agriculture in Patagonia. As we mentioned, Núñez [9] has linked this process to the feminisation of the land, since the different development plans that are deployed as of these years, even to the present day, are supported by metaphors that associate the type of woman territory that justifies the type of exploitation. Each activity and recognised geography refer to a different woman so that the development process has an impact on the social order. In rural Patagonia, the woman's representation of the land refers to that of the prostitute, the Indian or the slave, with a penetrable body, argument not only of the denial of feminine work but also of the destructive and contaminating environmental intervention. Against this vision, local populations survive supported by their family and emotional ties. From there, they establish cooperatives and communities, which take the objective of commercialization and articulate it to their particularities [40]. This survival is configured from care brands that refer to female work. The denial of this work impacts the denial of the whole set of associated practices.

The plots investigated, impact on the possibility of linking the visibility of rural female work in Patagonia, as opposed to the possibility of establishing new development policies from the demand for promoting family farming. Basically, the first reference is the need to recognise exchange dynamics alternative to international trade. This does not mean denying or demonising the gaze of higher institutional weight in the region, but putting it in dialogue with the practices that made institutionalisation possible as long as they configured the subsistence pattern.

The potential of this plot, linked to family farming and mostly female work, is what stands out when recognising the Patagonic productive plot in gender. Nevertheless, not only stands out for its dimension but the social sense of production. The Patagonian region has been investigated thinking that the human is reduced to extensive livestock, the same recognised by censuses, plans and that refers to female labour.

One of the areas of study where this is clearer are soil studies [107,108], Del Valle, argues 'Land degradation is both a form of ecosystem self-regulation and a cause of ecosystem fragmentation. The ecological equilibrium of Patagonia is highly susceptible to man's impact and the present methods of natural rangeland management, based on an extreme overuse in space and time. Overgrazing and woodcutting result in a gradual degradation of vegetation, which causes a reduction of the total cover and of the number of plants, the disappearance of valuable fodder species, the invasion of undesirable species, and finally the decrease of available forage. The effects of overuse of this resource are also evident in soil erosion. The consequences of the anthropic impact are also aggravated by drought phases' [107], p. 118.

The anthropic impact is seen as masculine even in 2019 texts, which ignore other non-extractive uses. This is observed even in studies on sustainable developments that do not mention women [109]. Sustainability is run from the limits established for the activity. Sustainability is, in fact, linked to the international price of wool as an excluding factor. Against this, the analysis carried out calls attention to the multi-activity to think about the region, rethink land uses and even to establish suggestions for technical intervention.

The point we reached, in the light of the challenges of gender theory in dialogue with the results presented, is that we must modify even the knowledge matrix. There are no references to sustainable land uses, against the incipient mentions of pastoral women who find in the care and use of native plants the basis for claiming the visibility of their history and their practices [110].

4. Discussion

As we recognize in the first part of the results, in the origins of the Patagonian order, we find that, as economic capital has a greater recognition to the population, the most reasonable tenure is presented as large estate and work as masculine. As a consequence, women have more fixed and hidden obligations, because much of the effort they need for the same subsistence is opposed to the productive and commercial rationality.

The second section of results allows us to indicate that the censuses that are configuring the idea of female work allow to see female work as urban, perhaps as part of the inequality between rural and urban orders that are also cited as a reference for the development of census indicators. The partial recognition of the woman's work of the 1895 census and the direct denial of the 1914 census show the basis of invisibility. Rural female work is categorised as non-work.

The results of the third and fourth section allow us to argue that the indicators and variables of the censuses, as part of the State discourse, linked to the precepts of international, reassert logics of appropriation and extractive capitalist construction, updating racial biases, sexist and xenophobes of space.

Thus, what is 'seen' is the dynamism of the market. Among people, the one 'seen' is the capitalised man, who is the owner of land and productive resources. From here, success is the trade (ideally international) which is presented as the answer to the rural development and sustainable development. Self-consumption is not present, nor are networks of cooperation and solidarity. There is not the territory of the feminine, despite many decades of reclamation and declamation to its valuation.

The third section of the results reinforces this. Production that does not target trade is omitted even in local studies. This is reinforced in the way that overall production is measured. The invisibility of what does not go on the market has a marked depth. This comes together in the fourth section, where family farming is shown in a paradox. Despite having a high symbolic recognition, it faces a recognition structure that denies the bases of its structure as part of the productive system.

Above all, we recognise the practices of women and men, linked to the construction of everyday life, as the basis of the meaning of local rural production. This production and these people print with their practices, specific senses to their environments which discuss the exit to the market as a solution and which demand new ways of considering what exists. From the speeches of the family farmers, the land is still a woman, but it is presented as a peasant, a fighter and demanding to be seen for what she decides, and not for what they say it is. She is herself plus her environment and her social relationships, she is not an individual, she is a network of relationships.

This needs a look anchored in territorial and ethnic particularity. Strong women are not occidental models; it does not matter if they belong to European migration; they are outsiders in their own culture. This has been endorsed by some lines of feminism.

We reflect on how the land organisation justified the rural production that eclipsed female work. So, the discussion about women rural work needs to discuss land organisation, rural production, history and censuses. This work sought to expose the most problematic issues about that in later integrated territories.

5. Conclusions

The results presented seek to show the historical weight of the invisibility of rural female work. They also endeavour to expose the plot of variables where invisibility is built, where the lack of recognition of women's work is associated with the shortage of recognition of characteristics of the territory. From this point, the potential impact on regional development plans implies the set of new approaches. Census indicators are an actual problem that could be solved by the design of new ones. Nevertheless, the problem of recognition is much more profound. Relationships have been established from plots that have denied this work from a vision of the territory that only recognises export activities. In this way, there is a plot between territory and settlement that is questioned. The cited soil studies show how the biased recognition of activities mediates the physical characterisation of the space itself.

Somehow, we go back to the beginning of the conclusions. What does it mean to recognise? What is recognised when survey tools deny existence? The first point is that in order to not recognise something so relevant, we have to slide the recognitions and set up an association of unequal recognition forms of many elements. In this case, it is linked to orders that support the subalternity of both the population and the territory.

Among the results that allow us to think about alternatives is the fact that rural women in Patagonia begin to be more visible from their organization and not so much from their individuality. Perhaps this organising and reclaiming strategy of the women of the steppe allows us to think of alternatives for the general recognition of rural women's work and family farming.

All this leads us to think that recognition denies. To recognise what work looks like meant to deny women. The consequences of this 'decision not to see' were not only reflected in the social order but also the territorial one. In discrimination, space is ordered, which brings us to a very delicate point if we think about recognition. Seeing outside that order is messy. Thus, integrating family farming in the terms that are intended, as if it were merely something that was missing, ignores the enormous constitutive contribution present that operates from not seeing those dimensions. The problem that links female rural work with family farming is that we do not need to add something, but look from a different perspective. This is certainly one of the challenges of feminism, we mean to think that new orders are possible. To relocate, which recognised marginal territories such as the Patagonian, allow us to think about future comparisons of a broader framework.

Author Contributions: Conceptualisation, P.G.N.; P.A.L.T. and C.L.M.; methodology, P.G.N. and C.L.M.; software, C.L.M.; validation, P.G.N. and C.L.M..; formal analysis, C.L.M.; investigation, P.G.N.; M.A.N. and C.L.M.; resources, P.G.N.; P.A.L.T. and C.L.M.; data curation, P.A.L.T.; writing—original draft preparation, P.G.N.; writing—review and editing, P.A.L.T., M.A.N., P.G.N. and C.L.M.; visualisation, P.G.N. and C.L.M.; supervision, P.G.N.; project administration, P.G.N.; funding acquisition, P.G.N. and C.L.M. All authors have read and agree to the published version of the manuscript.

Funding: This research was funded by Universidad de Los Lagos and PIP CONICET 0838.

Acknowledgments: To Carina Llosa, Fiorella Laco Mazzone, Emily Austin, and Jessica Hite for language revision. To Daniela Heim for her comments. To María Luisa Femenías for her guide. We especially appreciate the work of the three reviewers who helped us to improve the article.

Conflicts of Interest: The authors declare no conflict of interest. The funders had no role in the design of the study; in the collection, analyses, or interpretation of data; in the writing of the manuscript, or in the decision to publish the results.

References

1. Mellor, M. Ecofeminist Economics. *Women Work Environ.* **2002**, *54*, 7–10.
2. Chigbu, U.E. Anatomy of women's landlessness in the patrilineal customary land tenure systems of sub-Saharan Africa and a policy pathway. *Land Use Policy* **2019**, *86*, 126–135. [CrossRef]
3. Leon, M. Measuring women work: Methodological and conceptual issues in Latin America. *Inst. Dev. Stud. Bull.* **1984**, *15*, 12–17. [CrossRef]
4. Iuorno, E.; Crespo, E. *Nuevos Territorios, Nuevos Problemas*; UNCo: Neuquén, Argentina, 2008.
5. Nuñez, P.; Lema, C.; Michel, C. La animalidad patagónica y la modernidad marginal. *Tabula Rasa* **2019**, *32*, 81–101. [CrossRef]
6. Nuñez, P.; Lema, C. "Cipres, the triumphant". The andean patagonian forest, the science, the moral and the social health in Argentina between late XIXth century and the '30s decade. *Asclepio* **2019**, *71*, 258. [CrossRef]

7. Vallejo, G.; Miranda, M. Evolución y revolución: Explicaciones biológicas y utopías sociales. In *El Pensamiento Alternativo en la Argentina del Siglo XX: Identidad, Utopía, Integración (1900–1930)*; Biagini, H., Roig, A., Eds.; Biblos: Buenos Aires, Argentina, 2004; pp. 403–418.

8. Abramo, L. *The Social Inequality Matrix in Latin America*; United Nations ECLAC: Santo Domingo, Dominican Republic, 2016.

9. Núñez, P. The "She-Land", social consequences of the sexualized construction of landscape in North Patagonia. *Gend. Place Cult.* **2015**, *22*, 1445–1462. [CrossRef]

10. Benería, L.; Feldman, S. *Unequal Burden. Economic Crises, Persistent Poverty, and Women's Work*; Westview Press: Oxford, UK, 1992.

11. Zusman, P. Prólogo. In *Paisajes del Progreso: La Resignificación de la Patagonia Norte, 1880–1916*; Navarro Floria, P., Ed.; EDUCO: Neuquén, Argentina, 2007; pp. 7–11.

12. Nouaeillez, G. Patagonia as a borderland: Nature, culture and the idea of State. *J. Lat. Am. Cult. Stud.* **1999**, *8*, 35–49.

13. Sochen, J. Frontier women: A model for all women? *S. D. Hist.* **1976**, *7*, 36–56.

14. Núñez, P. Feminismo de frontera. La construcción de lo femenino en territorios de integración tardía. *Rev. Fem. S* **2018**, *31*, 205–230. [CrossRef]

15. Puleo, A. *Ecología y Género en Diálogo Interdisciplinar*; Plaza y Valdés: Murcia, Spain, 2015.

16. Ramilo, D.; Prividera, G. *La Agricultura Familiar en Argentina. Diferentes Abordajes Para su Estudio*; Ediciones INTA: Buenos Aires, Argentina, 2013.

17. Girbal Blancha, N. État, savoir, pouvoir et bureaucratie: Le déséquilibre régional agraire argentin 1880–1960. *Économies Sociétés* **2011**, *44*, 1601–1626.

18. Moreno, F. *Viaje a la Patagonia Austral 1876–1877*; Hachete: Buenos Aires, Argentina, 1969.

19. Ebelot, A. Introducción. In *Informe Oficial de la Comisión Científica Agregada al Estado Mayor General de la Expedición al Rio Negro (Patagonia). Realizada en Los Meses de Abril, Mayo y Junio de 1879, Bajo las Órdenes del General, Julio, A. Roca*; Entrega I—Zoología; Doering, A., Ed.; Imprenta de Osvaldo y Martínez: Buenos Aires, Argentina, 1881; pp. VII–XXIV.

20. Sarobe, J. *La Patagonia y Sus Problemas. Estudio Geográfico, Económico, Político y Social de Los Territorios Nacionales del Sur*; Aniceto López: Buenos Aires, Argentina, 1935.

21. De La Fuente, D. *Primer Censo Argentino, 1869*; Ministerio del Interior: Buenos Aires, Argentina, 1872.

22. De La Fuente, D. *Segundo Censo de la República Argentina, 1895*; Ministerio del Interior: Buenos Aires, Argentina, 1898.

23. Martínez, A. *Tercer Censo Nacional. Levantado el 1° de junio de 1914*; Ministerio del Interior: Buenos Aires, Argentina, 1917.

24. INDEC. *Manual del Censista, Censo Nacional Agropecuario 2002*; Ministerio del Interior: Buenos Aires, Argentina, 2002.

25. INDEC. *Manual del Censista, Censo Nacional Agropecuario 2008*; Ministerio del Interior: Buenos Aires, Argentina, 2008.

26. INDEC. *Cuestionario, Censo Nacional Agropecuario 2008*; Ministerio del Interior: Buenos Aires, Argentina, 2008.

27. Manzano, F. *El mercado de Trabajo Femenino en Argentina. Evolución de Sus Principales Indicadores Desde el Año 1869 al 2010*; XI Jornadas de Sociología: Buenos Aires, Argentina, 2015.

28. National Congress. *Daily Sessions of the Congress of the Argentinean Nation, 1853–1904*; National Government, Archivo General de la Nación: Buenos Aires, Argentina, 1904.

29. Daus, F. Trashumación de montaña en Neuquén. *CONI* **1947**, *8*, 383–426.

30. Raone, J.M. *Fortines del Desierto. Mojones de Civilización*; Editorial Lito: Buenos Aires, Argentina, 1969; Volume 2.

31. Silla, R. Identidad, intercambio y aventura en el Alto Neuquén. *Intersecciones Antropología* **2009**, *10*, 267–278.

32. Michel, C.; Núñez, P. Planificación y cambio en áreas rurales norpatagónicas. In *Reconfiguraciones Territoriales e Identitarias. Miradas de la Historia Argentina Desde la Patagonia*; Moroni, M., Funkner, M., Ledesma, L., Morales, E., Bacha, H., Eds.; Publicaciones UNLPam: Santa Rosa, Argentina, 2017; pp. 258–269.

33. Conti, S. Procesos Psicosociales de Subjetivación en Experiencias Asociativas y Autogestivas Rurales. Casos Recientes en la Zona Andina y en la Línea sur Rionegrinas. Ph.D. Thesis, Socio-Community Psychology, Buenos Aires University, Buenos Aires, Argentina, 2015.

34. Cobelo, C. Transformaciones Territoriales en los Andes Patagónicos. El Caso de las Zonas Rurales de El Bolsón, Río Negro. Ph.D. Thesis, Agricultural Sciences, Buenos Aires University, Buenos Aires, Argentina, 2016.

35. Valtriani, A. Modelos de Desarrollo Forestal, sus Conflictos y Perspectivas en el Sector de Mirco Pymes Forestales; Estudios de Caso en la Región Noreste y Centro de la Provincia de Chubut. Ph.D. Thesis, Economic Sciences, Buenos Aires University, Buenos Aires, Argentina, 2008.

36. Méndes, J. Sociedades del Bosque. Espacio Social, Complejidad Ambiental y Perspectiva Histórica en la Patagonia Andina Durante los Siglos XIX y XX. Master's Thesis, Social Sciences, CLACSO, Buenos Aires, Argentina, 2010.

37. Danklmaier, C.; Wienke, H.; Riveros, H. *Activación Territorial con Enfoque de Sistemas Agroalimentarios Localizados (AT-SIAL)*; IICA: El Bolson, Argentina, 2013.

38. Coronato, F.; Fasioli, E.; Scheitzer, A.; Tourrand, J. Rethinking the role of sheep in the local development of Patagonia, Argentina. *REMVT Revue D'élevage et de Médecine Vétérinaire des pays Tropicaux* **2016**, *68*, 129–133. [CrossRef]

39. López, S. *El INTA Bariloche. Una Historia con Enfoque Regional*; UNRN Editorial: Viedma, Argentina, 2016.

40. Biaggi, C.; Canevari, C.; Tasso, A. *Mujeres que Trabajan la Tierra. Un Estudio Sobre las Mujeres Rurales Argentinas*; Secretaría de Agricultura, Ganadería Pesca y Alimentos: Buenos Aires, Argentina, 2007.

41. Papuccio, S. *Mujeres, Naturaleza y Soberanía Alimentaria*; Librería de Mujeres Editoras: Buenos Aires, Argentina, 2011.

42. Siliprandi, E.; Zuluaga, G. *Género, Agroecología y Soberanía Alimentaria. Perspectivas Ecofeministas*; Icaria Editorial: Barcelona, Spain, 2014.

43. Siempre Vivas. *Las Mujeres en la Construcción de la Economía Solidaria y la Agroecología. Textos Para la Acción Feminista*; SOF-Fundación Heinrich Böll Cono Sur: Sao Paulo, Brazil, 2015.

44. Nogueira, M.; Urcola, M. La jerarquización de la agricultura familiar en las políticas de desarrollo rural en Argentina y Brasil (1990–2011). *Rev. Ideas* **2013**, *7*, 96–137.

45. Feito, M. Agricultura familiar para el desarrollo rural argentino. *Avá* **2013**, *23*, 1–18.

46. Carrapizo, V.; Speranza, M.; Ganduglia, F. *Nos Juntamos? Facilitando Procesos Asociativos a Partir de las Experiencias de la Agricultura Familiar*; IICA-INTA-Min, De agricultura: Buenos Aires, Argentina, 2016.

47. Muzi, E. *Atlas de la Población y la Agricultura Familiar en la Región Patagonia*; INTA: Buenos Aires, Argentina, 2013.

48. Trpin, V.; Rodriguez, M.; Brouchoud, S. Desafíos en el abordaje del trabajo rural en el norte de la Patagonia: Mujeres en forestación, horticultura y fruticultura. *Trab. Y Soc.* **2016**, *28*, 267–280.

49. Sánchez, S. Mujeres Artesanas y Sentidos que Hilan su Quehacer en Prácticas Cooperativas. La Experiencia de la Cooperativa Artesanal Zuem Mapuche (Río Negro, Argentina). Ph.D. Thesis, Psychology, Universidad Nacional de Córdoba, Córdoba, Argentina, 2018.

50. Raguileo, D. Trayectoria Socio-Ecológica en Valles Bajo Riego: El Caso de Sarmiento en la Provincia de Chubut. Master's Thesis, Producción de Rumiantes Menores, Universidad Nacional del Comahue, Universidad Nacional De Rosario & INTA, Bariloche, Argentina, 2020.

51. McDowell, L. The baby and the bath water: Diversity, deconstruction and feminist theory in geography. *Geoforum* **1991**, *22*, 123–133. [CrossRef]

52. Nzegwu, N.; Bockover, M.; Femenias, M.; Chaudhuri, M. How (if at all) is gender relevant to comparative philosophy? *J. World Philos.* **2016**. [CrossRef]

53. Rao, N. Caste, kinship, and life course: Rethinking women's work and agency in rural South India. *Fem. Econ.* **2014**, *20*, 78–102. [CrossRef]

54. Agarwal, B. *Are We Not Peasants Too? Land Rights and Women's Claims in India*; The Population Council: New York, NY, USA, 2002.

55. Agarwal, B. Gender and land rights revisited: Exploring new prospects via the state, family and market. *J. Agrar.* **2003**, *3*, 184–224. [CrossRef]

56. Baydar, G. Sexualised productions of space. *Gend. Place Cult.* **2012**, *19*, 699–706. [CrossRef]

57. Shiva, V. *Staying Alive: Women, Ecology, and Development*; North Atlantic Books: Berkley, CA, USA, 2016.

58. Plumwood, V. *Feminism and the Mastery of Nature*; Routledge: London, UK, 1993.

59. Lee, M.; Madden, E. *Irish Studies: Geographies and Genders*; Cambridge Scholars Publication: London, UK, 2009.

60. Lewis, P.; Simpson, R. Hakim revisited: Preference, choice and the postfeminist gender regime. *Gend. Work Organ.* **2017**, *24*, 115–133. [CrossRef]

61. Pullen, A.; Lewis, P.; Ozkazanc-Pan, B. A critical moment: 25 years of gender, work and organization. *Gend. Work Organ.* **2019**, *26*, 1–8. [CrossRef]

62. Ferraro, F.; Pfeffer, J.; Sutton, R. How and why theories matter: A comment on Felin and Foss. *Organ. Sci.* **2009**, *20*, 669–675. [CrossRef]

63. Femenías, M.L. *Itinerarios de la Teoría Feminista y de Género. Algunas Cuestiones Histórico-Conceptuales*; Universidad Nacional de Quilmes: Buenos Aires, Argentina, 2019.

64. Svampa, M.; Viale, E. *Maldesarrollo. La Argentina del Extractivismo y el Despojo*; Katz Conocimiento: Buenos Aires, Argentina, 2014.

65. Balibar, E. *Violencias, Identidades y Civilidad*; Editores Gedisa: Barcelona, Spain, 2005.

66. Barrancos, D. Gender and citizenship in Argentina. *Iberoam. Nord. J. Lat. Am. Caribb. Stud.* **2011**, *XLI*, 11–14.

67. Barrancos, D. La puñalada de Amelia (o cómo se extinguió la discriminación de las mujeres casadas del servicio telefónico en la Argentina). *Trabajos y Comunicaciones* **2008**, *8*, 111–128.

68. Lewis, R.; Mills, S. *Feminist Postcolonial Theory. A reader*; Routledge: New York, NY, USA, 2003.

69. Sourisseau, J. *Las Agriculturas Familiares y los Mundos del Futuro*; IICA-AFD: San José, Costa Rica, 2016.

70. Guereña, A. *Unearthed: Land, Power and Inequality in Latin America*; OXFAM International: Oxford, UK, 2016.

71. López, E.V. *El Movimiento Feminista. Tesis Presentada Para Optar por el Grado de Doctora en Filosofía y Letras*; Facultad de Filosofía y Letras-Imprenta Mariano Moreno: Buenos Aires, Argentina, 1901.

72. Femenías, M. Notas acerca de un debate en América del sur sobre la dicotomía «feminismo: ¿Igualdad o diferencia?». *Feminismo/S* **2010**, *15*, 193–219. [CrossRef]

73. Abarzua, F.; di Nicolo, C. Extractivismo en territorio del norte de la Patagonia. Frutihorticultura en los valles de Río Negro y turismo en Villa Pehenia-Moquehue, Neuquén. *Rev. Del Dep. Geogr. UNC* **2018**, *6*, 18–345.

74. Núñez, P.; Michel, C. Territorios conquistados y trabajos invisibles. Las mujeres en el ordenamiento territorial patagónico. *Pilquén* **2019**, *22*, 2.

75. Hooks, B. *Feminist Theory from Margin to Center*; South End Press: Boston, MA, USA, 1984.

76. Easdale, M.H.; Rosso, H. Dealing with drought: Social implications of different smallholder survival strategies in semi-arid rangelands of Northern Patagonia, Argentina. *Rangel. J.* **2010**, *32*, 247–255. [CrossRef]

77. Azcuy Ameghino, E. La cuestión agraria en Argentina. Caracterización, problemas y propuestas. *Revista Interdisciplinaria Estudios Agrarios* **2016**, *45*, 5–50.

78. Barsky, O.; Posada, M.; Barsky, A. *El Pensamiento Agrario Argentino*; CEAL: Buenos Aires, Argentina, 1992.

79. Sarmiento, D. *Facundo ó Civilización I Barbarie en las Pampas Argentinas*; Hachette: Paris, France, 1874.

80. Lema, C.; Núñez, P. Destruir para desarrollar. El rol de cienc. en la desigual. Del ordenamiento patagónico. *Cuadernos de Geografía: Revista Colombiana de Geografía* **2019**, *28*, 255–270.

81. Broguet, J.; Corvalán, M.; Drenkard, P.; Mennelli, Y.; Rodriguez, M. Argentina, where are you from? Performative and pedagogical strategies to tackle racism. *Revista Brasileira de Estudos da Presença* **2019**, *9*. [CrossRef]

82. Hull, G.; Scott, P.; Smith, B. *All the Women Are White, All Blacks Are Men, but Some of Us Are Brave*; The Feminist Press: New York, NY, USA, 1982.

83. Gilliam, A. Black feminist perspective on the sexual commodification of women in the new global culture. In *Black Feminist Antropology. Theory, Politics, Praxis and Poetics*; McClaurin, I., Ed.; Rutgers University Press: Utopia, NJ, USA, 2001; pp. 150–170.

84. Neely, B.; Samura, M. Social geographies of race: Connecting race and space. *Ethn. Racial Stud.* **2011**, *34*, 1933–1952. [CrossRef]

85. Giordano, V. Ciudadanía universal/derechos excluyentes. La mujer según el Código Civil en Argentina, Brasil y Uruguay (c. 1900–1930). *E-l@tina* **2003**, *1*, 10–28.

86. Rankin, K. Cultures of economies: Gender and socio-spatial change in Nepal. *Gend. Place Cult.* **2003**, *10*, 111–129. [CrossRef]

87. Picone, M. Landscaping Patagonia: A Spatial History on Nation-Making in the Northern Patagonian Andes, 1895–1945. Ph.D. Thesis, History, Emory University, Atlanta, GE, USA, 2019.

88. Rocheleau, D. Maps, numbers, text, and context: Mixing methods in feminist political ecology. *Prof. Geogr.* **1995**, *47*, 458–466. [CrossRef]

89. Cerdá, J.M. Los Censos históricos como fuente para el estudio de la participación femenina en el mercado. El caso de la provincia de Mendoza a comienzos del siglo XX. *Mora* **2009**, *15*, 53–72.

90. Quay Hutchinson, E. La historia detrás de las cifras: La evolución del censo chileno y la representación del trabajo femenino, 1895–1930. *Historia* **2000**, *33*, 417–434. [CrossRef]

91. Michel, C. Institucionalización del desarrollo territorial en la región de la Norpatagonia: Una mirada desde lo rural. In *Araucanía-Norpatagonia II: La fluidez, lo Disruptivo y el Sentido de la Frontera*; Núñez, N., Ed.; UNRN Ed: Viedma, Argentina, 2017; pp. 264–284.

92. Conti, S.; Núñez, P. La violencia del silencio, las mujeres de la estepa. *Revista Polémicas Feministas* **2013**, *2*, 67–76.

93. Gras, C. Changing patterns in family farming: The case of the pampa region, Argentina. *J. Agrar. Chang.* **2009**, *9*, 345–364. [CrossRef]

94. Núñez, P.; López, S. The North Patagonia territorialization, the Río Negro case in the second half of twenty century. *Cuad. Geográficos* **2015**, *54*, 38–66.

95. Lanari, M.; Reising, C.; Monzón, M.; Subiabre, M.; Killmeate, R.; Basualdo, A.; Cumilaf, A.; Zubizarreta, J. Recuperación de la oveja linca en la patagonia Argentina. *Revista Aica* **2012**, *2*, 151–154.

96. Femenías, M. 'How (if at all) is gender relevant to comparative philosophy?' A response to Nzegwu. *J. World Philos.* **2016**, *1*, 93–96.

97. Lattuada, M. Políticas de desarrollo rural en la Argentina. Conceptos, contexto y transformaciones. *Temas Y Debates* **2014**, *27*, 13–47.

98. Michel, C.; Núñez, P. La globalización en la norpatagonia andina desde la agricultura familiar. *Rev. Austral Cienc. Soc.* **2020**, *38*.

99. Debenedetti, S.; Acebal, M.; Abad, M.; Rosso, H.; Suarez, A. Patagonian mohair: Angora goat production in a really harsh environment. *Angora Goat Mohair J.* **2010**, *52*, 40–43.

100. Rodríguez Flores, L. El enfoque de género y el desarrollo rural: ¿Necesidad o moda? *Revista Mexicana Ciencias Agrícolas* **2015**, *1*, 401–408.

101. Rojo, F.; Blanco, V. *Guía Práctica Para Técnicos y Técnicas Rurales el Desarrollo Rural Desde el Enfoque de Género*; Ministerio de Agricultura, Ganadería y Pesca: Buenos Aires, Argentina, 2014.

102. Paz, R. Agricultura familiar y sus principales dimensiones: La pampeanización del término. *Revista Interdisciplinaria Estudios Agrarios* **2014**, *41*, 5–33.

103. Civitaresi, M.; Colino, E.; Landriscini, G. *Territorios en Transformación en la Norpatagonia. Análisis Comparado del Impacto de Procesos Globales en Ciudades Intermedias*; I Congreso Internacional De Geografía De La Patagonia Argentino-Chilena; Comahue National University: Neuquén, Argentina, September 2018.

104. Schiavoni, G. Describir y prescribir: La tipificación de la agricultura familiar en la Argentina. In *La Agricultura Familiar del MERCOSUR. Trayectorias, Amenazas y Desafíos*; Manzanal, M., Neiman, G., Eds.; Ciccus: Buenos Aires, Argentina, 2010; pp. 43–60.

105. Domínguez, D. Territorialidades campesinas entre lo heterónomo y lo disidente: Formas de gestión de la producción y tenencia de la tierra en el campo argentino. *Política Trabalho Revista Ciências Sociais* **2016**, *45*, 67–84.

106. Galer, A.; Manavella, F.; Bottaro, H.; San Martino, L.; Casiraghi, L. *Aportes al Desarrollo Rural en Patagonia Sur*; INTA: Trelew, Argentina, 2017.

107. Del Valle, H. Patagonian soils: A regional synthesis. *Ecol. Austral* **1998**, *8*, 103–123.

108. Marcos, M.; Carrera, A.; Bertiller, M.; Olivera, N. Grazing enhanced spatial heterogeneity of soil dehydrogenase activity in arid shrublands of Patagonia, Argentina. *J. Soils Sediments* **2019**, *20*. [CrossRef]

109. Villagra, E. *Does Product Diversification Lead to Sustainable Development of Smallholder Production Systems in PATAGONIA, Northern Argentina?* Cuvilier Verlag: Göttingen, Germany, 2018.

110. Conterno, C. La extensión rural del INTA con la comunidad Nehuen-CO, El Chaiful. Nuevas experiencias de trabajo comunitario junto a INTA Jacobacci. *Presencia* **2017**, *5*, 14–21.

 *land*

Article

Community Development through the Empowerment of Indigenous Women in Cuetzalan Del Progreso, Mexico

Pamela Durán-Díaz [1], Adriana Armenta-Ramírez [2], Anne Kristiina Kurjenoja [2] and Melissa Schumacher [2,*]

[1]   Chair of Land Management, Technische Universität München, 80333 Munich, Germany; pamela.duran@tum.de
[2]   Department of Architecture, Universidad de las Américas Puebla, 72819 Cholula, Mexico; adriana.armentarz@udlap.mx (A.A.-R.); annek.kurjenoja@udlap.mx (A.K.K.)
*   Correspondence: melissa.schumacher@udlap.mx

Received: 23 April 2020; Accepted: 17 May 2020; Published: 20 May 2020

Abstract: Women are an underappreciated economic force who, when empowered by association with a female organization, can be a catalyst for development. To assess the status of Indigenous rural women, as well as the mechanisms and impacts of their empowerment, this paper presents a case study of a community development approach based on the Masehual Siuamej Mosenyolchicacauani organization in Cuetzalan del Progreso, Puebla. The methodology used is a mixed-methods approach involving a literature review of two regional instruments: The Federal Program "Pueblos Mágicos" and the Land and Environmental Management Program "POET" for Cuetzalan. It also includes geo-data collection from public sources, empirical data collection from open-ended interviews, and focus group discussions with key informants from the Indigenous organization. The research found that, despite an inclusive legal and institutional framework, weak policy implementation and certain federal programs tend to segregate Indigenous communities. Mechanisms such as cultural tourism and inclusive land management programs, capacity building initiatives, and female associations have proven useful for empowering women and have had positive socioeconomic impacts on the community. The research concluded that female Indigenous associations are a tool to empower rural women, grant them tenure security, strengthen their engagement in decision making, and consolidate them as key stakeholders in community development.

Keywords: community development; Indigenous women organization; empowerment

1. Introduction

The identity of Mexico can only be understood through the connection of Indigenous ancestry and the rural world. "The children of corn", as Mexicans describe themselves, have strong bonds with the land that go beyond time, borders, languages, nations, and political parties. Nevertheless, despite the general perception of an identity deeply rooted in pre-Hispanic culture, Indigenous ethnicities are widely discriminated against in modern urban societies. In a country with more than 25 million Indigenous inhabitants [1] (one-fifth of the total Mexican population) and 68 local languages, Oxfam reports that 70% of these groups live in poverty, 43% of them face discrimination, only 8.5% reach higher education, 10% pay into social security, and 40.5% work informally [2]. Additionally, 51% of the Mexican Indigenous population live in rural areas, where, according to CONEVAL (Consejo Nacional de Evaluación de la Política de Desarrollo Social), poverty diminished from 76.05% in 2008 to 74.9% in 2018. The same census recognized that 41.4% of Indigenous men have better quality of life than Indigenous women [3]. In addition, CONAPO (Consejo Nacional de Población) reported that 23.2% of

Indigenous people that speak a native language are illiterate, and 43.95% drop out of school between ages 6 and 25. As a comparison, one in four Indigenous women has no formal education at all, and one in five finishes secondary school; however, one in six men has no formal education and one in four finishes secondary school [4].

As an outcome, the typical portrait of a Mexican individual living in poverty is presented as a young Indigenous woman who lives in a rural area and speaks a local language. This portrait covers the five dimensions of discrimination—education, employment, gender, ethnicity, and material wealth [2]. Hence, there is a need to conduct a study to describe the status quo of Indigenous rural women in Mexico and the community development achieved by Indigenous female organizations.

Female peasants are responsible for 50% of food production in Mexico, albeit having unequal access to land and a consequential lack of access to public facilities, infrastructure programs, loans, and economic support essential for agriculture [5]. Women's labor is, thus an undervalued catalyst for economic and community development. Throughout the rural Mexican world, female farmers, artisans, and peasants participate as key stakeholders that amalgamate the sense of community and increase social awareness to improve local well-being. Notwithstanding the obstacles to improving the quality of life of Indigenous females in the Northern Sierra of Puebla, Nahua women stand out as triggers of community development.

This paper is a case study research project with a community development approach that uses spatial planning instruments to carry out gender studies on Masehual Siuamej Mosenyolchicauani ("Nahua women that work together" in Nahuatl, hereinafter Masehual), an Indigenous female association located in, and operating from Cuetzalan del Progreso (hereinafter Cuetzalan), Puebla. Masehual originated in 1985 and is one of the first social organizations created and operated by Indigenous women.

The relevance of the case study is based on the following factors: first, Cuetzalan is one of few Mexican municipalities that has achieved the integration of Indigenous autonomy with a land-use planning instrument for community development based on the Ecologic and Land Management Committee of Cuetzalan (Comité de Ordenamiento Ecológico Territorial Integral de Cuetzalan, hereinafter COTIC). This works as a counterpoint to homogeneous village renewal policies as "Pueblos Mágicos" [6]. Second, social organizations such as Masehual, created by female partners to support women's well-being rather than by institutional mandate, have been empowering Indigenous women for more than 30 years. The Indigenous common good and self-determination are achievable when they operate from biocultural roots, in contrast to the commodification of the Indigenous cultures that keep these people poor and illiterate but are attractive for tourism and social policies.

Indigenous organizations in Cuetzalan face multi-dimensional challenges, such as predatory capitalist neoliberal trends that transform identity and culture into merchandise for the global market. Accordingly, this paper questions the benefits of village renewal policies based mainly on competitive tourism and the homogenization of Mexican culture through the Federal Program "Pueblos Mágicos", which perpetuates the condescending approach of social policies. The result is a policy that segregates a vulnerable local population in their own environment. On the other hand, Masehual and its women have appeared as key detonators of socio-political and community development. Additionally, the community complements common good organizations making use of instruments for communal spatial planning, such as the Environment and Land Management Program for Cuetzalan (Programa de Ordenamiento Ecológico y Territorial, hereinafter POET) while dissenting land and tourism policies.

Therefore, this research contributes to the body of research on the ethnographic study of female Indigenous organizations, on the basis that Masehual is a key stakeholder that engages in the generation of socio-spatial cohesion and community development in Cuetzalan.

The municipality of Cuetzalan is used as a case study that presents distinctive conditions worthy of analysis—socio-spatial isolation due to its location, and Indigenous knowledge and biocultural heritage to kick-start the project, embedded in a context of marginalization and poverty. This follows the four dimensions of community and land development—social (community), economic (local activities), political (government), and spatial (land). The following research questions consider these dimensions in the study area of Cuetzalan: (1) What is the status quo of Indigenous rural women in Mexico? (2) Which mechanisms are used to empower Indigenous women in the rurality of Cuetzalan? (3) What are the socio-political and economic impacts of empowered Nahua women in Cuetzalan? (4) What is the role of village renewal policies in community development?

To answer these research questions, the methodology of this case study research project consists of a mixed-method approach for geo-data and empirical data collection conducted in the study area of Cuetzalan. Specifically, we used a literature review of spatial planning instruments, geo-data from public sources published within a time framework from the origin of Masehual (1985) to date, and fieldwork consisting of observations, semi-structured in-depth interviews, and focus group discussions with key stakeholders. This paper begins with a problem statement to contextualize Indigenous self-determination as an asset for social cohesion and community development. Secondly, it describes the chosen data collection methods and introduces the Masehual organization in Cuetzalan as the case study. Thirdly, we analyze the impacts of the legal framework of the Federal Program "Pueblos Mágicos" and the regional spatial planning program on the Indigenous socio-spatial organization. The results and discussion describe Cuetzalan's development in terms of the female approach and explain the role of the Masehual organization in redefining the role of women in attaining a productive economy. By focusing on three statements linked to the research questions, this study aims to highlight the importance of female social organizations to preserving local identity in the face of global hegemonic culture. Finally, the research questions are answered, and in accordance with the example of Masehual in Cuetzalan, we conclude that the role of public policies should serve as facilitators of Indigenous female autonomy, for which we provide recommendations.

2. Indigenous Autonomy as an Asset for Social Resistance and Advancements of Land Rights

2.1. Problem Statement: The Struggle of Indigenous Groups to Secure Land Rights

"This community and the school could be explained through the concept of well-being, which for the Nahua people is not a matter of 'having' rather than 'being and existing'. Nahuas from here cannot exist and cannot live without the other. The exaltation of the individual is made with the other: I am who I am while I am with you" [7].

Indigenous groups in Mexico are usually perceived as unruly social groups. Benton Zavala [7] states that, on the one hand, Indigenous people represent the glorious past of ancient civilizations; on the other, they are recognized as an inconvenient presence that is subject to marginalization, discrimination, exclusion, and poverty. The paradox lies in a morphing identity that is based on a venerated past but is forced to adapt to a standardized modern society ([8] p. 6). For that reason, López Bárcenas ([9] p. 117] stated that one of the historical claims of Indigenous people is the right to self-determination as a form of autonomy and self-government, grounded on the pre-condition inherited by all Indigenous groups in the American continent of being subject to colonial rulers. According to Bonfil Batalla [10], the colonists and rulers (from Europeans to *criollos* to modern governments) subjugated the natives by exploiting them and subduing them to the lowest strata of viceregal society.

It is worth mentioning that the concept of private property was introduced to Indigenous territories through the colonial process. For most native people, the land is conceived as a collective good attached to water and earth. For López Bárcenas ([9], p. 118), the idea of a collectively owned land infringes on the principles of private property, according to Viceroyalty and the State who, over centuries, introduced and formulated several laws and regulations as instruments to divide and grab the land.

In this sense, the *ejido* system of communal land has been an asset for Indigenous social resistance, especially after the Mexican Revolution of 1910 and the Agrarian Reform of 1920. *Ejido* land, as described by Schumacher et al. [11], is a land tenure system based on peasants' and Indigenous communities' land rights to farm and protect land resources through the collective ownership of land. Afterward, the consolidation of neoliberal policies in 1990, together with the liberalization of the *ejido* and communal lands [11], led to a rise in Indigenous resistance movements such as the National Zapatist Liberation Army (EZLN). Similar to the goal of many other Indigenous organizations, the movement pursues the right to self-determination and to organize and rule the land as an integral common good.

This is the key factor to understanding the struggle of Indigenous people in being considered subjects of the law, a status that was not granted in the Mexican Constitution until 1992 with the modification of Article 4, which finally acknowledged the multi-ethnic conformation of the country. Since then, the Mexican Constitution has recognized and respected local languages, culture, traditions, resources, and social organization of autochthonous groups. According to López Bárcenas, "Indigenous movements exist for resistance and emancipation: resistance in order not to stop being *pueblos*, emancipation in order not to remain colonies" ([9] p. 166).

In addition to collective land possession, Indigenous organizations are based on the spirit of their culture, like Nahuas. In rural areas, communities are organized through rigorous disciplines based on agricultural activities and spiritual connection with *Tlali Nantli*, Mother Earth [7]. Hence, organizations such as Tosepan and Masehual are not only Indigenous but are grounded in peasant and autochthonous conditions that are essential for the production of foodstuff and social resistance. Consequently, the process of decision-making is based on consensus and the active participation of community members [9]. Furthermore, Aparicio Wilhelmi [12] stated that Indigenous organizations have a legitimate social claim and need in regions where the presence of the State weakens in the face of neoliberal and globalization trends. Moreover, the importance of the political autonomy of native people or *pueblos originarios* through the empowerment of collective groups lies in the likelihood of improving regional administrations.

In this context, the Nahua women's association Masehual Siuamej Mosenyolchicauani was created in 1985 to generate jobs through the fair trade of Indigenous textile handicrafts that would mitigate migration to big cities and to the United States [13]. Ten years later, they started operating the hotel Taselotzin, a tourist infrastructure project to create employment and promote Indigenous culture, as well as to stimulate actions for environmental care [13]. The socioeconomic impact of Masehual in the community of Cuetzalan is a remarkable feature that will be further described and analyzed in this research paper.

2.2. Contextualization: The Bond of Women With Land and Water According to Nahua Cosmogony

The origin of Cuetzalan as a settlement dates back to 1475, in pre-Hispanic times, as an *altepetl* (water-hill in Nahuatl, referring to a socio-political entity) tributary of the Great Tenochtitlan, capital of the Aztecan empire (1428–1521). According to Reyes García [14], the importance of *Altepetl* as an urban organization was defined by its symbolic significance: "the sea-soaked the soil through its veins and pipes, it roamed under the ground and the hills. The hills were full of water, as plenty of vessels or houses. Here, people have the custom of calling the villages where people live '*Altepetl*', which means hill of water, or hill full of water, or water-hill. The *Altepetl* used to be represented by the pyramids, which emulated the hills, the 'water generators'. For this reason, 'artificial hills' were 'worshipped, and in their construction, all the subordinates participated". Thus, as water is the origin of life and the mountains were seen as being full of water, they symbolized a connection with the gods and the origin of agriculture. Consequently, temples were constructed as artificial hills representing the links of life with water (subsoil), land (agriculture), and sky (the surrounding community life).

For the Nahua people of Sierra Norte of Puebla, native cosmologies are still present and confer a spiritual framework to community life. Although the earth is conceived to be female and the sky is male, there has always been a balanced composition and symmetry in the presence of men and women, revealing complementary organizing principles of Indigenous social and gender relations. Men and women are expected to live their lives as partners, much more so than as individuals ([15] pp. 1,4). Guided by this cosmological complementarity, women continued to own land, pay tribute, participate in the local economy, and possess certain legal status during the Spanish Colony, although their rights were certainly affected by the Catholic Church and colonial principles of gender relations. Thus, complementarity is the concept to use when describing a system in which men and women hold distinct roles and responsibilities considered necessary for the well-being of their households and communities ([15] pp. 12–14).

Among the Nahua people of Sierra del Norte, social construction is based on understanding people as interdependent entities. Such interdependency is a continuous and permanent framework for mutual action that is not based on power relations or imposition. Social relations and values do not exclude non-human entities sharing the cosmos with human beings as they are an active part of its material and transformational cycles triggering effect ([16] pp. 75–76).

According to Nahua cosmogony, agricultural cycles of corn are framed by rituals dedicated to *yeyekatlame* or "airs", who provide the weather conditions for crop growth. "The *yeyekatlame* carry out differentiated, complementary, and interdependent actions in order to achieve shared well-being with human beings. From a local stance, humans and different *yeyekatlame* inhabiting the universe help each other through the active exchange of labor and its products, with these being signs of mutual respect and effect. The *yeyekatlame* helps people survive, and, in exchange, the human community carries out the task of taking care of them" ([16] pp. 78,79).

Thus, the Nahua cosmic vision of life is connected with the surroundings, in which the sun, the moon, the earth, stars, mountains, valleys, caves, plants, animals, stones, water, and air are sacred and living beings [17]. Based on the concept of complementarity and community, the dual unity of feminine and masculine is fundamental for the creation and maintenance of the cosmos. Even today, the fusion of feminine and masculine in one balanced bipolar principle is a common feature of almost every Mesoamerican community ([17] p. 35). For instance, agricultural cycles are understood within the concept of time-duality of feminine and masculine, and the way that Nahua women deal with life and conflict can be described as "[t]heir philosophical background allows them both to resist impositions and to appropriate modern elements into their spirituality. Fluidity and selectivity in adopting novel attitudes and values speak of the ongoing reconfiguration of their world of reference" ([17] pp. 35–36]).

From the Indigenous women's point of view, places are connected to broader social and power relations based on the principles of respect, reciprocity, and obligation ([18] pp. 4,8). In this context, "the norms, laws, and systems of governance that guide these relationships at the level of the family, community, and human and non-human interactions are specific to place. […] Men and women inhabit, experience, and belong to the world and, as such, are holders of knowledge. This knowledge and belonging to the world are based on the interconnections among the social, political, spiritual, economic, and natural spheres" [18]. Rocheleau et al. [19] suggested that despite men having privileged access to resources, Indigenous women have specific knowledge of resources vital for the survival of the household. Indigenous women's knowledge expands beyond the daily activities commonly done by women to involve a system of inquiry involving processes of observing and understanding the protocols for being and participating in the world ([18] p. 9).

Nahua people from Sierra del Norte interact with, and relate to, water in an active interdependent way as actors in a system of reciprocity in terms of affection and respect ([20] pp. 75–76). According to McGregor [21], Indigenous people value water greatly, as it is considered to play the role of a source and supporter of life, and as such, it mediates interactions between living beings on Earth. Within the Nahua context of interdependence, women assume the responsibility of taking care of water, reinforcing water's life-giving force, as water is perceived to be the blood of Mother Earth, a living entity with the same rights to live as human beings. Moreover, among Indigenous people, women have multiple and specific responsibilities toward the protection of water and its quality, and the development and dissemination of knowledge related to water and water management to younger generations. This framework denotes a "cultural understanding of one's responsibilities to the Earth's living, non-living, and spiritual beings and natural interdependent collectives" ([21] p. 606).

Nahua people from Sierra del Norte have linked water with femininity since the origin of water resources. The myth transmitted by oral tradition indicates that "the airs [*yeyekatlame*] were creating the world and playing. 'We challenge you to break that hill', they said to [the female air named] *Sipaketle*, as they shaped a hill with a cloud. Then they said, 'Now, break it'. *Sipaketle* felt strong, and she went back to gather strength and threw herself to the hill to break it. However, being a cloud, she could not break it, she only went through it and crashed her head on the ground, making a big hole. In that hole, the sea was formed with her body. The elders say that each strand of her hair opened breaches on the ground when she sprouted at the top of the hill. That is where *ameles* come from—*ameles* are all the little water springs that we can find around here—forming streams. How many hair strands did *Sipaketle* have? How many water streams run through the surface of the Earth?" ([16] pp. 79–80].

There are four different expressions related to water in Nahua cosmogony: rain, spring, sea, and river; each of them is related to specific social practices of resource and heritage management. Therefore, productive activities are carried out within the framework of traditional knowledge of interpretation of nature [16].

McGregor states that "water finds significance in the lives of First Nations people on personal, community, clan, national, and spiritual levels. Water is understood as a living force that must be protected and nurtured: it is not a commodity to be bought and sold" ([21] p. 27). According to the Indigenous cosmological concept of harmony between the natural elements, the balance between men and women, and the collective responsibility to care for resources, women have a special relationship with water. Women, the Earth, and water have life-giving powers, which grants them special places in the order of existence ([21] pp. 27,28). For instance, when the government of the State of Puebla intended to build water supply pipelines to Sierra del Norte, the eldest members of the Nahua population resisted. They stated that the water pipeline would make people lose respect for water because of the lack of direct contact with its natural origins: streams, springs, wells, and women taking care of them. Nahua people argued that "water is sacred. Why would we want to pollute it? They [the government] tell the young people that they will install a water treatment plant later for the water to be clean again, but why? It is better not to pollute it directly; otherwise, the animals that live there will die, and there will be no more little frogs nor turtles, nothing! Water is not harming anybody, on the contrary, *she* helps us, and thus we should respect *her*" ([16] p. 86).

Nahuas from the Sierra del Norte fear that water would be spilled, polluted, and trifled away. They believed that water would abandon them, and they would, therefore, have no means to live. In this context, Indigenous autonomy is an asset for social resistance and is crucial for subsistence. Autonomy, understood as the power of ethnic groups to design their integral project of life, is the right to decide one's own fate, taking into consideration the cultural past and current reality to foresee a sustainable future in accordance with customary practices [22]. Given the struggle of Indigenous groups to secure land rights, the suitability of their autonomy within the constitutional and institutional framework is indispensable for the survival of their culture and community.

3. Methods and Case Study

3.1. Data Collection Methods

Mixed methods were used in this research for data collection. As for qualitative methods, the field research consisting of surveillance of the preservation state of Cuetzalan's urban image and observation of the interactions of Indigenous people with the *tianguistas* (stallholders) of the Sunday market, was useful for describing the environment and interacting with the participants. A semi-structured, open-ended interview with Rufina E. Villa Hernández, co-founder of Masehual Siuamej Mosenyolchicauani organization and former Secretary of the COTIC Land Management Committee for Cuetzalan was conducted. This interview provided an insight into Masehual's structure, vision, and role in women's empowerment and improvement of their quality of life, as well as the role of women in the preservation of Nahua traditional knowledge and natural resources. In order to assess the impact of the organization in Cuetzalan's community development, to identify the social and environmental roles of Nahua women in the community, and to understand the Nahua cultural standpoint, we carried out a remote focus group discussion based on an open-ended questionnaire with five female members selected from the Masehual organization and a single moderator.

Regarding the ethical standards required for research work involving Indigenous women, this research project was based on Anderson's [23] notion of the problem, in which the researcher, as an outsider, assumes the locally prevailing principles of respect, communication, and reciprocity. In this context, the Masehual organization, as a community of Indigenous Nahua women, regulated its communication with the media, local stakeholders, and policy makers. In our attempt to make the voice of Nahua women heard, while understanding and respecting their cultural practices and background, this research project complied with Masehual's ethical standards and procedures as signaled by their spokeswoman, Rufina E. Villa Hernández. She selected the participants for the focus group discussion and approved the questions asked in advance. As pointed out by Castle [24], it might be gauche to face the reality of contemporary Indigenous communities, particularly regarding the role of women as a part of such communities. In this paper, the voice of the community spokeswoman was taken as the guideline for the development of the theoretical framework in order to avoid the use of persistent stereotypes about women in Nahua communities and Indigenous beliefs.

A discursive analysis of public policy documents, such as government reports and programs, online newspaper articles, academic articles, and official government and community organization web pages, offered information on the objectives, vision, and challenges of the public and private initiatives implemented in Cuetzalan. The initiatives discussed and compared in this paper include those of Indigenous organizations, the Federal policy "Pueblo Mágico", and the COTIC Land Management Committee. Regarding quantitative data, raw geo-data from the governmental web pages SNIM (Sistema Nacional de Información Municipal), SECTUR (Secretaría de Turismo), CONEVAL, CONAPO, and INEGI (Instituto Nacional de Estadística y Geografía) were collected in order to ascertain the socio-economic status of Indigenous women in rural areas, contrast public and private economic investments in Cuetzalan after "Pueblo Mágico" certification, and assess the impact of the studied instruments and organizations on the population's quality of life.

3.2. Case study: Masehual Siuamej Mosenyolchicauani, the Nahua Women that Work Together in Cuetzalan "Pueblo Mágico"

Cuetzalan del Progreso is a municipality in the State of Puebla with an area of 182 km^2 [25] that is located at the heart of the northern highlands of a mountainous landscape characterized by abundant endemic fauna and flora, scenic waterfalls, and humid weather. According to SNIM, 64% of the total Cuetzalan population is represented by Nahua and Totonaca Indigenous people, of which nearly 12% do not speak Spanish. Table 1 shows that women constitute 52% of the population, where the number of people who do not speak Spanish is nearly double. Moreover, they represent 68% of the illiterate population, and only 14.4% are economically active. Nevertheless, women are a hidden economic force that sustains a third of Cuetzalan's households [25].

Table 1. Overview of the status of women in Cuetzalan. Source: Adapted from SNIM (Sistema Nacional de Información Municipal) [25] and INMUJERES (Instituto Nacional de las Mujeres) [26].

Overview of the Status of Women in Cuetzalan		
Total population	47,983	100%
Female population	25,067	52%
Population that speaks a native language	32,132	
Women that speak a native language	16,428	65.5%
Women that speak a native language and do not speak Spanish	3721	22.6%
Total population illiteracy (*)	6230	
Illiterate women	4238	16.9%
Economically active population	16,623	
Economically active women	3617	14.4%
Matriarchal familiar and non-familiar households	27%	
Note: (*) Population of over 15 years of age.		

Cuetzalan is divided into eight districts, each with its own religious and civil authorities and community organization ([27] p. 70). The urban core houses 12.5% of the municipality's population and concentrates health services, government offices, infrastructure, hotels, tourist attractions, and trading activities [25]. As a result, it contains a level of social diversity that has enabled the emergence of strategic projects to benefit the Indigenous population, such as the establishment of the Indigenous Court in 2002 ([27] p. 1), cooperatives (e.g., Masehual), civil associations, and the Nahua traditional medicine module at the local hospital ([27] p. 67).

In terms of cultural diversity, the multi-ethnic interaction since the Spanish conquest in 1522 has been a driving force for the emergence of cultural hybridization between pre-Columbian and European-Catholic traditions [28]. This factor has led to unique cultural representations such as the church altars crafted with beeswax and flowers, awarded the *Tesoro Viviente Poblano* by the State of Puebla [29], and the *Voladores* ritual recognized as an Intangible Cultural Heritage by the United Nations Educational, Scientific and Cultural Organization UNESCO [28]. Other natural and cultural attractions are cave systems, Yohualichan archeological remains, traditional ceremonies to honor natural cycles, textiles elaborated with the use of the waist loom, traditional medicinal knowledge, and the preservation of Nahuatl and Totonaca languages, the most widely spoken languages in the area. Otomi, Zapoteco, Mazau, Mixteco, Maya, Huasteco, Tepehua, Cuicateco, Cinanteco, Mazteco, Chol, Mixe, and Popoloca are languages that are also spoken in the region [25].

As a shelter of the regional cultural heritage, Cuetzalan is an invaluable repository of Mexican history and traditions. Therefore, after being raised to the category of the city in 1986, its urban core became a protected monumental area ([29] p. 18), which enabled the "Pueblos Mágicos" certification to be granted in 2002 [30]. Cuetzalan thus stands out as the first community in the State of Puebla and one of the first in Mexico to have achieved this certification due to its historical and cultural richness, described as "symbolic attributes, legends, history, transcendental events, and everyday life. That is to say, the magic that emanates from each and every one of its socio-cultural manifestations, which nowadays represents a great opportunity for tourism" [31]. Jacobo indicated that Cuetzalan's nomination as "Pueblo Mágico" was possible because it offers sufficient hotel infrastructure and opportunities for adventure tourism, in addition to its cultural richness and historical sites from pre-Columbian and Colonial periods [27].

Moreover, it offers visitors the opportunity to get in close contact with "Nahua culture through gastronomy, handicrafts like the *huipil*, [a decorated blouse typical of Indigenous costumes], and multiple events such as the coffee trade fair and the Saint Francis of Assisi's festivities" ([27] p. 72). A major tourist attraction is the traditional Sunday market, which takes place every weekend at the central square of Cuetzalan and along the main street. In the morning, Indigenous—mostly elderly—people arrive carrying their goods on their backs, many wearing typical clothing from the

Colonial period with pre-Hispanic features. Women wear *manta* (ordinary cotton fabric) blouses with colorful embroidery representing birds, flowers, and geometrical patterns ([8] p. 87). They match them with long, wide, white skirts held in place by a cloth belt, and they go barefoot. For festivities, women wear a *quechquemitl* over the *huipil*, a pre-Hispanic white gauze rhomboid garment to cover the torso, and a headdress consisting of approximately twenty purple wool cords rolled up in their hair. Women use colorful ribbons that hang over their shoulders and back, indicating their marital status. "If the ribbons hang over the *huipil*, they are single, if they are underneath the *huipil*, they are married" [8]. Men wear white cotton shirts, white pants tied at the ankles, and the traditional *huaraches* (rough leather sandals) ([27] p. 71). "An important feature is that only men wear *huaraches* because they are the ones who walk the longest distances, carrying their *tecomate* [a dry calabash used as a bottle or vessel] on their shoulder, full of water to quench their thirst" [32].

Despite the unequal conditions inherited from the Colonial structure, rural women have always been agricultural producers, although they are not being recognized as such. Their participation in the productive system has, however, major individual differences depending on their social class, life situation, family relations, age, and ethnic group. In general, women of Indigenous communities have greater engagement in agricultural production than mestizo or white rural women, yet this involvement depends on the crops cultivated, the type of assigned task, the frequency and form of contracting labor force, and the grade of mechanization of the work ([33] p. 67). Deere and León pointed out that the emerging right to own land under certain organizational conditions may reinforce the capacity and position of women in family and community negotiations and decision making [34].

In the case of Cuetzalan, civil organizations and Indigenous associations have been deeply concerned about the wealth of the community since the 1950s ([27] pp. 74,75). The most important organization, Tosepan, was established in 1980 as an outcome of the 1977 Indigenous Cooperative Movement to respond to the need to supply regional families with basic alimentary goods [35], offering the ability to develop the knowledge, defense, and management of the rights of local Indigenous groups.

Figure 1 shows the presence of cooperatives throughout the northwestern highlands. This weighty network was achieved through actions, such as the "own communal policy", steered by their members to increase the quality of life of Indigenous people, which were widely adopted by several communities in the region [35]. Nowadays, Tosepan incorporates eight regional cooperatives and three associations, each specialized in different areas. Currently, a total of 34,000 families are affiliated with the organization, mostly Nahuas and Totanacas, and it encompasses 410 local cooperatives from 28 different municipalities of the State of Puebla [35].

Tosepan's philosophy highlights its commitment to increasing the quality of life of its members, preserving their cultural identity, and resources through the implementation of diverse working programs, according to the needs of each locality. It also promotes fair trade and organic coffee production, among other activities [35].

The aim of the cooperative is to contribute to a positive revalorization of the local Indigenous identity and culture, which is being reached through the implementation of projects such as Tosepan Kalnemachtiloyan, a trilingual Nahuatl-Spanish-English primary school for Indigenous children [7]. Albeit undertaking a significant role in increasing the common good of Cuetzalan communities, at first, Tosepan failed to generate equal opportunities for women.

Subsequently, in 1985, the organization Masehual Siuamej Mosenyolchicauani was founded by Nahua artisan women from different communities of Cuetzalan, with the aim of triggering actions to dignify the lives of Indigenous women. Nahua women initially approached Tosepan in 1985 to take out a loan to kickstart the Masuehual handicraft project. Masehual's spokeswoman stated, "we decided to organize because we needed economic resources. We already knew how to craft our embroidered garments and to use the waist loom. When tourists arrived, they showed interest in our handicrafts" [36]. They remained affiliated to Tosepan until 1991, when Masehual became an independent organization exclusively for women, as "we needed a space of trust, where women would feel entitled to speak out. If two *señores* were there, women did not feel confident enough to share their opinion" [36].

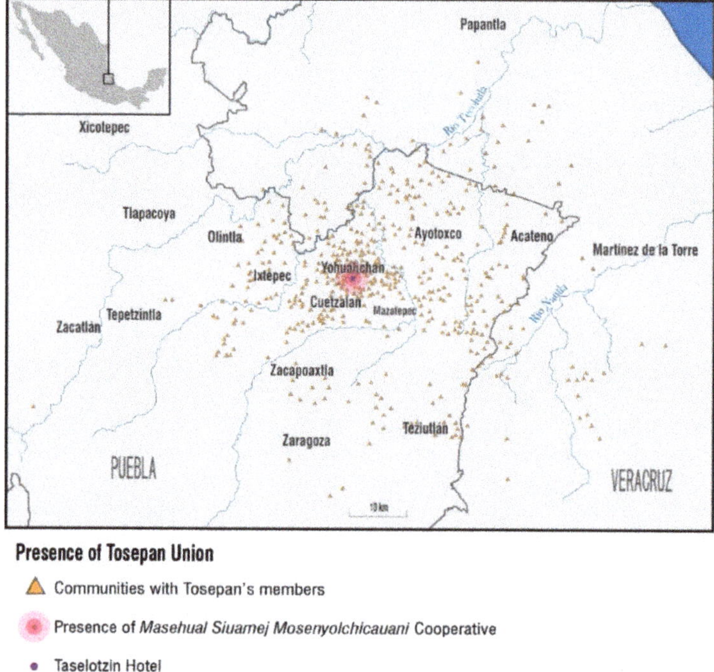

Figure 1. Presence of Tosepan's collectives and Masehual Siuamej in Puebla and Veracruz states. Source: adapted from Habitat International Coalition – Latin America HIC-AL (2018).

By 1989, Masuehual had grouped 300 women from eight communities in the form of a commission to train their practical agricultural skills, such as horticultural production and the use of organic production techniques in communal kitchen gardens. Later on, the participating women implemented the transferred techniques in their own backyards and combined them with their traditional knowledge of herbal and medicinal practices when working with seeds, plants, and tinctures ([37] p. 139).

Masehual separated from Tosepan in 1991 for multiple reasons: First, Tosepan was a mixed organization, while Masehual was a female-only organization with its own counsel. Second, although Masehual women had their own activities and procedures, they were also obliged to participate in Tosepan's meetings, which was overly demanding. Lastly, during the process of the formal registration of the Masehual Siuamej Mosenyolchicauani organization, they realized that they had not been formally associated with Tosepan since the beginning, which facilitated the process. However, the initiative of Masehual women to register their own organization shocked the *señores* of Tosepan [38]. The separation of these organizations generated the division of the Maseahual Siuamej into two associations and led to the creation of Tosepan's female cooperative under the name of Siuamej Sentekitini. Five communities and their 200 members continued to be affiliated to Masehual Siuamej, while the remaining three communities were absorbed by Siuamej Sentekitini ([37] p. 134).

In those years, exceptionally heavy seasonal rain impeded the sowing, harvesting, and trading of crops by men, and hence, family incomes shrunk alarmingly. Therefore, female members of the community began to produce and trade handicrafts through the newly founded Masehual Siuamej organization [39]. Besides solving economic emergencies, this collective of women soon engaged in providing literacy instruction, sensitizing about gender issues and rights, and including women in political affairs and community decision making through COTIC.

4. Literature Review: Institutional Framework for the Village Renewal and Indigenous Organization

4.1. Village Renewal through the "Pueblos Mágicos" Federal Program

In 2001, the Ministry of Tourism SECTUR created the "Pueblos Mágicos" Federal Program, a distinctive certification for the protection, renovation, and valorization of the history, architecture, and culture of traditional Mexican villages. One of the main objectives of the program is the advancement of tourism as an economic development strategy for the community to improve quality of life, increase employment, and stimulate investments through the development of a structured but diversified tourism market in the country. In 2011, tourism contributed 8.4% to the national Gross Domestic Product (GBP) equivalent of $728,186.5 million MXN pesos ([40] p. 15). Hotel activity grew by 6.8% in the period from January to June of 2013 compared with the same period in the previous year ([40] p. 17), and the national supply of rooms for tourists grew by 1.4% from 2011 to 2012, with Puebla being one of the states with the biggest contribution to the increase in rooms available for touristic purposes ([40] p. 16).

By 2018, a total of 121 towns and villages were included in the federal program and awarded $5.2 million MXN pesos each per year, meaning that there were a total investment of $6 billion MXN pesos by mid-2018 [41]. For SECTUR, "Pueblos Mágicos" is a strategic tool to mitigate poverty and favor the development of those communities into attractive areas for the tourist industry [42]. The awarding of "Pueblos Mágicos" certification to the first batch of communities improved the quality of life of the inhabitants in many ways, especially through their integration with tourist sector services and infrastructure. However, the most recent nominations seem to have implemented less inclusive public policies regarding the local population. The first group of 27 certifications was granted between 2000 and 2006; from 2006 to 2012, the number of towns in the program increased to 54, and finally, 40 new villages were included from 2012 to 2018. According to the statistics of the National Council for the Assessment of Public Policy and Development, CONEVAL, the communities that received a nomination in 2012 gained immediate economic effects—over three years, poverty reduced in 23 villages but increased in 11.

Until 2010, the criteria for the nomination and preservation of "Pueblos Mágicos" were notably stricter than in recent years. The conditions stated that the candidate villages had to install a tourist service directory, perform an inventory of resources, preserve their attractions and historical monuments, provide information about connectivity, communication, and infrastructure, and create a tourist services development plan. Towards the end of 2012, with the notable increment in the number of nominations through a more lax selection approach, the "Pueblos Mágicos" certification was jeopardized by the newly selected villages that did not comply with the original criteria [41].

Cuetzalan received the first of seven "Pueblos Mágicos" certifications awarded to the State of Puebla, ensuring a subsidy from the Federal Government to restore the façades, streets, and urban infrastructure of the urban core and to enhance economic tourism activities. Although Cuetzalan received $57.9 million MXN pesos from public funding from 2002 to 2012 and $42.5 million MXN pesos from private investment between 2011 and 2013 [42], the level of poverty increased from "high" in 2000 to "very high" in 2010 [25], despite the exponential growth of domestic and international tourism, as shown in Table 2.

Table 2. Public and private touristic economic investment vs. social welfare in Cuetzalan. Source: Armenta-Ramírez (2020) based on SECTUR (Secretaría de Turismo) [42], SNIM (Sistema Nacional de Información Municipal) [25] and SEDESOL (Secretaría de Desarrollo Social) [43]. CONAPO (Consejo Nacional de Población).

Public and Private Touristic Economic Investment vs. Social Welfare in Cuetzalan (*)		
Public funding 2002–2012	$57,918,208.35	
Private investment 4th trimester 2011, 1st trimester 2013 and 3rd trimester 2013	$42,580,000.00	
Total amount	$100,498,208.35	
	Social welfare indicators	
	2000	2010
Marginalization rank (**)	High	Very High
Migration (***)	648	1010
Population in poverty	NA	80.8%

(*) Amounts in MXN pesos (**) CONAPO classifies the marginalization rank as very high, high, medium, low, and very low (***) Population over five years by birthplace divided by gender, 2000.

This data confirms Riffkin's remark that only a small part of the economic wealth generated by tourism is evenly distributed among the local population ([44] p. 203). Jacobo revealed that the benefits derived from "Pueblos Mágicos" program mostly reach a limited group of Cuetzalan's population as the village's amenities, attractions, and infrastructure, such as hotels, restaurants, touristic events, and historical sites are concentrated at a few easily accessible sites, namely San Miguel Tzinacapan, San Andrés Tzicuilan, and Yohualichan ([27] pp. 72,73).

Hence, the federal program "Pueblos Mágicos" has been beneficial for Cuetzalan only in terms of the advertisement of touristic spots, but it has had multiple negative effects. A major issue is the non-continuity of urban and infrastructure projects. As each municipal administration manages the federal subsidy of 5.2 million per year [41] separately, unfinished works at the end of administrative periods are common. In addition, attempts to standardize "Pueblos Mágicos" have threatened endemic cultural values, as they attempt to turn each village into a generic place: "instead of enhancing the features of each site, it ["Pueblos Mágicos" Program] made them ordinary, making them lose their charm" [38]. In the case of Cuetzalan, the certification granted the cultural and environmental protection of the monumental area, but the village renewal program implied a homogenization of all nominated towns in order to unify the urban image, traditions, land uses, and socio-economic activities, even if it did not generate equal social and economic development.

Another deficiency of the program is that wealth is not perceived to be equitably distributed among the population. People who want to support their family economy but live uphill in remote communities with little or no access to infrastructure have to walk for hours each day to the local touristic sites in order to sell handicrafts and products ([27] p. 73). Although such efforts statistically indicate the activation of creative activity on the periphery of Cuetzalan, remote communities experience little or no enhancement in their quality of life. In this sense, "Pueblos Mágicos" and COTIC do not strive for a common aim.

Moreover, at some point, the policies of "Pueblos Mágicos" compelled local authorities to remove humble local Indigenous producers from touristic sites to soothe the urban image, adversely affecting the local family economy. " This is something we do not agree with. It is the people of the communities whom the visitors want to see in our Pueblo Mágico" [38].

The "Pueblos Mágicos" program, when adequately managed, could be an engine for a comprehensive community development plan with socio-economic impacts on the tourism sector, the community, and the local administration. However, when local authorities are corrupt, they tend to make use of the grant for the profit of their own circles at the cost of the privatization of cultural activities and built heritage, excluding local people—especially Indigenous women who have little or no representation in local committees—from the decision-making process.

4.2. POET: Land and Environmental Management Program for Cuetzalan

The *Programa de Ordenamiento Ecológico y Territorial*, or the Land and Environmental Management Program for Cuetzalan (hereinafter referred to as POET) [45] integrates land management tools with Indigenous cosmovision and the biodiversity of the region. The instrument also acknowledges the threats of illegal logging and the destruction of forests and rainforests to favor electric and mining companies. The preparative studies conducted to trace the POET detected that, 150 years ago, more than 40% of Cuetzalan's land used to be covered by cloud forest and 40% was covered by rainforest. In 2010, the cloud forest covered only 14% of the territory and the rainforest covered less than 0.81%. The instrument thus recognizes the vulnerability of ecosystems and culture to massive tourism policy, "Pueblos Mágicos", which may infringe and destroy them. It, therefore, aims to reinforce "tourism with identity", which benefits the local population rather than foreign capital. Strategically, several environmental policies for sustainable land use, forestry restoration, and environmental conservation were considered within the POET. In Cuetzalan, for instance, the land uses are divided into predominant, compatible, restricted, and incompatible land uses.

Endorsed by Mexican Environmental Law [46], POET established the following instruments for land use and ecological regulation: a spatial and ecological plan for Cuetzalan, a committee for spatial and ecological planning for Cuetzalan, administrative and technical working groups for the committee, and a commission for toxin and pesticide control. The main committee, COTIC, is secured by local authorities as an essential communication tool between the government and the community. Moreover, COTIC is a highly participatory committee in its conformation, as 80% of its representatives belong to the community [47]. Since POET's formulation in 2010, a wide range of academicians, planners, and locals have acted to protect Cuetzalan's ancestral living conditions, economics, natural resources, and landscape and territorial planning, with the aim of guiding land use towards sustainable development. COTIC has two working groups: one administrative and one technical. The Assembly elects the structure of the committee, as shown in Table 3.

Table 3. COTIC (Comité de Ordenamiento Ecológico Territorial Integral de Cuetzalan)/COEC (Comité del Ordenamiento Ecológico Territorial de Cuetzalan) Legal Structure. Source POET [46].

STAKEHOLDERS	LEADER(S)	ROLE
FEDERAL	Ministry of Environment and Natural Resources	Representatives
REGIONAL	Ministry of Environment	Representatives
LOCAL	Cuetzalan Major	Council President
	Committee Partner	Secretary
	Municipal officers for tourism, education, agriculture, economy, etc.	
COMMUNITY	Community members from each district	
	Citizen members of the rural development council	Representatives
	Citizens from each productive sector—tourism, coffee plantations, agriculture, agro-industry, cattle, handcrafts, health, infrastructure, etc.	
	Citizen members of social organizations registered at the committee (18)	
	Independent citizens elected by the committee (6)	
ACADEMY	Academics from the Autonomous University of Puebla BUAP	

COTIC is also the custodian of POET, and it is the strongest instrument that the community has, to regulate productive activities, environmental impact, and human settlements, according to sustainable land use.

5. Results and Discussion: Contextualization of Indigenous Rural Women's Rights and Roles

5.1. First Statement—Women Are Segregated in Rural Areas

When analyzing the position of women in Indigenous communities, we must consider that the pre-Hispanic principle of a complementary social organization was replaced in the 16th century by a Colonial patriarchal framework that designates the positions of Indigenous communities and women as being lower in the social hierarchy and *criollos*, the Catholic Church, and the State are the main rulers and landowners. "The sectors that are invisible to history are practically the same as those that are invisible to economics. These invisibles are of the greatest importance, and the fact that they remained unseen for such a long time is no accident. The reasons lie in our cultural traditions and evolution. That is to say, the evolution of the Western Judeo–Christian cultural branch. The undeniable fact is that humans—particularly men, as I also indicated by the account in Genesis—were placed above nature." ([48] p. 34). The effects of males' control over nature and women are evident in the exclusive male right to privately own land, the division of labor, and the high rates of domestic violence. "The mandate was not to integrate, which would have induced humility; the mandate was to subdue, and as such it could stimulate nothing less than actions and emotions of arrogance and disdain towards the environment, as well as towards those humans who were weaker or less prone to engage in games of power and domination" ([48] p. 36).

Under such conditions, it is difficult for women to be recognized as legal landowners; hence, the legal right to arable land has been one of the main issues of female inequality in Indigenous Nahua communities [34]. Geo-data indicates that 37.1% of women work between 40 and 48 h per week and 12.2% work more than 48 h per week, and even though they represent 29% of the national labor force and account for 50% of the food production, 40% are not able to report an income of their own. Moreover, "six out of 10 rural women live in poverty, the most lacerating expression of inequality" [26].

Since 1917, the Mexican Constitution has granted equal rights to men and women in Article 4: "all people, men, and women, are equal under the law. This article also grants all people protection to their health, a right to housing, and rights for children. Everyone has a right to an appropriated ecosystem for their development and welfare" [49]. However, evidence of this occurring is lacking. In customary practices, women's participation in decision making regarding land use and distribution is nearly null [34]. For instance, Mexican history provides evidence that the redistribution of land by the State in the 20th century through agrarian deals excluded women, mainly favoring male farmers, who were the heads of their families, with the right to vote in *ejido* and community assemblies. "The Agrarian Law is limiting because it is written in masculine terms. There is no specific article that states that women can hold property rights. It does not clearly establish how women can acquire land by simply living in a community or *ejido*. While it is true that women can inherit the land, it can only happen when their husbands die" [50].

In the case of Cuetzalan, in a focus group discussion, participants described the life of Nahua women before the creation of the female organization:

1. "Life used to be harder, with fewer opportunities, I worked in the fields" (Y.S.H., 48 years old).
2. "I used to work in the fields, I planted corn and beans, and could go to school. I learned to embroider at 11 years old by watching my mother doing it" (J.M.N.C., 51 years old).
3. "I was one of the few who got permission from my parents to continue studying until my teenage years" (C.A.L., 41 years old).
4. "Before Masehual, there were no opportunities for women, nor capacity building. We could not go out alone, although the community always supported us" (Y.C.S., 68 years old) [51].

These insights into the quotidian life of Indigenous women in rural areas in Cuetzalan portray the segregation to which they are subject to, despite being part of the workforce. The livelihood of barefoot economics, as described by Max–Neef [48], is a result of certain structural traditional inter-relations between work and the owners of the means of production.

5.2. Second Statement—The Feminization of Rurality through Social Organizations Empowers Women

Legal instruments such as the Convention on the Elimination of All Forms of Discrimination against Women (CEDAW, articles 14, 15 and 16) [52], signed and ratified by the Mexican state; the Mexican Constitution (articles 1 and 4) [49]; the General Social Development Act (article 3) [53]; and the Sustainable Rural Development Act (articles 6, 15, 118, 144, and 154) [54] establish criteria for gender, ethnic, and social equity, specifically dealing with the rights of rural women to ownership of land, inheritance, and property access, which are not strictly observed during the implementation of public policies. In effect, women are overlooked in rural areas in favor of customary practices. As a result, while 51.5% of the *ejido* population are females, only 21% are the *ejidal* landowners, and less than 11% are members of representation bodies [50], such as the *ejido* commissariat. This phenomenon is replicated across the country in different types of communal land tenure, where it is a specific requirement for women to be married in order to earn the right to request a plot of communal land. As an outcome, only 1.3 million of the 4.9 million agrarian landowners in Mexico are women; in other words, 73.46% of rural property is owned by men [55].

Still, female empowerment may occur as a result of the absence of a male partner [37]. In the case of Nahua communities, many of the women in a leading role are widows, single, divorced, or separated. Some of them also act as representatives of their absent husbands, who chose to migrate to Mexican metropolitan areas or to the United States ([56] p. 97). In the absence of men, some Nahua women tend to assemble in female cooperatives as a strategy to face economic crises that, after all, tender an opportunity to change female social roles at the community level.

In view of weak policy implementation, the female association is crucial for equal rights to be exerted and enforced, including land rights. "When I was 24 years old, I 'dared' to go to Cuetzalan, since women were not allowed to go alone, and I went there to sell my textiles. That is when I heard about the organization [Masehual], so I approached them to ask for support. Masehual changed my life because I gained independence as a woman" (J.M.NC., 51 years old) [51]. Hence, the emergent participation of women in economic activities, rural production, and land management generates transformational relational processes in the form of female collectives or groups of individuals. The women involved believe in their capacity to make favorable changes, in their right to make decisions about their own life, and in their potential to trigger wider social changes ([37] p. 110).

"Our organization has been our school, where we have learned the value of our Indigenous culture, to value ourselves as women, to feel that we are capable of achieving success, and to stop feeling inferior for being Indigenous women. Being Indigenous women is something to be proud of" [57].

Their increasingly visible participation in the community has also prompted changes in practices related to gender roles and hierarchy ([37] p. 111).

As a key stakeholder, Masehual is an exemplary case of what Marcuse [58] defined as strategic actions for social change and critical planning: expose (analyzing the problem), propose (planning and working with stakeholders for community development), and politicize (taking action, social organization, monitoring, and accountability) [59]. As represented in Figure 2, Masehual is the receptor of biocultural and local knowledge, in which women "water" the community and bring wealth through working with local organizations, investors, families, local producers, and key stakeholders. The role of Nahua women in taking care of water resources is essential for the sustainability of the community and their culture; thus, in this case, the empowerment of Indigenous women ensures environmental protection, land monitoring, and local economic development.

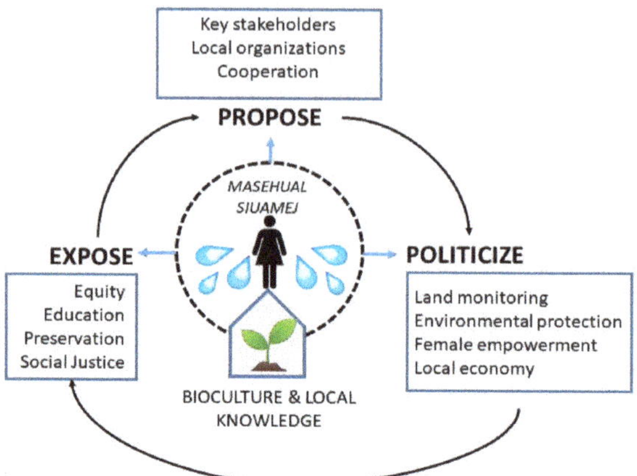

Figure 2. Transformation of gender roles through empowerment in the context of Masehual. Source: Schumacher (2020).

Tapia Villagómez affirms that the empowerment of rural Indigenous women in Mexico has been recently strengthened due to the urge to satisfy families' basic needs, thus activating a change in the traditional gender role that regards them more as caregivers rather than economic providers. This phenomenon is known as "feminization of the rurality" or "feminization of agriculture" [39], in which the attained economic security empowers women to make decisions regarding the future of family resources, namely land.

Furthermore, Tapia Villagómez discerns three dimensions in the processes of empowering Indigenous rural women: (1) progress in understanding the woman as a person, (2) progress in recognizing the organization to which they belong, and (3) transformation in the relations they establish with other members of their family [39]. "A significant achievement was convincing our husbands to allow our daughters to go to school, and to teach our sons to do the laundry, sweep the house, and clean their room. Therefore, the tasks are equally distributed" [36].

These three dimensions lie in the core of Masehual cooperative, where the dignity and plenitude of women go beyond providing economic security through the production of traditional handicrafts: the sorority envisions a global project that integrates culture, environment, health, human rights, and the sense of belonging [38]. What is initially understood as a small contribution to the family economy drives the emancipation and transformation of female roles in the Cuetzalan society and economy. The empowerment of Indigenous women of Cuetzalan through the holistic approach of Masehual has been perceived as successful by the female community because their voices are heard once their basic needs are met and surpassed, once they are educated on their rights, and once they recognize they are striving for a common cause.

5.3. Third Statement—The Empowerment of Indigenous Women Is a Catalyst for Development

To date, Masehual has gathered 100 women from six different communities to work on four main projects: the production and sale of handicrafts, herbalism, the Indigenous Women Care House (hereinafter, CAMI), and the management of the Taselotzin hotel. Each project contributes to the enhancement of members' quality of life through fair trade, housing improvement, and education on female and Indigenous rights, health care, environmental care, sustainable development, and the preservation, care, and dissemination of traditional ancestral knowledge [38].

In the focus group discussion, a 41-year-old craftswoman, an expert in *jonote* basketry weaving, stated, "When I joined Masehual, I lost the fear of speaking and participating" (C.A.L.) [51]. "I am proud of Masehual because we are women, and we are working together".

The co-founder of the organization stated, "I am proud of Masehual because women in the community have learnt to speak out, to participate, and to make decisions for the common good" (Y.S.H. 48 years old). A 61-year-old waist loom artisan who joined Masehual in 1985 pointed out, "I initially feared for the future of my children, we had no resources nor opportunities, but being part of the organization allowed me to sell my handicrafts and get ahead" (M.P.M.M.).

As shown in Figure 3, bioculture and Indigenous self-determination set the frame for equity among key stakeholders in Cuetzalan, in which Masehual is a nexus between local investments and community development.

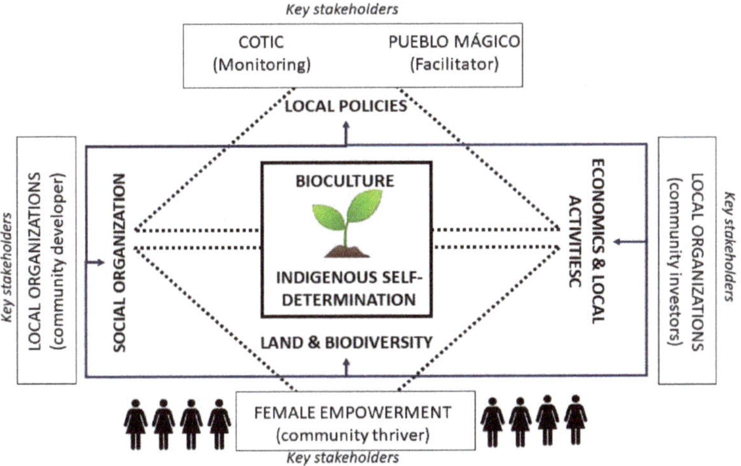

Figure 3. Bioculture and Indigenous self-determination frame for Cuetzalan. Source: Schumacher (2020).

CAMI evolved from initial handicraft workshops to training on human rights and how to support victims of domestic violence. CAMI's work in education about women's rights and health care is especially significant, considering that, although Spanish is the official language in Mexico, 72.51% of people in the state of Puebla speak Nahuatl, including 32,132 people in Cuetzalan [1]. The rate of teenage pregnancy is considerably higher among Indigenous language speakers than among non-Indigenous language speakers, as shown in Table 4. Mexico has the second-highest rate of teenage pregnancy (women under the age of 20 years of age) worldwide, with a rate of 74 out of 1000 women. Among Indigenous women in Cuetzalan, this rate reaches 85 out of 1000 [26]. This phenomenon is presumably linked to marginalization, and a lack of access to formal education in Indigenous communities.

Table 4. Fertility rate. Source: CONAPO [4], INMUJERES [26].

Female Fertility Rate in 2015 Indigenous Language Speakers vs. Non-Indigenous Language Speakers in Mexico	
Average number of live-born children for women who speak an Indigenous language	3.1
Average number of live-born children for women who speak a non-Indigenous language	2.2
Teenage pregnancy rate at the National level (15–19 years old)	74 births/1000 women
Teenage pregnancy rate in Indigenous language speakers (15–19 years old)	82.8 births/1000 women
Teenage pregnancy rate in non-Indigenous language speakers (15–19 years old)	61.4 births/1000 women

Masehual was the first organization in Cuetzalan to empower women through capacity development, encouraging their commitment, and active participation in family and community decisions. For the members of the organization, it was also an opportunity to learn to read, write, and speak Spanish, as a contribution to their empowerment. Consequently, women also engaged profoundly in defense of their territory, the preservation of their traditional knowledge, and the transformation of communal dynamics. Masehual has thus given women a voice to be heard in the regional decision-making process, engaging them in the defense against invasive infrastructure and mining projects in their lands. They have instead promoted the use of alternative renewable energy sources to reduce electric and mining mega-projects. Masehual, as a gender-capacity building project that enhances the creation of local businesses for women, is deeply rooted in Nahua's traditional vision of Mother-Earth. Likewise, it is embedded in the strong bond women feel with the land: "it is us, the women, who suffer the damage to land, the same land that gives us our daily sustenance. It harms us when water is polluted because it is us, the women, who collect the water […]. It is us, the women, who want to keep our lands clean, healthy, free of disease" [38].

The establishment of the eco-tourism hotel Taselotzin ("what the earth produces" in Nahuatl) in 1997 as a collective project by Masehual is an accomplishment that endured difficulties related to the rights of Indigenous women to own land [11]. The hotel operates both as a touristic enterprise and as a center for environmental education for the local population ([12] p. 141). In this context, the feat of acquiring the land to build the hotel could not have been possible as individuals—the land is owned by the organization. "This was a collective dream; we wanted to buy our land to build our huts" [36], stated a 51-year-old female peasant representative of the neighbor village Chicuelyaco who joined Masehual in 1991. The customary structures, endorsed by the groups in the power of the city council, challenged and slowed down the process towards tenure security but finally succeeded with the support of municipal authorities [60].

Hotel Taselotzin is a catalyst for the local economy through the consumption of products of 10 local organizations, sold in the compound. It also provides work opportunities to men and women who are not members of Masehual organization, for instance, through the maintenance work that is constantly needed at the hotel [38]. In 2001, the business model of Masehual was awarded the National prize for the most successful company led by women. It stated, "Transparency is very important to earn the trust of our members" [36].

Being initially formed as a cooperative of female artisans, Masehual uses Hotel Taselotzin as a platform to display and sell their handicrafts and to promote their traditions. "We have different types of handicrafts, like basketry made from the vegetable fiber *jonote*; regarding embroideries, we make napkin sets, tablecloths, blouses, *huipiles*, all hand embroidered. In the waist loom, we fabricate *rebozos* and scarves. Additionally, if anyone wants to see how we do our handicrafts, a group of associates from different communities will gladly come to show it. […] We tell the story of our traditions through the textiles and embroideries that are displayed everywhere at [Hotel] Taselotzin" [57]. Through the preservation of their language, garments, and know-how, Nahua women safeguard the permanence of their identity; the motifs that represent the biodiversity of the region through birds such as *elotot* (corn bird), *chiltotot* (red bird), and *huitzilin* (hummingbird), and geometric patterns that symbolize ferns and snakes, are deeply rooted in the Nahua cosmovision. Moreover, in 2017, the women of the association released a collectively written book, "*Hilando nuestras historias*" ("Spinning our stories" in Spanish) [61], where they tell their stories and experiences using embroidery as the narrative thread to explain Indigenous sorority.

In 2010, Masehual was listed in the Indigenous Touristic Network (RITA), a civil association that seeks the sustainable promotion of Indigenous touristic services as an effective way to preserve environmental and cultural heritage, driving community development with respect for, human rights and the specific rights and identity of Indigenous communities [62], which aims to accomplish the integral and dignified development of Indigenous touristic services.

The economic impact of Masehual's projects benefits 45 families in the region. "The objective of the organization is for women to have decent living conditions, and we are achieving it because we see beyond the economic part, we also consider our Indigenous culture, our environment, our health, our human rights" [36].

5.4. Fourth Statement—Village Renewal Policies Should be Facilitators of Community Development

The general aim of "Pueblos Mágicos" is to improve the quality of life of the local population through tourism, but the work of women in Cuetzalan suggests that this vision falls short in terms of equality, empowerment, and sustainable development. Masehual leverages the advertisement of Cuetzalan as a certified "Pueblo Mágico" and takes it to the next level to boost the local economy, empower Indigenous women, preserve traditional knowledge, and spread the local culture. They, nevertheless, defend their territory and counter questionable "Pueblos Mágicos" initiatives and mining and electric projects.

One of the main critiques of the Federal program is that it operates vertically, instead of transversally engaging the community in the decision-making process. Figure 4 portrays that cultural tourism programs such as "Pueblos Mágicos", when adequately managed, could be an engine for community development, impacting the local community and administration beyond the touristic sector. However, *Pueblo Mágico* seems to work against the COTIC since the Federal vision of a homogeneous Indigenous culture does not prevent excessive tourism and exploitation of natural resources.

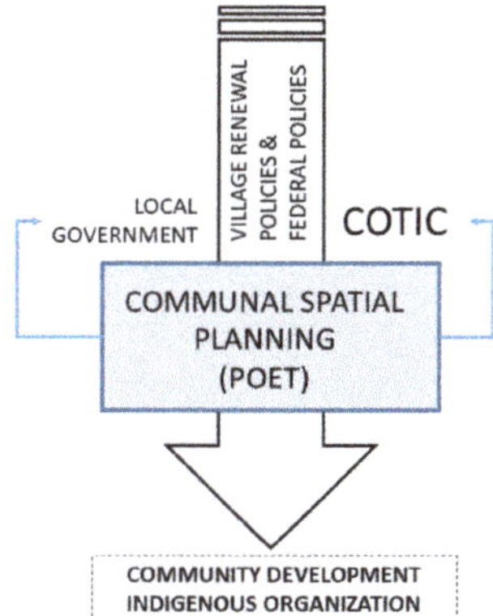

Figure 4. Transversal framework for village renewal policies in Cuetzalan. Source: Schumacher (2020).

COTIC, where the leaders of Masehual actively participate, is an exemplary case in which the impact of Indigenous female organizations goes beyond the aim of selling handicrafts to tourists, as the Federal Program "Pueblos Mágicos" narrowly intended. Nahua traditions and cosmovision are consolidated as the cultural, environmental, and social basis to counter the corporate vision of federal policies. Moreover, Tosepan and Masehual are key community developers that share a collective vision of well-being, and when their traditions and ecosystem are threatened, their actions are more socio-political than merely economic.

In Table 5, "Pueblos Mágicos" Federal Program, and COTIC, as the operative instruments in Cuetzalan, are described in order to visualize the spectrum of positive and negative issues of village renewal policies.

Table 5. Positive and negative aspects of the "Pueblos Mágicos" Program and POET (Programa de Ordenamiento Ecológico y Territorial). Source: Schumacher (2020).

ISSUES	"PUEBLOS MÁGICOS" PROGRAM		POET	
	POSITIVE ASPECTS	NEGATIVE ASPECTS	POSITIVE ASPECTS	NEGATIVE ASPECTS
ECONOMY	Federal Program that grants resources to the community and infrastructure investment.	Focuses mainly on attracting tourism.	Integration of different stakeholders and community members.	Does not have economic support from the Federal Government.
	Grants incentives for economic development.	The infrastructure investment is focused on touristic areas and not on other sectors of Cuetzalan.	Guides stakeholders in the decision-making process when new investment projects reach the community.	Does not provide direct economic incentives for the community.
PATRIMONY	Operates as a caretaker of built heritage.	The protection of built heritage is only granted when there are enough Federal resources.	Local caretaker of built and intangible heritage and natural resources.	Legal and death threats against members of COTIC, especially by mining, fracking, and infrastructure projects.
REGULATION	Regulates urban image.	The improvements to the urban image are only granted when there are sufficient Federal resources.	All new development and infrastructure projects must be approved by COTIC. The operation rules of the program are clear and comprehensive.	Regulations established by the POET are not always well received by majors and local authorities.
POLICY	Invest in the national and international advertisements to attract tourism.	The commodification of local culture and vertical decisions from the Federal Government to define Indigenous culture.	It incentive the creation of COTIC.	Some individual projects oppose to POET regulations, thus achieving a unanimous stakeholders' agreement with COTIC is challenging.
	The local government manages economic incentives.	Corruption and suspicious use of allocated economic resources.	POET is the only legal mechanism of land management and protection of natural resources.	It has not been updated since its publication in 2010.
SOCIO-SPATIAL	Delimitates an impact area of the program focused on the historical core.	The protection and delimitation of the "Pueblos Mágicos" area is not fit-for-purpose.	POET is the only permanent program that helps to improve local land management.	Insufficient instruments to integrate informal housing and urban planning over natural reserves.

Citizen participation is the key to identifying local needs and reacting accordingly through policy-making and implementation. Thus, clear, consolidated communication between stakeholders, including Indigenous and female organizations, is essential for village renewal policies to enhance community development.

6. Conclusions and Recommendations

In Mexico, Indigenous female peasants are the most vulnerable group to marginalization and poverty, despite their significant contribution to National food production and the workforce. The causes of this are the contextual inequalities in rurality, in which a fair legal framework clashes with weak policy implementation, a lack of enforcement, and customary discriminatory practices. Under such circumstances, Indigenous women's associations in Cuetzalan del Progreso empower women and strengthen community development and cultural identity through economic emancipation and education on women's rights. Moreover, they challenge the patriarchal scheme and thus change the role of women in society, which activates their participation in local committees and *ejido* Assemblies as stakeholders.

This case study of Cuetzalan provides an example of the community development that can be achieved through the empowerment of Indigenous women. The use of a mixed data collection method, involving both statistical and empirical material through a literature review and open-ended interviews and focus group discussions was adequate to understand and describe the process where community members take collective action to solve common problems and to answer the research questions depicted below.

1. What is the status quo of Indigenous rural women? Indigenous women face particular challenges in exerting their rights. This means that more than 10 million women live in conditions of poverty attributable to the socio-political and cultural context in which they are embedded and due to the patriarchal dependence networks, that they are forced to rely on for survival. A lack of access to land prevents them from producing food for self-consumption and trade, building a house, and achieving autonomy. It also obstructs their participation in decision making processes at the *ejidal* assemblies. This statement does not imply, however, that their contribution to farming is insignificant. On the contrary, women represent 34% of the workforce and are responsible for half of the country's foodstuff production.

2. Which mechanisms are used to empower Indigenous women in the rurality? There are legal instruments to empower Indigenous women in the rurality, such as the Mexican Constitution, the General Social Development Act, and the Sustainable Rural Development Act. They state that men and women have equal rights, recognize Indigenous autonomy, and seek social and gender equity throughout the development of rural actions and programs. Nonetheless, Nahua customary practices are weakly implemented, and thus, women's rights are under-rated. In Cuetzalan, under a patriarchal structure that neglects women's needs, worth, voice, and right to vote, Indigenous women have recovered their Nahua identity and cosmogony by uniting in the form of female associations. These mechanisms of self-empowerment, such as the Masehual organization, address the immediate needs of their members via capacity building, biocultural management, and tenure security with a sustainable rural development approach that has direct positive effects in the community. This vision ensures the sustainability of the project in the long run. As a result, the members participate more actively in decision making through COTIC. Cuetzalan is an exemplary case in which social resistance, respect for traditions, changes in roles, and the determination to succeed can result in inclusive programs, instruments, and mechanisms.

3. What are the socioeconomic impacts of empowered Indigenous women in Cuetzalan? In the case of Masehual, the empowerment of Indigenous women in the rurality has contributed to community development in terms of promoting female participation in the *ejidal*, communal, and land management assemblies, engagement in the COTIC, sustainable development, human rights, women's rights, economic opportunities, education, inclusiveness, equality, and identity. Training the local women to improve their agricultural, artisanal, and management skills is perceived among the participants as having direct positive effects on the household economy and social cohesion. Masehual stands as an exemplary case in which, despite social and legal

marginalization, the empowerment of Indigenous female farmers has led to social, economic, environmental, and cultural growth for the community.

4. What is the role of village renewal policies in community development? The "Pueblos Mágicos" program tried to homogenize Indigenous culture through urban image and economic incentives. The vertical approach of the Federal policy is not complementary to Cuetzalan's bioculture. Therefore, land management instruments like POET and groups like COTIC act as transversal monitors of the program. Therefore, the "Pueblos Mágicos" program fell by the wayside as it was limited to promoting touristic spots without facilitating community development. Cultural tourism could act as a catalyst for multi-dimensional development rather than bringing only economic growth if it is respectful towards the environment, cultural heritage, and local common good. Federal programs that foster cultural tourism projects, such as "Pueblos Mágicos" and POET, have had both positive and negative effects on the regional land management approach and economy. Although the programs were not formulated specifically to empower women, Indigenous female associations have taken part. As a prime example, the Taselotzin hotel successfully manages to integrate Indigenous traditional knowledge and global tourism demands through fair trade; moreover, its success proves that tenure security can lead to social cohesion and that sustainable projects can be profitable.

This research concludes that the empowerment of Indigenous women through female associations like Masehual boosts community development at multiple levels. Non-inclusive public policies, programs, and structures fail to identify women as key stakeholders who could introduce better agricultural practices, improve housing conditions, enhance environmental conservation, advance health coverage, and perform efficient resource management. In order to attain equal rights and preserve the authenticity of Indigenous rural communities, local authorities should protect cultural diversity and provide equal opportunities, conditions, and security for younger generations to preserve their ancestral roots, typical clothing, and local languages. Nevertheless, we should keep in mind that the function of Indigenous and female associations is not to substitute local authorities, but rather, to be the voice of peasant-Indigenous families.

Our recommendation for further research is to replicate the scenario of Cuetzalan in other contexts. A broader scope could be strategically used to understand the impact of female empowerment on Mexican rurality with a particular focus on Indigenous women, based on the indicators of Goal 5 of the Sustainable Development Goals, particularly Target 5.5—"Ensure full participation in leadership and decision making"; Target 5.A—"Equal rights to economic resources, property ownership, and financial services"; and Target 5.C—"Adopt and strengthen policies and enforceable legislation for gender equality" [63]. Hence, aligning the parameters of the global agenda with the geo-data of Mexican rurality could provide a wider picture of the current situation. To do so, we suggest the following:

(1) Study tenure security and landuse changes making use, among others, of the guidelines for Tenure Responsive Land Use Planning [64] and remote sensing;
(2) Conduct household surveys as a data collection method to determine and compare the household economy based on agriculture and cultural tourism managed by men and women;
(3) Carry out an assessment of the quality and quantity of the crops and hostelry services managed by men and women.

The role of social organizations is to provide a place of solidarity. Along the same line, community development should be grounded on inclusive mechanisms and policies, as the empowerment of women, Indigenous people, and other vulnerable groups will only lead to the improvement of the quality of life of all members of society.

Author Contributions: All authors have read and agree to the published version of the manuscript. Conceptualization, M.S., P.D.-D. and A.A.-R.; methodology, M.S., P.D.-D., and A.K.K.; software, A.A.-R.; validation, M.S., P.D.-D. and A.K.K.; formal analysis, M.S., P.D.-D. and A.A.-R.; investigation, M.S. and A.A.-R.; resources, M.S. and A.K.K.; data curation, P.D.-D. and A.A.-R.; writing—original draft preparation, M.S., P.D.-D. and A.A.-R.;

writing—review and editing, P.D.-D.; visualization, M.S. and A.A.-R.; supervision, M.S.; project administration, M.S. and P.D.-D.; funding acquisition, P.D.-D. and M.S.

Funding: This work was supported by the Dean's Office for Research and Graduate Studies from Universidad de las Américas Puebla (UDLAP) and the Technical University of Munich (TUM) in the framework of the Open Access Publishing Program.

Acknowledgments: The authors would like to thank Rufina Edith Villa Hernández, Yolanda Salvador Hernández, Cristina Álvarez López, María Petra M. Martínez, Juanita María Nicolasa Chepe and Yanira Contreras Segura for their time and openness regarding sharing their experiences at Masehual organization. We would also like to express our gratitude to Walter Dachaga, Lorna Pugh, and the MDPI English editing service for their comments and suggestions when proofreading this paper.

Conflicts of Interest: The authors declare no conflict of interest.

References

1. INPI. *Indicadores Socioeconómicos de los Pueblos Indígenas de México 2015*; Gobierno de México: Mexico City, Mexico, 2017.

2. Solís, P.; Güémez, B.; Lorenzo, V. *Por mi Raza Hablará la Desigualdad. Efectos de las Características Étnico-Raciales en la Desigualdad de Oportunidades en México*; Oxfam: Mexico City, Mexico, 2019.

3. CONEVAL. *Medición de Pobreza 2008–2018*; Consejo Nacional de Evaluación de la Política de Desarrollo Social: Mexico City, Mexico, 2018.

4. Consejo Nacional de Población. Infografía de la población indígena 2015. 2016. Available online: https://www.gob.mx/cms/uploads/attachment/file/121653/Infografia_INDI_FINAL_08082016.pdf (accessed on 13 December 2019).

5. Instituto Nacional de las Mujeres. *Las Mujeres Rurales Producen Más del 50% de la Producción de Alimentos en México*; INMUJERES: Mexico City, Mexico, 2019.

6. Tlali, T. El ordenamiento territorial ecológico de Cuetzalan, una herramienta para la defensa del territorio ante megaproyectos. *La Jornada de Oriente*. 17 June 2014. Available online: https://www.lajornadadeoriente.com.mx/puebla/el-ordenamiento-territorial-ecologico-de-cuetzalan-una-herramienta-para-la-defensa-del-territorio-ante-megaproyectos-el-caso-del-proyecto-de-pemex (accessed on 13 December 2019).

7. Benton Zavala, A.M. Paisaje Lingüístico en Tosepan Kalnemachtiloyan: 'Lecturas' sobre educación intercultural y revitalización. In Proceedings of the Conference "XIV Congreso Nacional de Investigación Educativa", San Luis Potosí, Mexico, 20–24 November 2017.

8. Bernkopfová, M. *La Identidad Cultural de los Nahuas de la Sierra Nororiental de Puebla y la Influencia de la Unión de Cooperativas Tosepan*; Karolinum Press: Prague, Czech Republic, 2014; p. 34.

9. López Bárcenas, F. *La Autonomía de los Pueblos Indígenas de México*; Universidad Nacional Autónoma de México, La Revista la Universidad de México: Mexico City, Mexico, 2019; pp. 117–122.

10. Bonfil Batalla, G. El concepto de indio en América: Una categoría de la situación colonial. *Boletín Bibliográfico Antropol. Am.* **1977**, *39*, 17–32.

11. Schumacher, M.; Durán-Díaz, P.; Kurjenoja, A.K.; Gutiérrez-Juárez, E.; González-Rivas, D.A. Evolution and collapse of ejidos in Mexico-To what extent is communal land used for urban development? *Land* **2019**, *8*, 146. [CrossRef]

12. Aparicio Wilhelmi, M. La libre determinación y la autonomía de los pueblos indígenas: El caso de México. *Boletín Mex. De Derecho Comp.* **2009**, *42*, 13–38.

13. Quién hay detrás del Hotel Taselotzin? Available online: http://taselotzin.mex.tl/frameset.php?url=/actividades.html (accessed on 25 February 2020).

14. Reyes García, C. *El Altépetl, Orígen y Desarrollo: Construcción de la Identidad Regional Náuatl*; El Colegio de Michoacán: Morelia, Mexico, 2000.

15. Sousa, L. *The Woman Who Turned into a Jaguar, and Other Narratives of Native Women in Archives of Colonial Mexico*; Stanford University Press: Redwood City, CA, USA, 2017.

16. Velázquez Galindo, Y.; Rodríguez González, H. El agua y sus significados. Una aproximación al mundo de los nahuas en México. *Antípoda. Rev. De Antropol. Y Arqueol.* **2019**, 69–88. [CrossRef]

17. Marcos, S. Mesoamerican women's Indigenous spirituality: Decolonizing religious beliefs. *J. Fem. Stud. Relig.* **2009**, *25*, 25–45. [CrossRef]

18. Kermoal, N.; Altamirano-Jiménez, I. *Living on the Land: Indigenous Women's Understanding of Place*; Athabasca University Press: Edmonton, AB, Canada, 2016.

19. Rocheleau, D.; Thomas-Slayter, B.; Wangari, E. *Feminist Political Ecology: Global Issues and Local Experience*; Routledge: London, UK, 1996.

20. Whyte, K.P. Indigenous women, climate change impacts, and collective action. *Hypatia A J. Fem. Philos.* **2014**, *29*, 1527–2001. [CrossRef]

21. McGregor, D. Anishnaabe-kwe, Traditional Knowledge and Water Protection. *Can. Woman Stud.* **2008**, *26*, 26–30.

22. Anaya Muñoz, A. Los Derechos de los Pueblos Indígenas, Un Debate Práctico y Ético. *Renglones* **2004**, *56*, 6–14.

23. Anderson, K. Notokwe Opikiheet-"Old Lady Raised": Aboriginal Women's Reflections on Ethics and Methodologies in Health. *Can. Woman Stud.* **2008**, *26*, 6–14.

24. Castle, E.A. Keeping one foot in the community: Intergenerational Indigenous women's activism from the local to the global (and back again). *Am. Indian Q.* **2003**, *27*, 840–861. [CrossRef]

25. Sistema Nacional de la Información Municipal (SNIM). 2015. Available online: http://www.snim.rami.gob.mx (accessed on 9 December 2019).

26. Instituto Nacional de las Mujeres. Sistema de Indicadores de Género. Gobierno de México. 2015. Available online: http://estadistica.inmujeres.gob.mx/formas/fichas.php?pag=2 (accessed on 13 December 2019).

27. Jacobo Herrera, F.E.; López-Levi, L.; Valverde, C. Cuetzalan del Progreso, Puebla. Un pueblo mágico organizado por sus habitantes. In *Pueblos Mágicos, Una Visión Interdisciplinaria*; Universidad Autónoma Metropolitana and Universidad Nacional Autónoma de México: Mexico City, Mexico, 2015; pp. 67–86.

28. Secretaría de Turismo. Cuetzalan del Progreso, Puebla. SECTUR Gobierno de México. 2014. Available online: http://www.sectur.gob.mx/gobmx/pueblos-magicos/cuetzalan-del-progreso-puebla (accessed on 28 October 2019).

29. Fundación Humbert para el Desarrollo Social y de la Biodiversidad A.C. *Asamblea Preliminar de Información a las Comunidades del Municipio de Cuetzalan de Progreso*; Municipio de Cuetzalan: Puebla, México, 2009.

30. CONEVAL. Lugares para visitar: Los Pueblos Mágicos. 2012. Available online: https://www.coneval.org.mx/Informes/boletin_coneval/marzoabril2012/pueblosmagicos.html (accessed on 28 October 2019).

31. Secretaría de Turismo. Pueblos Mágicos. Gobierno de México. 2014. Available online: http://www.sectur.gob.mx/gobmx/pueblos-magicos/ (accessed on 28 October 2019).

32. Escalante, P.; Staples, A. El Colegio de México. Sección de obras de historia. In *Historia de la Vida Cotidiana en México*; El Colegio de México: Mexico City, Mexico, 2005.

33. Sánchez Gómez, M.J.; Goldsmith, M.R. Reflexiones en torno a la identidad étnica y genérica, Estudios sobre las mujeres indígenas en México. *Política y Cult.* **2000**, *14*, 61–88.

34. Tovar-Hernán, D.M.; Tena Guerrero, O. Mujeres nahuas: Desapropiando la condición masculina. *Culturales* **2017**, *5*, 39–65. [CrossRef]

35. Unión de Cooperativas Tosepan. Breve reseña histórica. Tosepan. 2016. Available online: http://www.tosepan.com/ (accessed on 30 October 2019).

36. Alcalá, E. Masehual Sihuamej: Mujeres Indígenas Que Resisten, Trabajan y se Apoyan Juntas. 2018. Available online: https://luchadoras.mx/masehual-sihuamej/ (accessed on 11 May 2020).

37. Martínez Corona, B. *Género, Empoderamiento Y Sustentabilidad: Una Experiencia De Microempresa Artesanal de Mujeres Indígenas*; Universidad Nacional Autónoma de México, Centro Regional de Investigaciones Multidisciplinarias: Cuernavaca, México, 2016; pp. 109–150.

38. Villa Hernández, R.E. Semistructured interview about Masehual Organization. [interv.] M. Schumacher and A. Armenta-Ramírez. 2020.

39. Tapia Villagómez, I. *Emprendimiento Femenino Rural Indígena: El Hotel Taseoltzin, Cuetzalan, Puebla*; Universidad Iberoamericana Puebla: Puebla, México, 2017.

40. Secretaría de Turismo. *1er Informe de Labores 2012–2013*; Gobierno de México: Mexico City, Mexico, 2013.

41. Puga, T. Pueblos Mágicos… pero pobres. *El Universal*. 29 October 2018. Available online: https://www.eluniversal.com.mx/cartera/pueblos-magicos-pero-pobres (accessed on 29 October 2018).

42. Secretaría de Turismo. *Inversión Turística Privada Y Pública Cuetzalan*; Gobierno de México: Mexico City, Mexico, 2014.

43. Secretaría de Desarrollo Social. Pobreza y rezago. Entidad: Puebla. Municipio: Cuetzalan del Progreso. Clave 21043. Unidad de Microrregiones, Cñedulas de Información Municipal. 2010. Available online: http://www.microrregiones.gob.mx/zap/rezago.aspx?entra=zap&ent=21&mun=043 (accessed on 29 October 2019).

44. Rifkin, J. *The Hydrogen Economy: The Creation of the Worldwide Energy Web and the Redistribution of Power on Earth*; Tarcher/Penguin: New York, NY, USA, 2002.

45. Ayuntamiento del Municipio de Cuetzalan del Progreso. *Programa de Ordenamiento Ecológico Local del Territorio del Municipio de Cuetzalan del Progreso*; Municipio de Cuetzalan: Cuetzalan, Mexico, 2010.

46. Cámara de Diputados. Ley General de Equilibrio Ecológico y la Protección al Ambiente. *D. Of. De La Fed.* **1998**, 18.

47. González, A. El ordenamiento de Cuetzalan, una herramienta de defensa comunitaria. *La Jornada del Campo.* 17 February 2018. Available online: https://www.jornada.com.mx/2018/02/17/cam-cuetzalan.html (accessed on 2 May 2020).

48. Max-Neef, M. *From the Outside looking in: Experiences in Barefoot Economics*; Förlaget, N., Ed.; Dag Hammarskjöld Foundation: Stockholm, Sweden, 1982.

49. *Constitución Política de los Estados Unidos Mexicanos, Artículos 1 Y 4 [Capítulo I]*; H. Congreso de la Unión 25 Legislatura, Mexico: Mexico City, Mexico, 1992.

50. FAO. Mujer Rural y Derecho a la Tierra. In *Situación General de las Mujeres Rurales e Indígenas de México*; FAO: Mexico City, Mexico, 2017.

51. Masehual Women. Focus group discussion with five representatives from Masehual Sihuamej Mosenyolchicauani. [interv.] M. Schumacher and P. Durán-Díaz. [trans.] Pamela Durán-Díaz. Cuetzalan. 9 April 2020; Telephone interview.

52. United Nations Human Rights. Articles 14, 15 and 16. In *Convention on the Elimination of All Forms of Discrimination Against Women*; United Nations Human Rights: New York, NY, USA, 1979.

53. *Ley General de Desarrollo Social, Article 3*; H. Congerso de la Unión 25 Legistatura: Mexico City, Mexico, 2018.

54. FAO. *Ley del Desarrollo Rural Sustentable*; FAO: Mexico City, Mexico, 2017.

55. Registro Agrario Nacional. *Datos Geográficos de las Tierras de Uso Común, Por Estado*; Gobierno de México: Mexico City, Mexico, 2018.

56. Rodríguez Blanco, E. Género, etnicidad y cambio cultural: Feminización del sistema de cargos de Cuetzalan. *Política y Cult.* **2011**, 35, 87–110.

57. Villa, H.; Rufina, E. *Mujeres Masehual, Deshierbando el Machismo*; La Coperacha: Mexico City, Mexico, 2017.

58. Marcuse, P. Whose right(s) to what city? In *Cities for People, not for Profit*; N. Brenner, P.M., Mayer, M., Eds.; Routledge: New York, NY, USA, 2012; pp. 36–53.

59. Schumacher, M. Peri-urban Development in Cholula, Mexico. Ph.D. Thesis, Technische Universität München, Munich, Germany, 2016.

60. Hernández-Loeza, S.E. La participación en los procesos de desarrollo. El caso de cuatro organizaciones de la sociedad civil en el municipio de Cuetzalan, Puebla. *Econ. Soc. Y Territ.* **2011**, XI, 95–120.

61. Masehual Siuamej, M. *Hilando Nuestras Historias. El Camino Recorrido Hacia Una Vida Digna/Ibero Puebla*; Cecilia Ramón, F., Ed.; Instituto de los Derechos Humanos Ignacio Ellacuría: Puebla, Mexico, 2017.

62. Red Indígena de Turismo de México (RITA). 2013. Available online: http://www.rita.com.mx (accessed on 10 February 2020).

63. United Nations Development Program. *Transforming Our World: The 2030 Agenda for Sustainable Development. Goal 5*; United Nations: New York, NY, USA, 2015.

64. Chigbu, U.E.; Chigbu, U.E.; Haub, O.; Mabikke, S.; Antonio, D.; Espinoza, J. *Tenure Responsive Land Use Planning: A Guide for Country Level Implementation*; UN-Habitat: Nairobi, Kenya, 2016.

 *land*

Article

The Nexus between Peri-Urban Transformation and Customary Land Rights Disputes: Effects on Peri-Urban Development in Trede, Ghana

Barikisa Owusu Ansah and Uchendu Eugene Chigbu *

Chair of Land management, Faculty of Aerospace and Geodesy, Technical University of Munich (TUM), 80333 Munich, Germany; barikisa.owusu@tum.de
* Correspondence: ue.chigbu@tum.de; Tel.: +49-(0)89-289-22518

Received: 6 May 2020; Accepted: 28 May 2020; Published: 5 June 2020

Abstract: Typically, peri-urban areas are havens and vulnerable receptors of customary land rights (CLRs) disputes due to the intrusion of urban activities or an uncoordinated mix of both. Although it is a dictum that CLRs cause setbacks to socioeconomic and spatial development, there seems to be a paucity of empirical studies on the effects of the CLRs disputes on the development of peri-urban areas, especially in developing countries, such as Ghana. This study addresses this issue by establishing a link between peri-urban transformation and emerging CLRs disputes, while assessing the effects of these disputes on the development of peri-urban areas. The study adopted a problem-centered mixed methods approach with a focus on the case of Trede, a town in Ghana transitioning from rural to urban status. Findings reveal that the changes leading to enhancing of peri-urban transformation are also the same changes inducing CLRs disputes in the area. It was found that the implementation of a local land use plan is a critical driver of CLRs disputes in Trede. A land-use plan implemented as a major step in converting rural lands into urban plots, triggered tenurial changes, land market development, high land values, loss of agricultural land, etc., which become recipes for the CLRs disputes in the study area. These CLRs disputes have hatched detrimental consequences on the economic, social, and physical developmental trajectories of Trede. As a way forward, the study proposes measures for peri-urban land management and CLRs dispute prevention.

Keywords: Ghana; peri-urban area; peri-urban development; customary land rights disputes; land rights; land-use plan; tenure security; Trede

---

1. Introduction

The challenge posed by customary land rights disputes (defined as any claim, disagreement, or contestations over the right to use, control, and management of customary lands) in developing countries has been well documented in literature [1–4]. The dispute-affected areas and countries are known to suffer detrimental consequences in their economic, social, and physical developmental trajectories. However, the effects of customary land rights (CLRs) disputes vary across the different spatial continuums—from rural to peri-urban to urban areas. Typically, peri-urban areas are more gravid with the effects of customary land disputes due to the frequent and enormous changes that occur to their land use patterns. There is increasing evidence that peri-urban areas have become contested zones when compared with rural and urban areas, mostly because it is a geography where rural and urban functions occur side-by-side [1,4]. Particularly in peri-urban areas, there is the blur of rural and urban boundary, as such boundaries of traditionally urban land and rural land remain a core debate in peri-urban studies. Similarly, the Global Land Tool Network (GLTN) [3] noted that the uncertainty of 'what is rural' and 'what is urban' land in peri-urban areas has caused the regulation of different land tenure rights to become subjected to different land management regimes and administration system which may come in conflict leading to disputes, contestation, and violence. These have resulted in significant peri-urban development issues.

Ghana is one of the countries, at least within the Sub-Saharan African region, facing a lot of challenges in its peri-urban development. Legal pluralism in land tenure is recognized and practiced in Ghana. Consequently, customary land laws and Statutes contemporaneously exist in regulating land administration and governance in Ghana. In peri-urban areas of Ghana the management of land tenure operates within a plural environment. The construction of the peri-urban land involves two major actors; the customary authority and the statutory authority. Despite the pluralistic nature of land tenure system in Ghana's peri-urban areas, most developments are undertaken within the customary land tenure regime. Therefore, any barrier confronting customary land rights have detrimental implications on the transition from rural to peri-urban. The management of the transition of rural areas solely by customary authorities without statutory intervention affect the transition process [4,5]. According to Mends [4], the inability of customary authorities to manage changing customary tenure systems (defined as the traditional rules and regulations that govern the holding, use and transfer of interest and rights in land) have resulted in tenure insecurity, land ownership conflicts, and boundary disputes.

Several factors have resulted in significant changes in land tenure practices and land rights in peri-urban areas. Chauveau [6] noted that urbanization and rapid population have given rise to the high demand for rural land. Pressure from demand has caused the rapid transformation of customary lands to peri-urban lands to engender changes in the customary tenure system [4,7,8].The transformation manifests mainly in the physical development and conversion of agricultural land uses to non-agricultural land uses. Communal land rights, commonly practiced in rural areas, are gradually changing to individualized land rights in peri-urban areas [6,9,10]. Wehrmann [11] also observed that the emergence of peri-urban areas has come with new forms of land tenure systems, which may not be compatible with the customary land tenure system and, as such, contradict customary rules.

It is worth noting that changes in customary land tenure system move in tandem with changes in ownership, control, and land uses. Consequently, new actors, such as urban elites and foreign investors, move into peri-urban areas to compete for land for residential, commercial, and speculative purposes [12,13]. The change in land uses, ownership, and control of land in peri-urban areas have therefore been characterized by intense competition engendering exclusion to access to land leading to CLRs disputes [2,11,14,15]. For instance, the conversion of agricultural lands has made local farmers in peri-urban become subjects of expropriation. Many farmlands are lost to residential and commercial developments without fair compensation to farmers [16]. Some local farmers, in their attempt to defend their rights, enter direct struggle with chief, whiles others physically attack new buyers from entering their lands or sometimes go to the extent of destroying newly built-structures [13,17]. These are some of the undesirable consequences of customary land rights disputes.

The problem of customary land rights disputes thwart the development potentials of peri-urban areas in Ghana. Although policymakers do realize that CLRs disputes pose serious setbacks to development, the underlying effects of CLRs disputes on the development of peri-urban areas remain largely unexplored. It is within this context that the study seeks, by way of its purpose, to examine the effects of CLRs disputes in the context of peri-urban areas through a case study of Trede, a town which is currently undergoing a transition from rural to urban. This study is critical to informing policy for an efficient peri-urban land management and CLRs dispute prevention.

The remaining section of this paper is structured as follows: first, we introduce the conceptual review of peri-urban, peri-urban transformation, development, and customary land tenure. Next, we discuss the data collection and analysis methodology which were employed to investigate the problem. We then present and discuss the results in respond to the study objectives. In the last section, we draw logical conclusion on the analysis and make recommendations towards policy reforms in managing peri-urban lands and CLRs dispute prevention.

2. Peri-Urbanization and Land Tenure: A Conceptual Review

The interrelated concepts of peri-urban, peri-urban transformation, development, and customary land tenure are examined in this section of the paper.

2.1. Putting Peri-Urban and Peri-Urban Transformation into Context

The term peri-urban is multi-dimensional and has been defined differently by various scholars. It has been described as a place, process, or concept. As a concept, peri-urban is when rural and urban activities meet; as a process, peri-urban is the gradual transformation of rural areas as they attain more urban characteristics; and as a place, it is the region between rural and urban zone [18]. In addition, studies, such as Reference [19–21], considered peri-urban as a geographical location. Some groups of researchers also considered peri-urban as a mental state of mind [22,23]. Similarly, the definition of peri-urban has been conceived differently across discipline and professions. The sociologists view peri-urban as areas of social compression or intensification where social stratification increases leading to conflict and social evolution [23]. The planners consider peri-urban as a transitional zone between urban area and rural area predominately under urban influence [24,25]. The environmentalist conceptualizes peri-urban as an interface of heterogeneous mosaic of natural ecosystems, agro-ecosystems, and urban ecosystems affected by rural and urban interactions [26].

Even though there is still no universal definition for peri-urban, preponderances of urban researchers and policy analysts place higher priority on the place-based definition of peri-urban. However, Iaquinta and Drescher [23] argued that defining peri-urban as a geographical location is not satisfactory. They highlighted the need to incorporate socio-psychological and institutional components that characterize the reality of peri-urban areas in any definition. Meanwhile, Ravetz et al. [21] proposed that arriving at consensual definition of peri-urban was not realistic due to the disparate perspectives and contexts in which the concept is deployed and described. They suggested that it will be better to view peri-urban as a spatial phenomenon in continuous flux and transition. Nonetheless, researchers have recognized the complexity and diverse nature of the concept [26,27]. In this study, peri-urban area is regarded as an area of "tenurial transformation" [19] where customary lands transition into urban lands. This conceptualization is useful in providing an appropriate context for examining the land dispute types that come along with the tenurial transformation in peri-urban areas.

Peri-urbanization is a global phenomenon; however, the trend of peri-urbanization varies across globe. While some researchers have viewed peri-urbanization as harmful to physical environment, others have associated it with socioeconomic development [28,29]. According to Chiris [28], peri-urbanization in developed countries is characterized by industrial development, whereas, in most developing countries, the process has resulted to sprawl endangering the physical environment. On the other hand, Setterthweite et al. [29] argued that land problems, such as sprawl, are not the resultant effect of peri-urbanization but, rather, the lack of appropriate management of the rural-urban interface by planning authorities. Peri-urban areas in Ghana consist of mainly customary lands; as a result, the management of these areas are largely under the control of customary authorities, with limited interventions from state actors. Due to the nature of land management in peri-urban areas of Ghana, the process of transformation is, to a large extent, unplanned, leading to land disputes.

2.2. The Interface between Land Tenure and Peri-Urban Development

To appreciate the implications of land tenure on peri-urban development, it is useful to demystify the terms 'land tenure' and 'development'. Land tenure has been defined as the laws, rules, and obligations governing how people hold land [30]. The Food and Agriculture Organization [31] considered land tenure as an institution invented by society to regulate human-land relationship. These suggest that land tenure provides an institutional structure (customarily or formally) for people to exercise their rights and interest in land.

On the other hand, development is the process of societal change that brings improvement in the quality of lives of people [32]. It symbolizes "the attainment and maintenance of a reasonable and acceptable standard of living, based on the traditions, institutions, values and standards of a given society" (Reference [30], p. 20). Development is usually equated with economic growth. While economic growth is one of the important components of development, other components, such as social and physical factors, exist. Development can be in the form of physical action, for instance, implementation of physical projects, like road infrastructures, community hospitals, etc. It could also come in the form of process change not involving any physical intervention. Regardless of the form, development affects people beliefs, traditions, and values [30].

Land tenure is an important aspect of institutional context of development in many developing countries [33]. Issues of land tenure have become a global concern. In May 2012, the Committee on World Food Security (CFS), initiated the "Voluntary Guidelines on the Governance of Tenure of Land, Fisheries and Forests" at the Global Food Security Symposium. These guidelines were adopted with the aim of providing an overarching framework for tenure systems capable of ensuring tenure security and equitable access to land for all. In addition, the GLTN through UN-Habitat has been working on developing pro-poor land use planning tools to secure tenure in most developing countries. Other laudable initiatives, such as the Sustainable Development Goals (SDGs) adopted in 2015 at the UN Sustainable Development Summit (Agenda 2030), are all aimed at reducing poverty and ensuring sustainable development through secure land tenure systems.

Land tenure is crucial for socio-economic development. Kasanga [30] asserted that land tenure forms the basis not only for agricultural production but also for social and economic development. In a similar view, De Soto [34] and Payne [33] emphasized that tenurial problems have adverse effects on economic development. Other studies have equated land tenure to economic development [8,17]. However, USAID [35] reported that estimating the impact of land tenure on economic development is a difficult and complex process to determine. Yet, outcome variables, such as poverty rates, land values, and agricultural productivity, can provide the bases for measuring the impact of land tenure on economic development. Nevertheless, land tenure can act as a considerable constraining factor impeding development. Nkwae [27] showed that tenure problems, such as tenure insecurity, litigations, land disputes, and loss of agriculture land, could compromise food security, land rights, and agricultural productivity, with detrimental effects on economic growth, and may act as a recipe for poverty. It is, therefore, necessary to conclude land tenure has a correlation with the development of any nation.

2.3. The Dynamic Customary Tenure System Versus Transformative Peri-Urban Area

Land tenure as part of the institutional context of peri-urban is subject to change as rural areas transform into peri-urban areas. In effect, there is an interrelationship between changes occurring in the peri-urban areas and changes in the land tenure. It is essential to note that the transition in peri-urban areas corresponds to changes in the customary tenure; thus, a sudden change in the social or economic fabric of peri-urban areas results to changes in customary tenure [27]. Twaib (1996), as cited in Nkwae [27], observed that "any change in the pattern of land ownership automatically carries with it changes in economic, family and social relationships". This assertion reflects the views of Iaquinta and Drescher [23], who noted that peri-urban is reciprocal. This simply means that changes in peri-urban areas change people's land relations, and vice versa.

The main change arising from customary land rights in peri-urban areas is the shift from collective ownership to individual rights. This view is echoed by Mends [4], who noted that the concept of customary land rights has evolved from the notion that land which was collectively owned by the community now belongs to the person who initially cultivated it. Thus, the communitarian approach gradually moves toward individualization. Wehrmann [11] observed that change from communitarian approach to individualization would not follow a smooth process but, rather, one that is characterized by tenure insecurity, contestations, and disputes. Boserup (1965), as cited in Wehrmann ([11], p. 85), also warned that "each new step on the road to private property in land may create less and not more security of tenure, and a vast amount of litigation is the obvious result". It is therefore necessary to understand the changes in land tenure as customary land transition to peri-urban land, the land disputes that comes along with, and how these disputes can be managed.

The customary tenure is regarded as flexible, dynamic, and can respond to changing socio-cultural and economic conditions [4,10,17]. However, Chauveau [6] found that the way in which customary tenure responds to changes varies substantially depending on the diversity of the area in context. In some cases, customary tenure still maintains some resilience despite the urban pressure. Notably, some traditional authorities have been steadfast in maintaining the customary land management practices. On the contrary, the customary tenure could be entirely eroded due to changes in the composition of the indigenous population. Chauveau et al. [5] suggest that where customary land tenure has eroded, government intervention is required for effective land management. In the peri-urban areas where this problem is common, Reference [26,36] claim that policymakers are not capable of addressing land management problems because of the increasingly complex nature of peri-urban areas. To propose a solution to this quandary, this paper makes recommendations towards policy reforms in managing peri-urban lands.

3. Methodology

3.1. Description of Case Study Area

Trede is situated south of Ashanti region and is about 16 km from the regional capital, Kumasi (see Figure 1). Trede has an estimated land size of about 1200 acres with a population of 5438 people according to the 2010 population Census [37]. The town lies within the huge conurbation area on or near the road from Kumasi to Obuasi. Kumasi is fast growing, and this has resulted in the emergence of urban sprawl at the fringes of the city. The sprawl of Kumasi has fast encroached its neighboring rural communities, including Trede. The proximity of Trede to Kumasi and the consequences of urban sprawl facilitated the transformation of the town from rural to per-urban. However, on the other side of proximity are some shortcomings. As an emerging peri-urban area, the issue of how to meet the demand for rural lands for urban development and how to manage the rural-urban interface have presented serious land problems in the area. The choice of the above case study area was based on the persistence of land disputes in the area. Hence, the selection of the area for the study was done purposively.

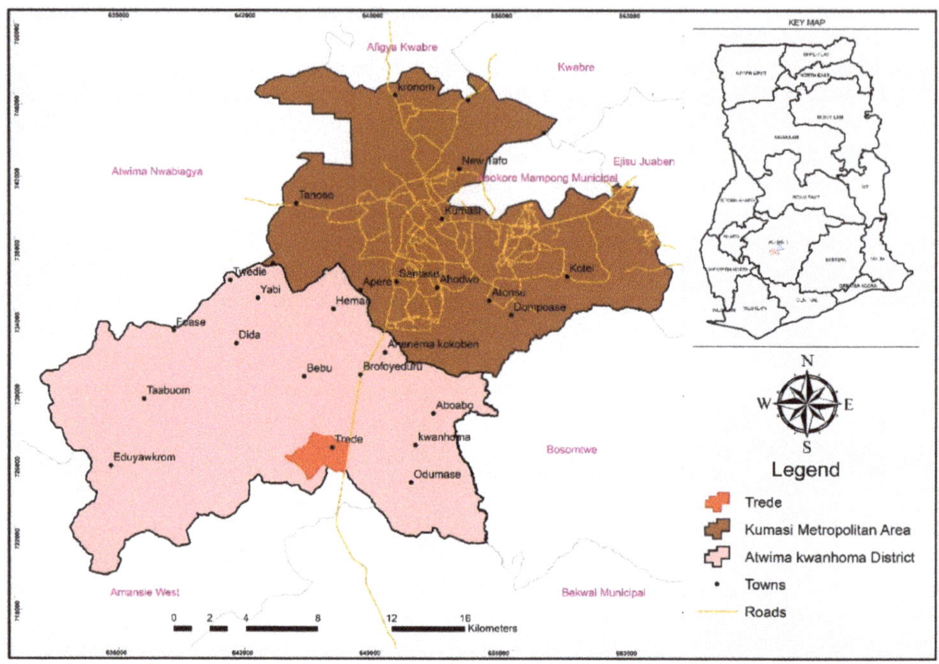

Figure 1. Location of Trede in conurbation of Kumasi Metropolitan Area. Source: Authors.

3.2. Research Methods

The present study sought to examine how the study area, Trede, transformed from rural to peri-urban status and how the problem of customary land rights disputes has affected the development in the area. The study approaches this problem by striving to answer three key research questions. These questions include: (1) What process did Trede undergo in its transformation from rural land to peri-urban land? (2) Did this process lead to increased customary land disputes, and if so, what are the characteristics of these disputes (including their types, drivers, effects, and consequences) on the development of the area? (3) In what ways can the situation be improved? By investigating these questions, it was possible to grasp the past and the present situations of land related problems in Trede, as well as discern ideas to improve the situation going forward.

Considering the complex and dynamic nature of peri-urban areas and land disputes (and the nature of the questions under investigation), it will be challenging to explore the associated issues using a single method. Thus, a mixed methods research approach was adopted to investigate the problem. This approach is more problem-centered and is suitable for gaining a deeper understanding of complex problems [38]. It offers the opportunity to counterbalance the weaknesses of each method. The requirement of using mixed method renders it appropriate to collect and analyze both qualitative and quantitative data either at the same time or different stages. Therefore, data collection and analysis for the study was done simultaneously. A case study analysis through key informant interview, field observation and document review have been used to generate qualitative primary and secondary data. Quantitative descriptive data was also generated from household survey. In the context of the research problem under investigation, the qualitative data emanated from the case study analysis and quantitative data resulted from the households' survey. In effect, a case study analysis of Trede was done using key informant interview, field observation document review, and maps to investigate the problem.

The case study was used to gain an in-depth understanding of the historical and present development in the town that influences its transition from rural to peri-urban, the customary tenure arrangements in the town, existing land disputes, evidences of the effects of these disputes, and the way forward in solving them. In-depth interviews were undertaken using semi-structured interview with traditional leaders, family heads, and state officials. These key informants were selected by purposive sampling technique since they are known to be involved in dispute resolution process and are in better position in answering questions on the causes of these disputes and interventions in strengthening the local disputes resolution mechanisms. In all, a total of six (6) traditional leaders and three (3) family heads were interviewed. Traditional leaders here constitute members from the palace who play key roles in the customary land dispute resolution, as well as members who have enough knowledge on the history of the town and can give an account on the transformation process of the town. Officials, including three (3) Senior Technical Officers from the Public and Vested Land Management Division (PVLMD), the District Planning Officer from Physical Planning Department (AKDA), Secretary of the Trede Stool lands, the District Land Surveyor from Physical Planning Department (AKDA), and three (3) personnel from the Otumfuo Land Secretariat (CLS), were interviewed. In all, nine (9) state officials were interviewed.

To bring to light the effects of land disputes on the spatial development of the town, some forms of observations were carried out on land parcels in the study area. This approach facilitated in identifying underdeveloped land within the area due to land disputes. Again, observations of the customary land dispute resolution process on two different occasions were undertaken. This was necessary to gain insights into how land dispute resolution procedures were organized and how efficient (process oriented) and effective (goals oriented) was the system. The Household survey has been used to collect information about demographic distribution, factors influencing change, and types of customary land disputes from landowners, developers, and users in the study area. The target population for the survey included landowners, developers, and land users from the study area who were mostly the victims of customary land disputes and are best placed to offer their lived experiences and responses on how land disputes impinge on their livelihood's and the community at large.

A total of 165 households were covered for data in the survey. The simple random sampling technique was used in selecting respondents; however, the snowballing technique was further employed in the selection of respondents. For household survey, simple random sampling technique, alongside snowballing sampling, was used to select a total of 165 households for the survey. Since issues of land disputes cases are delicate and sensitive, and victims of land disputes might not be ready to discuss their bitter lived experiences, the snowballing sampling was the only way to get in touch with some of these land dispute affected victims. Data from the 165 households were collected using semi-structured questionnaire comprising open-ended questions and some multiple choices closed-ended administered face-to face to the respondents. The analysis of data for the study involved quantitative, qualitative, and spatial analysis. Responses to the open-ended questions from the household survey and interviews were analyzed descriptively manually through content and narrative analysis. Responses to the closed-ended questions from questionnaires survey were coded with numerical values and keyed into the computer, which was analyzed using the Statistical Package for the Social Sciences (SPSS). Descriptive statistics were used to generate frequencies and percentages of the various units of analyses. Spatial data, including topographical map, as well as aerial and satellite images obtained from Google map, were analyzed using ArcGIS to visualize the changes taken place in the transformation process of the study area. The study was conducted over a period of eight (8) months, starting from July to February. Table A1 in the Appendix A section presents a matrix of the research methods adopted for the study.

4. Results and Discussion

4.1. The Transformation of Trede from Rural Land to Peri-Urban Land

Findings from all data sources in the study area revealed that the transformation of Trede from rural to peri-urban took four different dimensions: institutional, economic, social, and physical.

4.1.1. Institutional Transformation

Institutional transformation entails the changes in the customary land tenure system in response to urban demand. Trede town practices customary land ownership system where all lands are vested in the stool and held in trust for the community by the paramount chief *(Tredehene)* and other sub-chiefs. The Allodial title is vested in the paramount chief who exercises ultimate authority over all land in the town. According to interviews conducted, the history of land ownership and use rights in Trede traces back to the first settlers of the community. One of the sub-chiefs recounted that when *Nana Nuben* (the first settler and founder) settled in Trede, other families later joined him and requested land for settlement and farming. Each family had the freedom to choose any part of the forest to farm. The portions they cleared and cultivated became family lands, and these families were identified as the first landowners in Trede.

During this period, access to or the right to use land was based on social relations, either by family ties or by acquiring membership status in a community. Interviews with family heads revealed that family lands acquired by the first ancestors have passed on from generation to generation. Individual landowners could also transfer their land rights through gift or inheritance. Land acquisition by then was simple and cheap. One only needed to declare his/her intentions to the chief or family head, and, upon approval, the person presents a token (drinks); from then, one can start using the land. Both natives and strangers could enjoy free access (usufructuary rights) to and use of land.

From the mid-1960s, Trede started evolving from rural to peri-urban. Since then, the town has witnessed significant alterations in its customary land tenure practices. Some of the changes in Trede's customary land tenure include change in the process of land acquisition, change in the transfer of land, recordation of land transactions, and change in inheritance rules on land transfer. Interviews conducted show that these changes are occurring in the wake of population growth, shortage of land, and the high demand for land in the area. According to chiefs, the era of clearing uncultivated forest to access land is no more, and, as such, access to land is now dependent upon transfers of land from kins or on market transactions, such as leasing or renting.

Table 1 below present a cross tabulation of age by mode of land acquisition in Trede. It was found that majority of the landowners 41 (50%) acquired land through inheritance. However, further analyses showed that, in contemporary Trede, access to land is largely dependent on purchase. This was because most of the respondents who acquired land through inheritance were the older generation of ages 60 years and above.

Table 1. Mode of acquisition of land by age of respondent.

Age		How Did You Acquire the Land?				Total
		Inheritance	Purchase	Gift	Rent	
Under 30 years	Count	4	5	2	0	11
	(%)	(36)	(46)	(18)	(0)	(100)
Between 30–60 years	Count	16	21	2	2	41
	(%)	(39)	(51)	(5)	(5)	(100)
Over 60 years	Count	21	3	6	0	30
	(%)	(70)	(10)	(20)	(0)	(100)
Total	Count	41	29	10	2	82
	(%)	(50)	(35.4)	(12.2)	(2.4)	(100)

Source: Field survey, 2018.

The majority of the 21 respondents (51%) who purchased land were between the ages of 30–60 years. According to the Ghana Statistical Service (GSS), the age group of 30–60 years is economically active and has the highest working-class population. This means that people from this age bracket have got a high purchasing power, hence the reason for having the high rate of land purchasers from this age group. Census data from GSS also shows that people under 30 years are mostly unemployed and, as such, may be the reason for the low purchase land from this group. Meanwhile, other modes of acquisition, like gifts and rent, are gradually extinguishing from the town.

The investigation revealed that communal land rights in Trede are gradually giving way to individual rights under urban pressure. It was discovered from the survey that, out of the 82 landowners in Trede, 32 (39%) respondents hold a leasehold interest in their land, and 50 (61%) respondents have freehold interest. From these leaseholders, ten respondents (32%) purchased land from the chief and sub-chiefs, 8 (32%) purchased from private landowners, and 14 (44%) respondents purchased from family heads. Figure 2 below represents respondents' rights in land and represents grantors of leasehold interest.

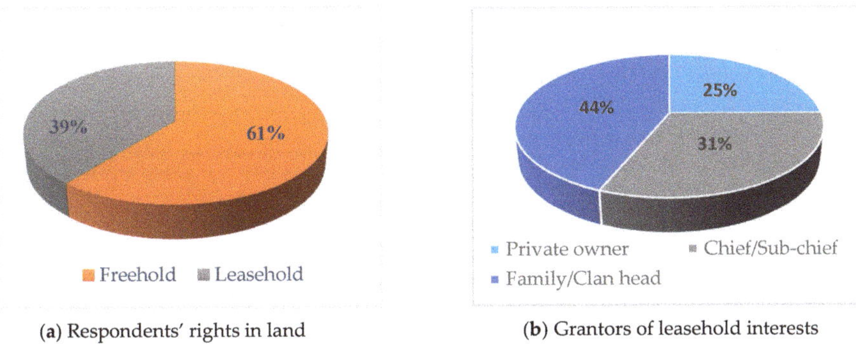

(**a**) Respondents' rights in land   (**b**) Grantors of leasehold interests

Figure 2. Respondents' rights in land (**a**) and grantors' leasehold interest (**b**) (source: Authors).

The results show that family heads have granted most of the leasehold interest in land. Investigations show that some leaseholders were also natives who had to enter into leasehold agreement to secure their land rights. Currently, in Trede, the ordinary usufruct right can only be secured if family members entered into lease agreement with family heads. Therefore, poor natives whose family lands have been disposed of are put in the same position as migrants to compete for land. This has rendered many poor native landlessness in peri-urban areas. This finding supports other research findings [5,7] on peri-urban land market, where access to land is dependent on one's purchasing power rather than socio-cultural characteristics, such as on gender, age, and social status. This means the usufruct rights of free access to land do not apply in peri-urban areas.

4.1.2. Economic Transformation

Economic transformation considers the changing economic importance of land (that is, land as a social commodity changing to an economic commodity) and the influence of urban development on the economic activities of peri-urban dwellers.

According to the chiefs, prior to the intrusion of urban activities, land in Trede was under rural production and solely used for farming and housing. Farming was mainly on a subsistence basis. The inhabitants of Trede cultivated crops, such as cassava, plantain, okra, and pepper, to feed themselves and their families. Monetary payments for the use of land was not a common practice. The common form of leasing land was by *abunu* and *abusa* (sharecropping arrangement). Sharecropping was mainly transacted between native landowners and migrant farmers. In a sharecropping arrangement, the land was given to a migrant farmer on the basis that the landlord was entitled to a third (1/3) share of the farm produce and the farmer received two-thirds (2/3) (*abusa*), or each party was entitled to half (1/2) share *(abunu)*. Now, these forms of leasing are gradually fading out, and landowners prefer to give land out on long-term leases because of the changing economic importance of land.

Active land sales became much more evident in the 1960s when the first local plan (layout) was prepared for Trede. According to one sub-chief, the first commercial sale of land in Trede was in the year 1969, when about 2560 acres of land was sold to a poultry farmer. This did not come as good news to the youth of Trede. They opposed the sale and lamented that such a sale would deprive some inhabitants of their farmlands. The chief ignored their concern and went ahead with the sale. This resulted in a conflict between the chief and youth of Trede. This conflict has been recorded as the most violent in the history of Trede. Interviews therefore suggest that the initial local plan was the main driver of transformation from rural lands to peri-urban lands, as it leads to the subdivision of rural land with unclear boundaries to urban plots. The resultant effect of the local plan was an increase in demand for urban plots and increasing land values.

Economic transformation also resulted in the livelihood diversification of rural dwellers in Trede. Some farmers reported that they have lost their farms without compensation as a result of urban expansion. This was because the chief demanded that, as the town expands and gets to your farm, you must sell it for development. If you delay in selling, and he sells it on your behalf, you may not get a good share of the sales. The fear of many farmers losing their farmlands caused them to sell their farms and switch to non-farm activities in the town. Indeed, those still practicing farming have other non-farming activities as their secondary occupation.

Even though many economic activities continue to spring up, the survey revealed that farming remains the dominant occupation of the people in Trede, despite the urban pressure. A total of sixty-five (39%) respondents out of the 165 respondents are into farming (small scale farming), such as crop farming and poultry (chicken) rearing. Forty-Two (25%) respondents were engaged in artisanship (the artisans included carpenters, masons, dressmakers, tailors, and local mechanics, among others). Artisanship accounted for the second largest occupation of the respondents, particularly the youth. The youth in the town were either engaged in masonry, carpentry, plumbing, and other construction works because of the many residential and commercial on-going development in the study area. Thirty-one respondents (19%) were into trading activities, such as shop keeping, food vending, and selling agricultural commodities, because Trede has a vibrant daily market for both external and internal trading activities. Other respondents included 4 students (2%); 6 civil servants (4%), and 5 drivers (3%). Twelve respondents (8%) had no occupation, and these were people who had gone on pension, those who could no longer work on their farms, and those who had lost their farms through customary land disputes. Table 2 shows the occupation of the respondents.

Table 2. Primary occupation of respondents.

Occupation	No. of Responses (N = 165)	Percent (N = 100%)
Farming	65	31
Trading	31	19
Civil Service	6	4
Artisan	42	25
Student	4	2
Driver	5	3
None	12	8

Source: Authors.

The analyses revealed that the new economic development in land have resulted in the shift from traditional land-based activates to more urbanized jobs. This finding confirms Tacoli's [39] conclusion that economic diversification is high in peri-urban areas as people diversify a range of farm and non-farm income earning activities to supplement their household income. Many farmers explained the reason for diversifying livelihoods as the fear of losing farms and the need to supplement household income.

4.1.3. Physical Transformation

Physical transformation relates to the changes that have occurred in the physical configuration of the study area, which is concomitant with the transformation process.

A common physical change in Trede is the conversion of agricultural land to non-agricultural use. Interview with the District Surveyor (DS) of Atwima-Kwanwoma District revealed that, since the implementation of the first planning scheme, the built-up area and population size have both expanded tremendously. Table 3 below indicates the proportion of developed and undeveloped land in Trede. The period of study runs from 1975 to 2018. As of 1975, the total developed area of Trede constituted 145.30 acres of developed land and 1091.61 acres of undeveloped land. Between 1975 and 2018, there has been a rapid increase in developed lands by 68.66% and a reduction in undeveloped land by –41.83%. This can be accounted for by the high demand for land for residential and commercial uses. The rapid expansion of the built-up area played a significant role in the current peri-urban status of Trede.

Table 3. Proportion of undeveloped and developed land in Trede (source: Field survey, 2018).

Land Use	1975 Area (acres)	%	2000 Area (acres)	%	% Change	2018 Area (acres)	%	% Change
Developed	145.30	11.75	379.45	30.70	19.00	782.09	63.60	32.9
Undeveloped	1091.61	88.26	856.47	69.30	18.96	447.71	36.41	–32.9

Additionally, Figure 3 below visualizes the proportion of developed and undeveloped land in Trede over time.

Figure 3 above revealed that the town of Trede has witnessed changes in its land use after the implementation of the local land use plan. The pattern of change shows a continuous decrease in agricultural lands and a rapid increase in the built-up area. Since physical development in the town follows the requirements of the existing land use plan, it can be deduced that the land use plan is monolithic as it gives priority to urban land uses than agricultural uses. In terms of population growth, the town recorded a 31% population increase between 2010 and 2018 (that is, from 4065 to 7730) according to the Ghana Statistical Service [37]. Observation and interviews revealed that the high rate of population growth in the town resulted from the natural increase and the influx of migrants from other towns.

Local government intervention has also contributed to the physical transformation of Trede. In Ghana, the provision of local government services is mostly skewed towards urban settlement. As an emerging peri-urban area, the provision of services and infrastructural facilities provided by the government has facilitated the transformation process. The following facilities were provided by government intervention: schools, postal service, agriculture extension service, daily market, electricity, district administration block, area council, police post, health center, public toilet, and a borehole. According to the interviewees, the construction of the first-class road connecting Kumasi to Trede in 2018 was the turning point for the town to become more urban.

(**a**) 1975.

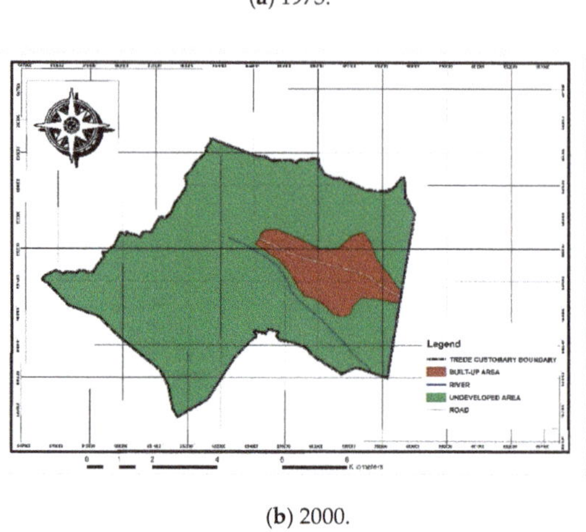

(**b**) 2000.

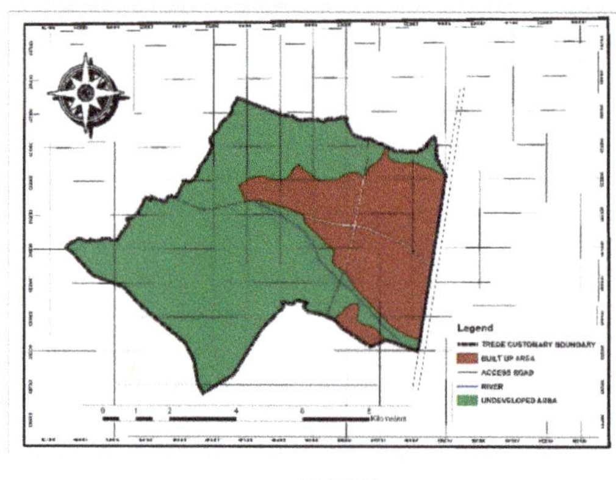

(**c**) 2018.

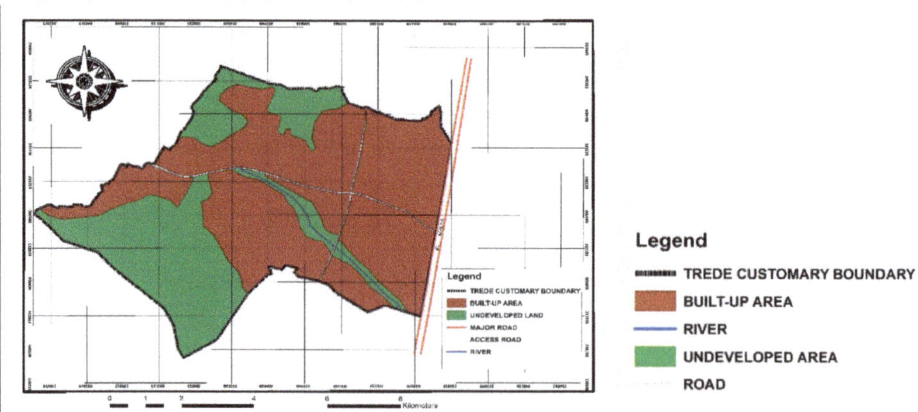

Figure 3. Maps of built-up area of Trede between 1975–2018, (**a**) 1975, (**b**) 2000, and (**c**) 2018 (source: Authors (based on Geographic Information System Analysis).

4.1.4. Socio-Psychological Transformation

Socio-psychological transformation entails the change in ideas, perception, values, and the lifestyle of rural dwellers in response to urban pressure. It constitutes the transformation in the mental orientation of rural dwellers. It is mostly caused by the infusion of urban values in rural areas. The proximity of Trede to Kumasi and the influx of migrants from diverse culture into the town have contributed to the change in the lifestyle of the people in Trede contributing to the transformation process. For instances, change from communitarianism approach of land tenure to individualization. Interviews revealed that, in the past, people preferred to pool their resources together to cultivate and share the farm produce after harvest. Back then, land was a collective resource, not an individual property. Another aspect of change observed during the survey was that people's understanding and perception on the family system has taken new forms in peri-urban areas. Interviews revealed that the solidarity that existed in family relations is no more. A woman commented: *"In the olden days when we say family we meant our parents, siblings, grandparents, uncles, (…), but today family means you and your children, nothing more"*.

It was noted that this new perception had engendered a deviation in the inheritance rule leading to many intra-family land disputes. The people of Trede practice the matrilineal system of inheritance. The rules of this inheritance system stipulate that children shall not inherit from their father but, rather, their uncle (mother's brother). According to interviewees, people now find the need to keep their responsibility towards their children other than their nephews due to economic hardship. The transfer of land from father to son rather than nephew has broken down many extended families and promoted intra-family land disputes in the area. It was found that the traditional family system and lifestyle have gradually been eroded by modern and urban lifestyle as the town is increasingly becoming peri-urbanized. This confirms other findings [23,40], which noted that, as an area transforms from rural to peri-urban, family and social relationships are modified, created, and sometimes discarded.

The below diagram Figure 4 shows how the various dimensions of the peri-urban transformation are interlinked.

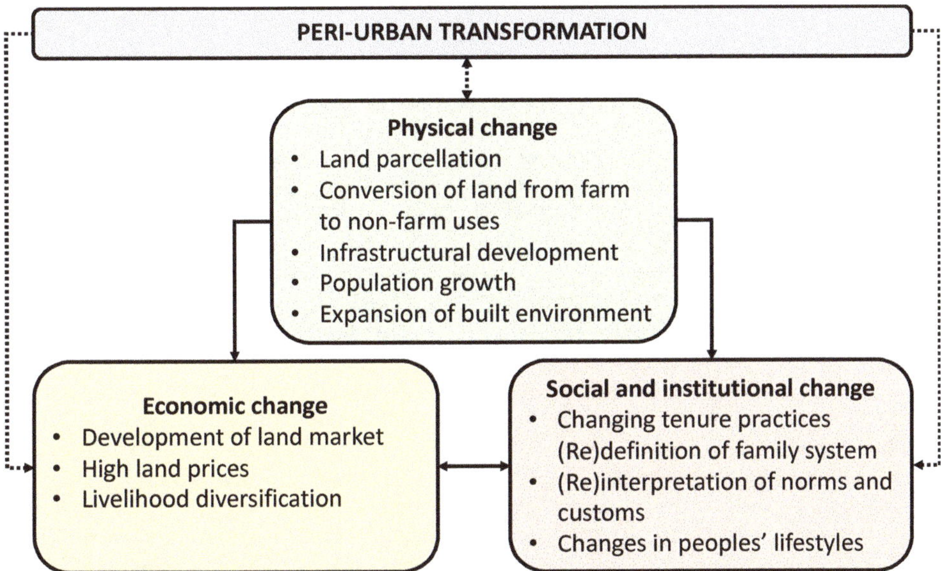

Figure 4. Structural model of the interrelated transformation process of Trede (source: Authors).

It is worth noting that the peri-urban transformation is an interrelated process. Thus, the changes taking place are interactive and mutually supportive. This was evident in the case of Trede, as the change in the physical composition accounts for both changes in the social and economic composition of the town. For instance, land use change under the physical dimension is an enabler of change in livelihoods of peri-urban dwellers. Market development as an element in the economic change induced changes in the customary tenure rules, resulting in the transition from communal to individualized rights. Socially people's new perception about what constitutes family has implications on inheritance rules to land transfer. The emergence of Trede from rural to peri-urban demonstrates that the pattern of change in the peri-urban environment are interconnected and a change in one aspect can trigger a change in another aspect. In this study, the interaction between social, economic, and physical elements influenced the transformation from rural to peri-urban. The drivers and the elements of changes that occurred in the peri-urbanization of Trede collaborates with that of Rauws and De Roo's [40] study, which analyzed peri-urban transformation in various European cities (including Leipzig, Warsaw, and the greater Hague region). Although similar characteristics exist in peri-urbanization in developed and developing countries, the resultant effect of the process may differ. Whereas, in developed countries, the implementation of planning schemes guiding peri-urban transformation have yielded some positive outcomes, the situation is different in most developing countries [28]. In the case of Trede, the implementation of a land use plan to facilitate the transformation of the area became the source of many land disputes.

4.2. Customary Land Rights Disputes: Types, Drivers, Effects, and the Way Forward

Specifically, respondents were asked to mention and relate any land disputes encountered with respect to their land rights over the years. Out of the 165 respondents, 52 (32%) respondents indicated boundary dispute as the most dominant type of CLRs dispute in the area. Based on the opinions of the respondents, the study noted different forms of boundary disputes. It is found that boundary disputes arose if some people gradually encroach on other people's land or try to forcefully extend their boundaries into other people's land. This situation arises when one party tries to take advantage of another for failure to identify exact location of his/her boundary, absence of another party, or physical or financial weakness of another party.

Not only did boundary disputes occur between individuals or families but between Trede town and other neighboring towns. At the time of the field survey, Trede stool was in boundary dispute with Sabin Akrofrom, Pakyi, and Nkoransa, three neighboring communities. An interview with the District Surveyor revealed that the neighboring communities have encroached about 2 km beyond the boundary originally delineated between them and Trede. A further investigation into the boundary disputes revealed that some landowners in the disputed area purchased land from chiefs from the various communities. As such, landowners bemoaned that the boundary dispute have made it difficult to ascertain documentation on the land. This was a challenge during the survey whether to include these landowners as inhabitants of Trede (because they bought the land from other communities) as some of these respondents could not actually confirm whether they were in Trede or otherwise. To avoid doubt, the landowners in the disputed area were excluded from the survey.

The second prevalent land rights disputes relate to intra-family disputes over land ownership, of which 44 (23%) respondents disclosed. Intra-family disputes relate to disputes between close relatives with equal chances of inheriting or having access to the family land. Most of these disputes were related to disputes over the inheritance or gift of land and typically arose between deceased persons' nephews and their children. Intra-family dispute was found between elders and younger family members (elders versus youth). This form of land dispute was largely driven by plot division or sharing of proceeds from the sale of land. In Trede, many farmers depend on family lands for farming. According to the District Planning Officer, about 60% of the farmers depend on family lands as their source of livelihoods. Family disputes over land ownership, therefore, present difficulties for farmers in accessing family lands for farming.

The third major type of land dispute discovered during the survey is multiple sales of land, which accounted for 23% of land disputes in Trede. Multiple sales of land in this instance denotes a situation where the same parcel of land is sold to different persons. It is often encountered by private developers, who acquire land for residential, commercial, and industrial uses. Geographically, these disputes were prevalent in new areas where development was now springing up in the town. It became evident that multiple sales of land in the study area emerged from two forms. It resulted from the deliberate act of unscrupulous landowners selling the same piece of land to different buyers, as well as the absence of a land secretariat office to keep a reliable database of land transaction. In response, a land secretariat office was recently set up in Trede to keep records of land transactions.

The last type of land dispute identified during the survey was the disposition of rights in land by traditional leaders. It was found that the disposition of rights in land by authorities have often resulted from the demand for land for community projects, such as schools, clinics, and other infrastructures. The community members never disputed against such communal decisions in using their lands for developmental projects, like schools, and, in fact, always found pleasure in these beneficial projects when implemented in the town. However, it is a common practice for traditional authorities to take more land than is needed for such projects and later selling the remaining land for personal gains. In such circumstances, expropriated members agitated and expressed dissatisfaction through violence over such attitudes, which respondents described as uncultured and uncouth on the part of the chiefs. Indeed, some farmlands were forcefully sold by traditional authorities because farmers had delayed in either converting or selling their farmlands for residential development. Currently, it is a requirement in the town that, when the town expands unto areas used for farming, the affected landowners must convert their lands for their own residential use or sell to other persons for similar uses. Notwithstanding, where farmers delay in converting their lands, the chief will order his elders to sell and compensate the affected person with a small amount. These undemocratic depository acts of the chiefs are often weakly justified by the superficial entrenched custom that chiefs are the sole owners of land and choose to do whatever they want with the land. As one sub-chief reiterated that:

"farms were given to natives in the past to 'eat from' but not to own it forever. So, if the time comes for the chief to take back land for development works nothing can stop him. If you are respectful enough and beg him, he can find you an alternative place to farm. The chief is determined that farmlands cannot be a barrier to the development of the town".

This pronouncement of chiefs confirms the current emerging trend where some chiefs view themselves as owners of communal land rather than trustees. Figure 5 below shows the predominant type of land disputes in the study area.

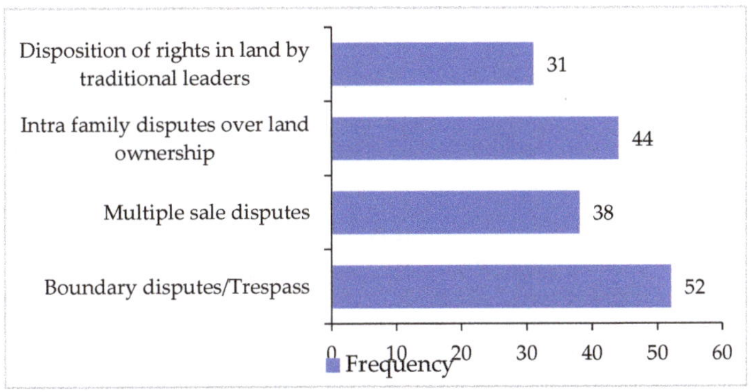

Figure 5. Types of customary land rights disputes in Trede (source: Field survey, 2018).

A further deliberation revealed that customary land disputes happening in Trede could be broadly categorized into three groups: land ownership dispute, land use dispute, and land value dispute. In the case of Trede, boundary disputes between neighboring families/farmers and intra-family disputes over land ownership, multiple sales, and the disposition of rights in land are well-situated within the category of land ownership dispute. Disposition of rights leading to dispute over compensation value shall fit under land value dispute, whereas intra-family disputes over the use of land between family heads and members are classified under land use disputes.

4.3. Drivers of Customary Land Rights Disputes

4.3.1. Deviation from Inheritance Rule

It was found that the transformation of Trede from rural to peri-urban is causing inheritance related disputes. The people of Trede are classified as matrilineal, and, therefore, land ownership in the town is based on matrilineal descent. The matrilineal system plays a key role in who gets access to and control over land. For instance, the matrilineal system of inheritance allows nephews to inherit land from their uncles. According to the matrilineal principle, children particularly, the son do not inherit from their father but, rather, their maternal uncle (mother's brother). The matrilineal inheritance rule also stipulates that self-acquired property can be passed on to children, but property inherited from the matrikin will forever remain a family property. However, information obtained from some of the respondents revealed that, due to population increase and economic hardship, successors of family property are either selling family lands for personal gain or transferring to their children rather than nephews. Some respondents attributed the land inheritance related disputes to changing social relations. They complained that the matrilineal system of inheritance is changing as some family members have begun redefining the family relationships. Fathers now prefer bequeathing their property to their children even when such property was inherited from the matrikin. The inheritance right to land generates tension between the nuclear family and the extended family, and this has been the source of many intra-family land disputes in the town.

4.3.2. Chiefs Using the Right of Re-entry to Resell Lands Already Allotted

It was found that chiefs use their right of re-entry to resell land they have already allocated when allottees fail to develop such lands within a specified period. As part of the covenants stated on land allocation notes, chiefs require allottees to commence development on the land within three years from the date of allocation. Any failure by the allottee (be individual or investor) to develop the land within the stated period entitles the chief to re-entry. According to respondents, most people had lost their land through this practice where chiefs resold their lands when they failed to develop the land within the three years. For developers who lost their lands to the chief, it is due to financial constraints that hindered them from developing the land. One developer commented:

> *"Building a house is not a one-day job. It is a step by step process. The plot of land itself is expensive. After working and getting money to buy the land, you need time also to work to accumulate money before you can start the building project. But, by the time you get the money to start building, your land will be gone - thus sold by the chief".*

The narration above shows that the chiefs are not considerate with the timeline within which land acquirers are expected to develop the land. The chiefs, however, justified their actions by stating it is a provision in the allocation note which ought to be followed, failure of which entitles them (the chiefs) to re-enter. It was found that many land purchasers, upon signing the allocation note, do not pay attention to this clause (3-year timeline). It was also found that those who are most likely to fall prey to the 3-year timeline condition are the illiterate land purchasers who cannot read and write to understand the terms of the purchase. However, interviews with land officers at the Lands Commission revealed that chiefs are taking undue advantage of the right of re-entry and allocating the same piece of land to different buyers, giving rise to multiple land sales. It is expected that, before re-entry, chiefs are required to give notice to landowners. However, most chiefs have failed to give prior notice which is a violation of Section 29 of the Conveyancing Decree, Act 1973 (N.R.C.D. 175), which specifies the restrictions on the right of re-entry as it states that:

(1) a right of re-entry or forfeiture under any provision in a lease for a breach of any covenant, condition, or agreement in the lease shall not be enforceable, by action or otherwise, until:

(a) the lessor serves on the lessee a notice: specifying the particular breach complained of; if the breach is capable of remedy, requiring the lessee to remedy the breach; and (except where the breach consists of a non-payment of rent) requiring the lessee to make reasonable compensation in money for the breach; and

(b) the lessee has knowledge of the fact that such notice has been served.

It is evident that the high demand for urban lands has caused land values to rise. These are motivating factors for chiefs to engage in multiple sales using the right of re-entry. The right of re-entry, though legal, has chiefs violating it as they fail to give due notice as required by law. This creates disputes between chiefs and land buyers.

4.3.3. The New Local Plan as a Recipe of Land Disputes

The growth of Trede has necessitated the creation of a new layout to regulate development and land use. According to the respondents, the new layout is causing land boundary disputes and evictions. Apart from the first layout for the town, which was implemented over 50 years ago, there had not been an updated layout to reflect the spatial transformation in the town. Indeed, the first layout only covered about 40% of the total area. Until 2018, there was no settlement layout guiding physical development in many parts of the town. This resulted in uncontrolled development and conflicting land uses in the town. In 2018, the new chief found the need to re-plan the town. The recent effort by the new chief to implement a new layout to properly regulate development has escalated the problem of land disputes. An interview with a District Surveyor revealed that the new layout is a continuation of the old plan, with slight changes. He explained that there was the need to demarcate the town again because several land uses and developments had sprung up since the first layout was prepared five decades ago. Other respondents also accused the surveyors of bias and bribery, indicating it was mostly the poor people whose lands were affected by roads, schools, and other projects. One interviewee complained;

> *"It was only the poor people's land which was earmarked in the layout to be used for the roads, schools and other projects. The rich people who could bribe their way through influenced the surveyors who designed the layout to favor them (the rich). If per the layout, a road was passing through your land/house and you could bribe the surveyors, it would be excluded".*

According to some chiefs, land uses that were found to be incompatible with the new layout were changed and owners of such land were given alternative land, but only if they approached the chief with respect and requested for another place. Other respondents lamented that the implementation of the new layout was a strategy adopted by the chief to reclaim land from individual owners. Since there was no reliable database for landowners in the town, all landowners were asked to register their land parcels at a fee with the chief. The amount to be paid is irrespective of your membership status: native or stranger; or mode of acquisition: purchase, gift, or inheritance. The refusal of some landowners to make payment has caused land disputes between the chiefs and landowners. Landowners did not understand why they should pay for land they have acquired long ago. Thus, they claim consideration for the land had been paid when they purchased the land. Others also complained that the registration fees were expensive and arbitrary. The chiefs, however, justified the charges, indicating it as meant for town development.

4.3.4. Rising Land Values

The high demand for land in the study area is accompanied by increase in the price of land. Interviews conducted show that there is no fixed price for land in the study area. Prices vary depending on the location and the purpose for which one is acquiring the land. High land values were a motivating factor for some landowners to indulge in double sales of land. The increasing land prices automatically squeezed out the poor from the land market. As one respondent noted:

> *"I bought land from my Abusuapanin (Family Head) in 2001 at a price for GH¢1500 ($750). I did not have the money to pay at once, so he agreed I pay him in installment. I had paid this amount over a period of 3 months with only GH¢100 (24$) outstanding. Before I could make my last payment, he told me he already resold the land to another person who could offer a better price than me. I believe if I had the money to pay him everything he would not have denied me the land".*

It was observed the financial incapability of some prospective buyers has led to disputes between landowners and prospective buyers. This finding is consistent with previous studies [11,13,15], who all reported that access to land in peri-urban areas is largely dependent on one's purchasing power, rather than membership status.

4.4. Effects of Customary Land Rights Disputes on Development in Trede

The study reports the effects of customary land disputes across the economic, social, and physical development trajectories of the study area.

4.4.1. Loss of Income for Local Farmers and Private Developers

Interviews conducted revealed that victims of land disputes suffer income losses. It was found that, when a land dispute is heard in the chief's palace or at the law court, an injunction is placed on the disputed land. In such a situation, conflicting parties are banned from using the land until the dispute is settled. Where the disputed land is farmland, farmers are not allowed to cultivate on the farm. Farmers expressed that they are prevented from generating income from their farms when the case is being heard at the palace or law court. One farmer recounts that: *"you will not be allowed to harvest your crops when your land is in dispute. I lost my seasonal harvest when I had dispute on my land".* Particularly in the law court where there is backlog of cases, it could take several years before a case is settled. This means that the disputed land remains unproductive for that period, this has led to the loss of income for farmers in disputes.

Disputants also indicated that even when the case is not heard in court, you may not want to use the land for the fear that your opponent will attack you on the farm. Conflicting parties were fond of issuing death threats or destroying each other's farm when a dispute arises. During the interviews, some disputants reported incidents where they were attacked or physically assaulted on their farms by 'gangs or *landguards*' sent by their opponents. For safety and security reasons, farmers left their farms idle without any productive activity carried out during the periods of decisions on CLRs disputes. Preventing farmers from working on their farms because of on-going CLRs disputes leads to reduction in farm production, which has implications for food security in the community and the nation at large. On the part of private developers, land disputes meant delay in construction works, as well as high transaction and construction costs. Some respondents confirmed that construction works came to a halt when cases were before the chiefs or pending in court. During field observations, it was common to find uncompleted structures that were abandoned due to land disputes. Disputants indicated that disputes imposed additional costs to construction when properties are destroyed. Disputants who were involved in multiple land sale disputes indicated they incurred additional cost to protect their structures from destruction. One respondent remarked: *"I hired someone at a high cost to watch over my building materials because I feared my opponent would take them away".* These expenses incurred by developers during disputes reduces their income or their return on investment.

It was also found that disputants during disputes resolution incur legal cost, transportation cost, and waste a lot of time that could be used on other productive ventures. It was noted that the cost of dispute resolution varied depending on the medium used to resolve the dispute. Disputants reported that a lot of costs are incurred if disputes are settled in the law court. This confirms why many land disputes in the study area were resolved at the local level by the customary court. However, it was found that the party who is sure of winning the case is ready to spend any amount in the resolution process. This was because, after passing judgment, any amount incurred by the winner during the dispute shall be reimbursed by the loser. Disputants complained that, where bribe is involved in the dispute resolution process, the amount is inflated. This condition provided the opportunity for the rich and powerful to take advantage of the vulnerable in the society as they used money and influence to take land from the poor. As one woman said: *"when I had dispute with Mr. A (not the real name) I did not litigate further, I left the land for him to take it forever. If I did not allow him to take it, he will take me to court, pay bribe and take the land while I reimburse him the cost of expenses"*. This finding corroborates the ideas of Reference [11,13,16], who found that the poor in the society fail to defend their land rights because institutions for dispute resolution have come under the direct influence of the rich and powerful in the society. As such, all types of cost—legal, transportation, damage, security, and lost time—results in reduction of household income, return on investment, and capital for victims of land disputes.

4.4.2. Unemployment

Many youths were either engaged in masonry, carpentry, plumbing, and other construction works because of the many residential and commercial development in the study area. As at the time of the survey, there was an injunction on some construction projects either by the court or chief. As a result, youth who were engaged in these construction projects as a means of earning a living, were jobless for the period of the injunction. A 24-year-old man commented that: *"I am a bricklayer and I earn GH¢50 per day, but for the past two weeks I cannot find a job to do"*. Many of the farmers who had lost their farms in the new layout project complained that they had remained unemployed since then. It was found that many of these farmers were already old, and it was difficult for them to find alternative jobs elsewhere. The unemployment situation has accounted for the high dependency ratio in the town. According to the 2010 population and Housing Census, the age dependency ratio in Trede is 80.8; this indicates that there is relatively high dependence on the working population. This finding matches the observation by Budiyantini and Pratiwi [41] that there is a relatively high dependency ratio in the peri-urban areas. In addition, the unemployment caused by the effects of CLRs disputes could lead to other problems in the town, such as crimes. To make ends meet, some unemployed, especially the youth, could resort to criminal acts, like theft or armed robbery.

4.4.3. Break Down of Family Relationship

Intra-family disputes have brought separation between the extended and nuclear family relationships in Trede. The survey results point to an emerging distortion of family ties in the community. As indicated by some respondents, the support from uncles to their nephews has ceased because of family disputes over ownership of land. As such, it is now common to find nephews and uncles fighting over ownership of land. One respondent remarked: *"I was arrested by my own uncle because I refused to leave my father's land for him"*. A good relationship between family members and solidarity in a community create room for a productive environment that enhances community development.

4.4.4. Discouragement of Communal Spirit and Solidarity

A direct effect from the breakdown of family ties is the lack of communal spirit in the town. Disputes arising between family members, as well as between chiefs and community members, have resulted in a low sense of commitment from community members to support developmental projects. One respondent mentioned that: *"The chief usually held durbar and invite community members to discuss community matters. But these days when he invites, nobody attends because the people are not happy with him"*. The unwillingness of community members to participate in community projects have negatively impacted in the development of the area. The involvement of community members in community decision-making creates room for problem-solving.

4.4.5. Emotional Distress and Loss of Lives

It was found that the only CLRs dispute that was ever violent in Trede was the dispute in 1968 between chiefs and community members over the sale of a large tract of land to a private investor. The youth took up arms to defend their land rights and use of the land. However, there was no reported case of death in this dispute, but it was discovered during the survey that even though current land disputes are not often characterized by many physical confrontations, yet some people have lost their lives. Respondents cited instances where disputants who have unjustly lost their lands in dispute committed suicide. This means that land constituted the last hope for livelihood for such people. Some disputants also experience emotional stress and trauma, which take a long time for them to recover from. One respondent commented that: *"my father is down by stroke because of land dispute, he spent all his life savings on this case, but he lost his land to another family. He cannot do anything as we talk now"*. Therefore, irrespective of the violence profile of a dispute, one can lose his/her life.

4.4.6. Limited Infrastructure Development

Survey results from both key informant interviews and semi-structured questionnaire administration unfolded that Trede has lost investment projects because of land disputes. It was found that, in 2008, the government was considering siting the Atwima Kwanwoma district capital in Trede but, due to on-going chieftaincy and land disputes cases, such decision could not materialize. It is currently the area council for the district. One sub-chief commented:

"We deserved to have the district capital in our town because Trede is big and more developed than the other towns in the District. However, matters of disagreements related to chieftaincy and land made us lose this great opportunity".

Trede currently has one clinic, and there has been a proposed hospital project for the town. However, this project has been pending for years because families whose lands were acquired for the project are disputing over the land since they were not compensated in any form either by money or alternative land. One sub-chief commented that: *"any time these investors come to inspect the land they find people cultivating on the land"*. Findings from this study show that land dispute adversely affected physical, social, and economic developments of Trede.

4.5. Conceptual Nexus between Peri-urban Transformation, CLRs Disputes, and their Consequences on the Development of Trede

In Figure 6 below, we show the relation between peri-urban transformation, types of CLRs their causes and the effects of these disputes on peri-urban development. It was found that, in the transformation process of Trede, the change in the physical composition accounts for both changes in the social and economic composition of the town. The physical transformation marked by population increase, expansion of the town, the creation of new layout, etc., have led to some changes in customary tenure practices, such as registration, recordation of land transaction, etc. The field survey suggests that these changes have been the sources of CLRs disputes in the area. CLRs disputes arising have been categorized in three groups: dispute over ownership, use, and value of land. The land ownership disputes include the struggle over the ownership right to land, as well the revenue that come with it. Land use disputes manifested in the disagreement over what use land should be put to. Land value disputes relate to disputes over the right to compensation. Physical changes have been a major source of all the types of CLRs disputes in the study area. Economic changes also accounted for land use and land value disputes. The consequences of land disputes can be broadly categorized into social and economic, as well as physical, consequences. For the local farmers in Trede, land disputes have led to economic consequences, such as reduction in farm profitability, loss of income, and unemployment. To the private developers, CLRs disputes result in a delay in construction, high transaction and construction cost, and reduced return on investment. For the community at large, land disputes lead to social consequences, such as lack of community spirit, breakdown of family relations, emotional stress for community members, and a critical barrier to infrastructure development in the community.

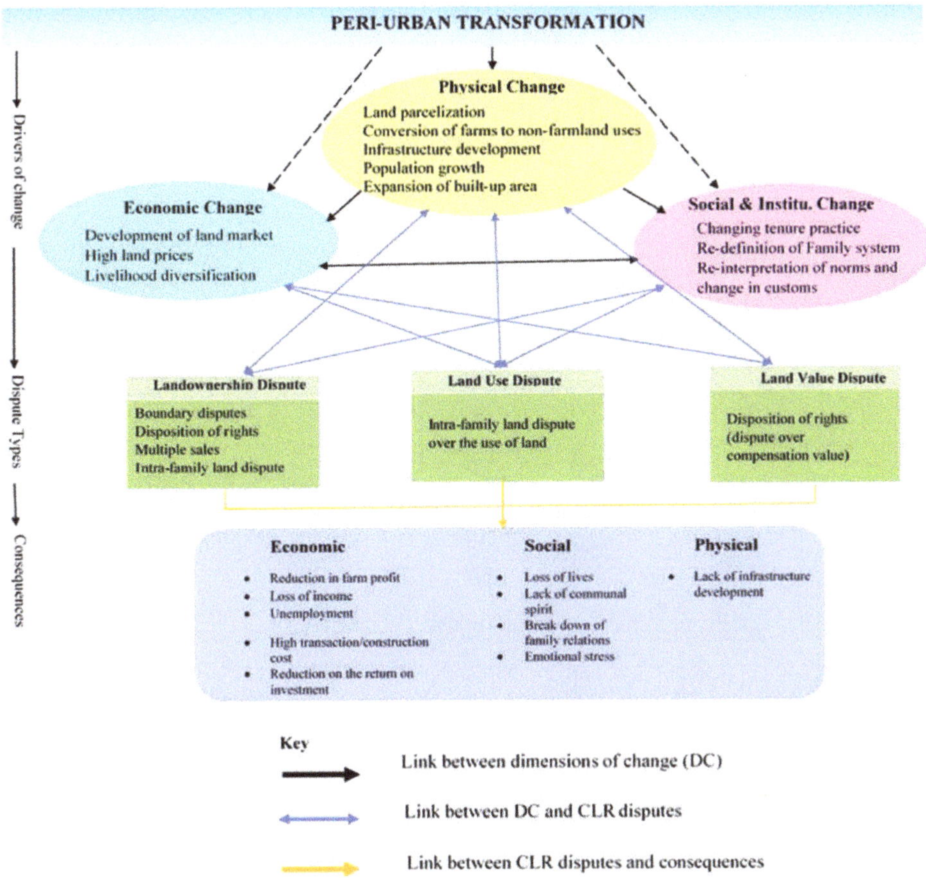

Figure 6. Linkages between peri-urban transformation, customary land rights (CLRs) disputes and their effects on peri-urban development (source: Authors).

4.6. Disputants' Perception of the Effectiveness of the Dispute Resolution Mechanisms

Respondents who have suffered CLRs disputes in the study area were asked to specify which mechanisms they adopted in resolving their land disputes. It was found that disputants usually first resort to the customary court before trying any other alternative means of resolution. The survey found that 41 (49%) disputants consulted the chief's palace for resolution, 19 (22%) disputants said they used the law court, and 13 (15%) disputants sought the intervention of the police. Alternative Dispute Resolution (ADR) was the least used by disputants. The survey shows that many who had used the customary court but were not satisfied with the outcome used other alternative mechanisms. Nine (9) disputants indicated that they did not employ any mechanism for resolution. Those who did not use any mechanism reasoned that land litigation was expensive and a waste of time. Others expressed that their land dispute involved the chief, and there was no way they could litigate with him. It should be noted that data in Figure 7 (below) involves 85 cases total, which exceeds that of the 71 sampled disputants because a disputant may have adopted multiple sources for resolving the dispute.

Disputants were asked to further elaborate based on their opinions of the effectiveness of the mechanisms they employed in attempting to resolve their disputes. Most of the disputants, 39 of them (55%), indicated that they did not find it effective, although many have used the customary court system. They reasoned that the customary court's decision is not as binding as the state courts. Respondents also tagged the customary court with mistrust and corruption and indicated they preferred to use the law court but were hindered by financial constraint. This finding agrees with Crook et al. [42] and Ubink [13], who noted that, despite the patronage of the customary dispute resolution institutions, many had accused the chiefs' court of being corrupt, lacking accountability, and being partial. On the contrary, 32 (45%) disputants regarded the customary court as effective. They explained that traditional leaders better resolve certain land cases, like boundary disputes, than law court. This is because the state court requires that you prove your boundary with a title document, which most landowners did not have. Unlike the state court, chiefs are the landowners, and they know the boundaries very well. They only ask for a witness to confirm, and, therefore, it was less stressful and relatively fast compared to the state court.

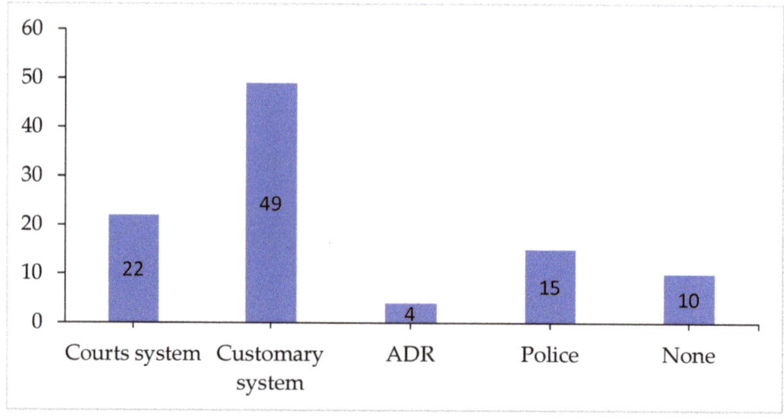

Figure 7. Land disputes resolution mechanism employed by disputants (source: Field survey, 2018).

4.7. The Role of State Land Sector Agencies in Customary Land Management

Interviews were conducted with Lands Commission (LC) and Town and Country Planning Department (TCPD) to assess their role in customary land management and dispute resolution. It was found that traditional authorities are required by government to seek statutory intervention in the management of peri-urban lands. It is expected that chiefs collaborate with the TCPD to survey, plan, and sub-divide rural land into urban plots as a town transforms from rural to peri-urban. The LC, on the other hand, shall serve as a regulatory body to check the activities of traditional rulers regarding customary land management. Clause 3 of Article 267 of the Ghana Constitution requires that there shall be no disposition or development of any customary land without approval of the LC [43]. However, it was observed that there were some restrictions as to how state land agencies interfere with customary land management. Personnel from the TCPD reported that they cannot plan a town without the consent of the chief. This means that the initiative to collaborate with the statutory body emanates from the chief. In the case of Trede, it was found that, after the intervention of TCPD in 1968, there was no collaboration with TCPD until 2018, when the area had already developed before the chief requested a new local plan. The delay in producing a local plan to regulate development accounted for some of the CLRs disputes in the area. It was further found that these statutory bodies did not play a direct role in customary land dispute resolution; however, they formulated measures that prevent land disputes.

5. Policy Implications and Recommendations

Based on the findings, the following recommendations are made in respect of land management and the control of CLRs disputes in peri-urban areas.

Tenure Responsive Land Use Planning (LUP) to Guide the Peri-Urban Transformation

Planning authorities, including local government, traditional authorities, and state land agencies, should adopt Tenure Responsive LUP in managing peri-urban lands. Ideally, the purpose of implementing a land use plan is to ensure the sustainable use of land to meet the needs and demand of the people [44]. However, in practice, the outcome of land use planning has often not yielded the expected results because it is often not linked to tenure security [45]. Evidence from this study demonstrates that the implementation of a land use plan to regulate development and prevent land disputes rather gave rise to more CLRs disputes and created tenure insecurity problems. The avoidance of such undesirable outcome of the well-intentioned plan could have been achieved by the adoption of a Tenure Responsive LUP to respond to the spatial requirement of the peri-urban growth. This is achievable by streamlining or integrating land tenure issues in the local land use planning. Adopting a land use planning that integrates land tenure is essential because it gives recognition to all forms of land rights (including CLRs), allows for participatory planning, and incorporates tenure security elements in its implementation [45]. Adopting such a measure will help to provide protection to locals from unwarranted disposition and reduce the occurrence CLRs disputes from arising. In any circumstance where a land use plan poses tenure insecurity (or causes the disposition of land rights), stakeholders should reconsider strategies or measures, such as providing alternative land or fair compensation to affected persons. Moreover, a land use plan for tenure security will help avoid monolithic planning where land uses are skewed towards residential and commercial uses only. The Local Government authorities, including Municipal, Metropolitan, and District Assemblies (MMDAs), who are responsible for approving the local land use plan, should ensure that there is an equitable distribution of land uses which cater for agricultural, residential, and commercial uses. This will ensure that farmlands are protected from being taken over by residential development. A participatory land use planning can serve as an effective tool to reconcile competing claims in land use rights between families, traditional authorities, and individual landowners. Chigbu et al. [46] reported that the flexibility of Tenure Responsive LUP allows for the incorporation of other land management tools (such as land mediation), hence its being an effective strategy to prevent CLRs disputes and for resolving existing land disputes. In this regard, mediation sessions can be held in the process of LUP for disputants to identify their positions, needs, and interests to reach a consensus. A common approach to participatory land use planning can be done through an open forum discussion organized for planning authorities, as well as land users, to ascertain existing farmland boundaries and land ownership prior to the demarcation of new land uses. In this regard, planning authorities can elicit for local knowledge and integrate it with the legal regulation to improve a local plan. The Land Use and Spatial Planning Act, 2016 (Act 925), emphasizes participatory approach when it states that at least three public hearings should be organized to ensure the participation of all community members before a local land use plan is approved. A Tenure Responsive LUP is a necessary precondition to facilitate peri-urban transformation and prevent CLRs disputes as it espouses the principles of equity, justice, and sustainability.

Participatory Land Management between State Land Agencies and Traditional Authorities

It is found that peri-urban transition is multidimensional and complex. Therefore, land management solely by the traditional authorities or the statutory agencies does not yield an effective land delivery system that can match changing trends in peri-urban areas. In Ghana, land management in peri-urban areas continues to be under the control of traditional authorities, with barely any state intervention despite urban intrusion. The absence or the delay of state action in land management in peri-urban areas is partly to blame for some of the land disputes problems that occur in peri-urban areas. To reverse this situation, there is a need for timely intervention by statutory agencies in the transformation process of peri-urban areas. By law, the Physical Planning Department (PPD) is responsible for the preparation of the local land use plan. However, the onus often lies with the chiefs to request for a planning scheme for their areas. This implies that the state land agencies wait on the local authorities' request before they can intervene. Most of the time, there is no involvement by the state land agencies in local land use planning at the early stages when rural areas begin transforming into peri-urban. A participatory land management approach where traditional leaders and state land agencies collaborate, consult, and cooperate with each other can provide a useful tool for peri-urban land delivery where CLRs disputes will be minimized [47]. For instance, in the early phases of peri-urbanization, the local authorities should seek the assistance of PPD to prepare a local plan to guide peri-urban growth. In addition, the PPD should be empowered to enforce the implementation of local plan to ensure that land development is in conformity with the local plan. The Lands Commission should prevent indiscriminate allocation or disorderly use of customary land by monitoring and ensuring that land allocation complies with approved local plans as emphasized in the constitution of Ghana.

Checks and Balances of Local Powers on Customary Land Administration

Findings from the study revealed that the statutory land agencies had limited control over the management of peri-urban lands. This is because land in peri-urban areas are largely customary lands and under the control of traditional authorities. The management and control of customary land by the traditional authorities are legally recognized and backed by the Constitution of Ghana. However, some traditional leaders abuse their powers and role as traditional land managers, often causing problems detrimental to the land rights and tenure security of natives in peri-urban areas. In recent times, some chiefs are reinterpreting their role as fiduciaries and engaging in corrupt and illegal customary land practices. Cases of how some chiefs in some peri-urban areas have engaged in illegal and multiple customary land sales and embezzlement of stool land revenue abound. To ensure chiefs exercise their authority, which is in conformity with customary laws and statutory provisions, there should be reasonable checks and balances on local land administration. Chiefs should not have the final say in customary land allocation for development. There should be a reformation in the customary norms and rules that allow community members to have a strong voice in customary land management decisions. With regard to change of norms, a central task for all stakeholders at the individual and household levels is to adhere to relations that limit land disputes. This is possible if peri-urban residents, real estate developers, and planning and customary authorities engage in new ways of living (concerning their behaviors, attitudes, ethics, and awareness towards land matters) [48]. Typical land matters that lead to disputes are gender-related biases, as well as excessive statutory and customary controls being exercised by district/municipal and customary authorities, respectively. Most important is that individuals should respect planning regulations where they apply.

In addition, Article 267 of the Ghana Constitution should not only focus on ensuring that land allocation at the community level is in conformity with planning regulations but also that the provision should be amended to include community members consent.

Institutional Capacity Building on the Part of Planning Authorities

Due to financial, human resource, and logistical constraints, state planning institutions are not pro-active enough to plan rural areas. Planning, in most cases, occurs after an area might have been fully transformed into peri-urban or urban town. The lack of planning schemes from the PPD for rural communities often creates the opportunity for chiefs to create their own local land use plans. This practice is fraught with problems, such as disagreement with later plans from the planning authorities, little or no regard for formal regulations, charging of exorbitant fees from landowners to cover cost of land use plan, etc. Consequently, this results in CLRs disputes because such land use plans often fail to address the land needs of the peri-urban dwellers. A redress to this problem demands that the local government provide the PPD at district levels with the adequate resources to enable them to initiate land use planning in rural areas before traditional authorities take over and manage it in their own interest. Capacity building should also aim at educating and training planning authorities on how they can integrate social norms with legal regulation in the planning peri-urban areas.

6. Conclusions

Using the experience from Trede (in Ghana), this study sought to create an understanding of how peri-urbanization and its associated CLRs disputes effects on the development of the peri-urban space. The study explored how CLRs disputes plague development with multiple effects in peri-urban areas because of the several socioeconomic and physical transformations that take place during the process of peri-urbanization. In this regard, the study identified CLRs disputes types evolved from the transformation of Trede, as well as their causes and effects on the development of the Town. The study has shown that the transformation of Trede from rural to peri-urban was influenced by social, economic, physical, and institutional changes. A major driver of the transformation process was the implementation of a local plan in 1968, which led to the conversion of rural lands to urban plots. A land-use plan implemented as a major step in converting rural lands into urban plots, triggered tenurial changes, land market development, high land values, loss of agricultural land, etc., which become recipes for the CLRs disputes in the study area. This situation is indicative that a mere implementation of a land-use plan is not enough to protect legitimate landowners from conflicts, unless the process of its implementation is designed to be tenure sensitive and capable of securing the tenure of all stakeholders involved. This scenario is important because it captures a situation that has not been well documented in urban literature. That reality is that peri-urbanization within customary jurisdictions comes with spatial transformations that lead to increased built-up areas alongside tenure transformations that cause increased land use (and land rights) conflicts that contribute to unorganized spatial development. In Trede, tenure related transformations engendered boundary, intra-family disputes, multiple sales of land, and disposition of land rights. CLRs disputes in Trede have significantly threatened livelihoods of the peri-urban dwellers, jeopardized physical development, increased tenure insecurity, and weakened social structure and relations. CLRs disputes have long-term implications on the development of peri-urban areas and therefore the need for responsive measures to curb their occurrence. Ways to improve this situation is what this study was able to investigate.

To ensure that the resulting CLRs disputes are tackled in a balanced form, the study outlined four specific policy related measures for action. These include: (1) the embrace of a tenure responsive LUP to guide the peri-urban transformation; (2) the engagement in participatory land management between state land agencies and traditional authorities; (3) introduction of checks and balances to ensure that the powers local authorities on customary land administration stay within the boundaries of good governance; and (4) to delve deeper into institutional capacity building to ensure that the planning authorities have the necessary capacities to administer peri-urban controls over land resources in ways that embrace responsible resolution of CLRs disputes. These measures are not only mere recommendations for improving peri-urban land administration in Trede (or any other small town in sub-Saharan Africa); they constitute the pre-conditions for responsible governance of tenure in peri-urban areas within land jurisdictions that are predominantly customary in nature.

Author Contributions: Conceptualization, B.O.A. and U.E.C.; Formal analysis, B.O.A.; Investigation, B.O.A.; Methodology, B.O.A.; Resources, B.O.A. and U.E.C.; Supervision, U.E.C.; Validation, B.O.A.; Visualization, B.O.A.; Writing—original draft, B.O.A.; Writing—review & editing, U.E.C. All authors have read and agreed to the published version of the manuscript.

Funding: This research received no external funding.

Acknowledgments: This work was supported by the German Research Foundation (DFG) and the Technical University of Munich (TUM) in the framework of the Open Access Publishing Program.

Conflicts of Interest: The authors declare no conflict of interest.

Appendix A

Table A1. Overall research matrix (source: Authors, 2018).

Research Variables	Information Required	Source of Data	Data Collection Method	Data Analysis Method
Transformation of Trede from rural to peri-urban town	History of Trede Past and future changes in development Changes in land tenure practices Factors influencing change Spatial maps visualizing changes in transformation process	Traditional leaders Landowners(natives) Government Officials Field observations	Key informant interviews Household survey Observations Governmental spatial data set review	Content and spatial analysis
Types of CLRs disputes	Land tenure arrangement in Trede Land ownership types Process of land acquisition Actors involved in disputes Common types of disputes in Trede Conflicting land ownership in Trede Description of the boundary disputes in Trede	Traditional Leaders and Family heads Landowners, developers, and users Government Officials Literature	Key informant interviews Household survey Observations Literature review	Descriptive statistics, content and spatial analysis
Causes of CLRs disputes	Changes in land tenure system Procedures of land acquisition Changes in land acquisition process Procedures for the sale of customary land Actors involved in disputes Actors interest and position in disputes Reasons for land disputes occurring	Traditional Leaders and Family heads Landowners, developers, and users Government Officials Literature	Key informant interviews Household survey Literature review	Descriptive statistics, content and narrative analysis
Effects of CLRs disputes on peri-urban development	Effect of CLRs disputes on member's livelihoods Effects on private investment Effects on physical development in the town Effects on economic development	Traditional Leaders and Family heads Landowners, developers, and users literature	Key informant interviews Household survey Literature review	Descriptive statistics, content and narrative analysis
Resolving CLRs disputes	Role of traditional leaders in dispute resolution Role of Government officials in disputes resolution Collaboration between Government officials and traditional leaders in dispute resolution Assessing the effectiveness of the customary court Existing mechanisms in resolving disputes in Trede Interventions needed in resolving land disputes	Traditional Leaders and Family heads Landowners, developers, and users Government Officials (land managers) Literature	Key informant interviews Household survey Observations Literature review	Descriptive statistics content and narrative analysis

References

1. Pradoto, W. Development Patterns and Socioeconomic Transformation in Peri-Urban Area.-The Case of Yogyakarta, Indonesia. Ph.D. Thesis, TU Berlin, Berlin, Germany, 10 February 2012. [CrossRef]
2. Adam, A.G. Urbanization and the Struggle for Land in the Peri-Urban Areas of Ethiopia. In Proceedings of the Annual Bank Conference on Africa, Lusaka, Zambia, 23–27 May 2016.
3. Global Land Tool Network (GLTN). *Bridging the Gap on Land Governance and Tenure along the Urban-Rural Continuum*; UN-Habitat/Global Land Tool Network: Nairobi, Kenya, 2018.

4. Mends, T.M. Customary Land Tenure and Urbanization with a Case Study on the Peri-Urban Area of Accra, Ghana. Master's Thesis, The Netherlands International Institute for Geo-information Science and Earth Observation Enschede, Enschede, The Netherlands, 2006.

5. Chauveau, J.P.; Colin, J.P. Customary transfers and land sales in Côte d'Ivoire: Revisiting the embeddedness issue. *Africa* **2010**, *80*, 81–103. [CrossRef]

6. Chauveau, J.P. *Changes in Customary Land Tenure Systems in Africa*; IIED: London, UK, 2007.

7. Ubink, J.M. *In the Land of the Chiefs: Customary Law, Land Conflicts, and the Role of the State in Peri-Urban Ghana*; Leiden University Press Van Vollenhoven Institute, Faculty of Law, Leiden University: The Netherlands, 2008; p. 256.

8. Kanji, N.; Cotula, L.; Hilhorst, T.; Toulmin, C.; Witten, W. *Research Report 1 Can Land Registration Serve Poor and Marginalised Groups?* International Institute for Environment and Development: London, UK, 2005.

9. Alden Wily, L.; Hammond, D. *Land Security and the Poor in Ghana. Is There a Way Forward? A Land Sector Scoping Study*; DFID Ghana Rural Livelihoods Programme: Accra, Ghana, 2001.

10. Amanor, K.S. Family values, land sales and agricultural commodification in South-Eastern Ghana. *Afr. South East. Ghana Afr.* **2010**, *80*, 104–125. [CrossRef]

11. Wehrmann, B. *Land Conflicts: A Practical Guide to Dealing with Land Disputes*; Deutsche Gesellschaft für Technische Zusammenarbeit (GTZ) GmbH: Eschborn, Germany, 2008.

12. Barry, M.; Danso, E.K. Tenure security, land registration and customary tenure in a peri-urban Accra community. *Land Use Policy* **2014**, *39*, 358–365. [CrossRef]

13. Ubink, J.M. Negotiated or negated? The rhetoric and reality of customary tenure in an Ashanti village in Ghana. *Africa* **2008**, *78*, 264–287. [CrossRef]

14. Akaateba, M.A. Urban Planning Practice under Neo-Customary Land Tenure: The Interface between Government Agencies and Traditional Authorities in Peri Urban Ghana. Ph.D. Thesis, TU Berlin, Berlin, Germany, 2008.

15. Owusu, G. Indigenes' and migrants' access to land in peri-urban areas of Accra, Ghana. *Int. Dev. Plan. Rev.* **2008**, *30*, 177–198. [CrossRef]

16. Peters, E. Contesting Land and Custom in Ghana. State, Chief and the Citizen. *J. Agrar. Chang.* **2010**, *10*, 604–608. [CrossRef]

17. Kasanga, K.; Kotey, N.A. *Land Management in Ghana: Building on Tradition and Modernity*; UK's Department for International Development (DFID): London, UK, 2001.

18. Narain, V.; Nischal, S. The Peri-Urban Interface in Shahpur Khurd and Karnera, India. *Environ. Urban.* **2007**, *19*, 261–273.

19. Kasanga, R.K.; Cochrane, J.A.; King, R.; Roth, M.J. Land markets and legal contradictions in the peri-urban area of Accra Ghana: Informant interviews and secondary data investigations. *Res. Pap.* **1996**, 1–90. [CrossRef]

20. Narain, V. Growing city, shrinking hinterland: Land acquisition, transition and conflict in peri-urban Gurgaon, India. *Environ. Urban.* **2009**, *21*, 501–512.

21. Ravetz, J.; Fertner, C.; Nielsen, T.S. The Dynamics of Peri-urbanization. In *Peri-Urban Futures: Scenarios and Models for Land Use Change in Europe*; Routledge Publications: New York, NY, USA, 2013.

22. Holland, M.D.; Kasanga, R.K.; Lewcock, C.P.; Warburton, H.J. *Peri-urban Baseline Studies: Kumasi, Ashanti Region, Ghana; Vol. 1: Executive Summary and Main Report*; Natural Resources Institute: Greenwich, UK, 1996; 86p.

23. Iaquinta, D.L.; Drescher, A.W. Defining peri urban: Understanding rural-urban linkages and their connection to institutional contexts. In Proceedings of the Tenth World Congress of the International Rural Sociology Association, Rio de Janeiro, Brazil, 1 August 2000; Volume 1, pp. 3–28.

24. Rakodi, C. *Poverty in the Peri-urban Interface. Natural Resources Systems Programme (NRSP) Research Advances No. 5*; DFID: London, UK, 1999.

25. Piorr, A.; Ravets, J.; Tosics, I. (Eds.) *Peri-urbanization in Europe. Towards European Policies to Sustain Urban-Rural Futures, Syntesis Report*; Forest & Landscape: Frederiksberg, Denmark, 2011.

26. Allen, A. Environmental planning and management of the peri-urban interface: Perspectives on an emerging field. *Environ. Urban.* **2003**, *15*, 135–148. [CrossRef]

27. Nkwae, B. Conceptual Framework for Modelling and Analysing Periurban Land Problems in Southern Africa. Ph.D. Thesis, University of New Brunswick, Fredericton, NB, Canada, 2006.

28. Charisa, I. Peri-urban Dynamics and Regional Planning in Africa: Implications for Building Healthy Cities. *J. Agric. Ext. Rural Dev.* **2010**, *2*, 15–26.

29. Satterthwaite, D.; McGranahan, G.; Tacoli, C. Urbanisation and its implication for food and farming. *R. Soc. Publ.* **2010**. [CrossRef]

30. Kasanga, R.K. *Land Tenure and the Development Dialogue: The Myth Concerning Communal Landholding in Ghana*; Department of Land Economy, University of Cambridge: Cambridge, UK, 1988.

31. FAO. *Gender and Access to Land*; FAO: Rome, Italy, 2002; Available online: http://www.fao.org/3/Y4308E/Y4308E00.htm (accessed on 15 February 2020).

32. Dale, R. *Development Planning: Concepts and Tools for Planners, Managers and Facilitators*; Zed Books Limited: London, UK, 2004.

33. Payne, G. Land tenure and property rights: An introduction. *Habitat Int.* **2004**, *28*, 167–179. [CrossRef]

34. De Soto, H. *The Mystery of Capital: Why Capitalism Triumphs in the West and Fails Everywhere Else*; Civitas Books: New York, NY, USA, 2000.

35. USAID. What is Land Tenure? 2014. Available online: https://www.usaid.gov/land-tenure (accessed on 15 February 2020).

36. Hudalah, D.; de Roo, G. *Transition: A Relevant Issue to Planning?* Urban and Regional Studies Institute, University of Groningen: Groningen, The Netherlands, 2007.

37. Ghana Statistical Service. *Population of Atwima Kwanwoma District in Ghana*; Atwima Kwanwoma District Assembly: Atwima Kwanwoma, Ghana, 2018. Available online: www.statsghana.gov.gh (accessed on 10 February 2020).

38. Teddlie, C.; Tashakkori, A. *Foundations of Mixed Methods Research: Integrating Quantitative and Qualitative Approaches in the Social and Behavioural Sciences*; Sage: Beijing, China, 2009.

39. Tacoli, C. Changing Rural-Urban Interactions in sub-Saharan Africa and their, Impacts on Livelihoods: A Summary. In *Working Paper Series on Rural-Urban Interactions and Livelihood Strategies*; Working Paper 7; IIED: London, UK, 2002.

40. Rauws, W.S.; De Roo, G. Exploring transitions in the peri-urban area. *Plan. Theory Pract.* **2011**, *12*, 269–284. [CrossRef]

41. Budiyantini, Y.; Pratiwi, V. Peri-urban typology of Bandung Metropolitan Area. *Procedia Soc. Behav. Sci.* **2016**, *227*, 833–837. [CrossRef]

42. Crook, R.C.; Affou, S.; Hammond, D.; Vanga, A.F. The Law, Legal Institutions and the Protection of Land Rights in Ghana and Coˆte d'Ivoire: Developing a More Effective and Equitable System. In *Final Report SSRU Project R 7993*; Institute of Development Studies UK: Brighton, UK, 2005.

43. Government of Ghana. *Constitution of the Republic of Ghana*; Assembly Press: Accra, Ghana, 1993. Available online: https://www.HomePage/republic/constitution.php?id=Gconst5.html (accessed on 10 February 2020).

44. GIZ. *Land Use Planning: Concept, Tools and Applications*; GIZ: Eschborn, Germany, 2012.

45. Chigbu, U.E.; Masum, F.; Leitmeier, A.; Antonio, D.; Mabikke, S.; Espinoza, E.; Hernig, A. Securing tenure through Land Use Planning: Conceptual framework, evidences and experiences from selected countries in Africa, Asia and Latin America. In Proceedings of the World Bank Conference on Land and Poverty, Washington, DC, USA, 23–27 March 2015.

46. Chigbu, U.E.; Haub, O.; Mabikke, S.; Antonio, D.; Espinoza, J. *Tenure Responsive Land Use Planning: A Guide for Country Level Implementation*; UN-Habitat: Nairobi, Kenya, 2016; 92p.

47. Bugri, J.T.; Yuonayel, E.M. Traditional Authorities and Peri-Urban Land Management in Ghana: Evidence from Wa. *J. Resour. Dev. Manag.* **2015**, *13*, 68–79.

48. Chigbu, U.E. Anatomy of women's landlessness in the patrilineal customary land tenure systems of sub-Saharan Africa and a policy pathway. *Land Use Policy* **2019**, *86*, 126–135. [CrossRef]

Article

Using a Gender-Responsive Land Rights Framework to Assess Youth Land Rights in Rural Liberia

Elizabeth Louis [1],*, Tizai Mauto [1], My-Lan Dodd [1], Tasha Heidenrich [1], Peter Dolo [2] and Emmanuel Urey [1]

[1]   Landesa, 1424 4th Ave, Seattle, WA 98101, USA; tizaim@landesa.org (T.M.); myland@landesa.org (M.-L.D.); tashaheid295@gmail.com (T.H.); emmanuelu@landesa.org (E.U.)
[2]   Development Education Network-Liberia (DEN-L), Dementa Road, Gbarnga, Bong Country, Liberia; dev_edunet@justemail.net
*   Correspondence: elizabethl@landesa.org

Received: 30 June 2020; Accepted: 15 July 2020; Published: 27 July 2020

Abstract: This article summarizes the evidence on youth land rights in Liberia from a literature review combined with primary research from two separate studies: (1) A qualitative assessment conducted as formative research to inform the design of the Land Rights and Sustainable Development (LRSD) project for Landesa and its partners' community level interventions; and (2) a quantitative baseline survey of program beneficiaries as part of an evaluation of the LRSD project. The findings are presented using a Gender-Responsive Land Rights Framework that examines youth land rights through a gender lens. The evidence highlights that female and male youth in Liberia face significant but different barriers to long-term access to land, as well as to participation in decisions related to land. Our suggested recommendations offer insights for the implementation of Liberia's recently passed Land Rights Act as well as for community-level interventions focused on increasing youth land tenure security in Liberia.

Keywords: youth land rights; gender-responsive land rights framework; Liberia Land Rights Act; land governance; tenure security

1. Introduction

Broad-based land tenure security and equitable land governance are pressing issues in Liberia. Land and natural resources have always been, and remain, crucial to Liberia's economy [1,2]. Seventy percent of the active population is dependent on agriculture for their livelihood and over half of the country's inhabitants live in rural areas [3]. Youth (ages 15–35) constitute approximately 34% of the population [1] and rural youth depend primarily on agriculture to support their livelihoods [4]. Many rural youth lack access to farmland or suffer from high levels of land tenure insecurity.

The viability of the youth demographic is crucial to Liberia's social, political, and economic future. The exclusion of youth from effectively accessing land to support themselves and their families could have negative implications for the transfer of knowledge and skills, as well as food security, youth employment, economic development, and security. The civil war arose from the systematic denial of land (and other economic assets) and exclusion from governance of the indigenous Liberians who constitute the majority of Liberia's population [5]. After the conflict, the country made strides

---

[1]   Although this age range is generally accepted as a definition of youth, in reality there are wide range of social factors that define youth and adulthood in rural communities in Liberia. The Government of Liberia in its National Youth Policy defines Liberian youth as being between the ages of 15 and 35 (Brownlee et al., 2012), a definition also used by the African Youth Charter (2006) and the National Federation of Liberian Youth (Nasser 2012).

toward peace, stability, and economic growth. However, poverty, food insecurity, inadequate human capacity and infrastructure, a high unemployment rate, particularly among youth, and land tenure insecurity threaten further progress.

The Land Rights Act, adopted in September 2018, was formulated to address several inequities in land access and land governance, giving communities ownership rights and empowering them to make decisions on the lands that they have customarily accessed for decades [6]. The LRA explicitly recognizes the rights of all community members so long as they meet the LRA's legal definition of community member. Through these provisions, women, youth, and "strangers" [2]—groups that have been traditionally marginalized within rural communities—enjoy land rights by operation of the LRA. Landesa worked with the Government of Liberia to provide technical input into the framing of the Land Rights Act. Additionally, through the Land Rights and Sustainable Development Program (LRSD), Landesa currently works with local CSO partners DEN-L and FCI to raise awareness around land rights for women and youth in order to foster sustainable development in rural communities.

This paper summarizes the evidence from a literature review combined with primary research from two separate studies conducted by Landesa, which focused on youth land rights in Liberia. The particular issues that youth face with respect to land vary considerably across Liberia. For example, there are wide disparities between the rights of male and female youth, as well as between the rights of youth who come from the original landholding families and those from land-poor households. Perhaps the most marginalized of youth are "stranger" youth who are not considered full community members. The extent to which Liberian youth face challenges in accessing land varies by area, though the scope of the problem is difficult to assess given the lack of available evidence.

The findings in this paper are presented using a Gender-Responsive Land Rights Framework [3], which is a conceptual tool that allows us to analyze and present nuanced information on the various dimensions of land rights, including how land rights are granted and realized by various actors in various contexts as well as how they are mediated by formal and informal institutions. The framework is used here to examine the rights of female and male youth in the context of rights to customary land in rural Liberia.

The findings highlight that youth face significant but different barriers to long-term access to land with female and "stranger" youth facing considerably more barriers to accessing land than male local youth. While the Land Rights Act has the potential to improve land access for youth and other marginalized groups, youth lack knowledge about land policies and traditional elders and elites still control land and resources in rural communities. Furthermore, Liberian youth are limited in their participation on land-use decisions and the crops they grow, and have limited access to credit and extension services, with female youth being less able to participate in land-use decisions than male youth. Youth's land tenure insecurity impacts their ability and desire to practice agriculture.

This paper is organized as follows: The next section reviews the data collection methods and explains the gender-responsive youth land rights conceptual framework used to present the findings. This is followed by a presentation of the findings using our conceptual framework. The next section discusses the findings and proposes recommendations to promote youth land rights in Liberia based on our findings.

---

[2] "Strangers" are residents who are not in the direct landowning lineages of communities in Liberia. Usually they have various degrees of land-use rights but not ownership rights. Some of them, especially youth, are born in the communities, but because their parents are not in the landowning lineage they are often referred as strangers and have less secure land access and land rights.

[3] The framework has gone through several iterations with contributions made by different stakeholders in the land sector. A version of the framework was used by Doss and Meinzen-Dick (2018), who cite Place et al. (1994) as originating the approach. An version of the framework was also used by Landesa to examine women's land rights in Myanmar, see Louis, Eshbach, Roberts and Htee (2018) "An assessment of land tenure regimes and women's land rights in two regions of Myanmar", Paper Presented at the World Bank Conference on Land and Poverty. Washington, D.C. March 2018.

2. Materials and Methods

The research discussed in this paper come from three separate research activities: (1) A review of evidence on youth and land access in Liberia; (2) a qualitative assessment on youth land rights conducted in August 2018 to inform the design of LRSD project for Landesa and its partners' community level interventions; and (3) a quantitative baseline survey of program beneficiaries conducted in October 2018 as part of an assessment of the LRSD program's outcomes on knowledge of community members' land right and attitudes toward women and youth land rights.

2.1. Literature Review

The literature review focused the available but limited evidence on youth access to land and the related obstacles they face in supporting their livelihoods in rural Liberia. The literature identifies several different obstacles that youth face in accessing land. These are discussed within the conceptual framework along with the findings from the primary research.

2.2. Formative Qualitative Research

The objective of the formative qualitative research was to understand youth and land issues in rural Liberia in order to inform the design of Landesa's Land Rights for Sustainable Development (LRSD) Project. The research focused on four major thematic areas: (1) youth livelihood activities, (2) youth access to land, (3) land-related disputes, and (4) youth land governance and community relationships. The research was conducted by Landesa in collaboration DEN-L, one of Landesa's civil society partners in Liberia.

Sixteen communities in Lofa and Bong County were included in the study and were purposively selected to represent the variety of land issues, cultural values, ethnic dynamics, and land tenure systems in rural Liberia. In addition, communities with proximity to concession activities as well as communities near urban areas were included, in order to better understand these dynamics.

A total of 16 focus group discussions (FGDs) and 46 key informant interviews (KIIs) were conducted. The FGDs included community members with a focus on youth, while KIIs included youth, government officials, customary authorities, and other leaders. Each FGD included nine to twelve community members and a total of 192 community members participated in the FGDs in both counties; 70% of the participants were "youth" (15–35 years old) and 44% were female. By occupation, 70% of FGD participants were farmers and 20% were teachers. FGD participants were recruited by asking elder and youth leaders to identify youth, using the community definition of youth. Some participants over 35 years self-identified as youth or participated because of interest in youth issues.

KIIs were conducted with national and regional government officials, land experts, and civil society organization leaders (CSO). In addition, at least two KIIs were conducted in each community, with traditional leaders, elders, CSO leaders, youth leaders, and other key land stakeholders. About 10% of KIIs were with youth stakeholders. By primary occupation, KII participants mirrored the FGD (majority famers, although 15% said CSO leader was their primary occupation). CSOs and opinion leaders were selected to balance information in the assessment, as many of them have no direct link or interest in community lands. KII participants were selected according to DEN-L's knowledge of the stakeholders in the communities. The KIIs generally discussed the same topics as the FGDs, but were geared to KII participants' unique knowledge, viewpoints, and unique roles/responsibilities in youth and land issues.

2.3. Baseline Survey

Landesa conducted a baseline survey with LRSD project beneficiaries located in seven communities in Bong, Lofa, and Rivercess Counties in October 2018 just before implementation of community-level activities through its CSO partners. These communities were targeted for land rights awareness activities implemented by Landesa's partner CSO organizations, DEN-L and FCI. The survey sought to establish a baseline on beneficiaries' knowledge of and attitudes toward community membership; land rights, including the land rights of women and youth; land laws and institutions; land governance; and Alternative Dispute Resolution (ADR).

The research sample was exhaustive and targeted all beneficiaries selected to participate in the trainings. In some cases, community members who were not selected to participate took part in training; these additional participants were not included in the baseline survey. The beneficiaries of the training programs were selected by DEN-L and FCI staff through an intensive community consultation process in each of the seven program communities, which took into account community members' ability and willingness to participate in the trainings. The participants represented a broad spectrum of community members, ranging from traditional elders, leaders (including religious, women, and youth leaders), teachers, as well as a diverse spectrum of community members at large (including women and youth). Table 1, below, illustrates the age and gender of the participants/survey respondents in the seven program communities.

Table 1. Age and Gender of Baseline Respondents, Percentage in Program Communities.

Community	No. of Respondents	Male	Female	Adult	Youth *
Lofa County	85	62%	38%	49%	51%
Pasama	36	47%	53%	42%	58%
Bardezu	24	29%	71%	79%	21%
Gbonyea	25	32%	68%	32%	68%
Bong County	73	41%	59%	47%	53%
Gbarnga Siaquelleh	36	50%	50%	44%	56%
Shankpowai	37	67%	33%	50%	50%
Rivercess County	50	30%	70%	82%	18%
Neezuin	25	32%	68%	80%	20%
Little Liberia	25	28%	72%	84%	16%
Total	208	43%	57%	57%	43%

* Youth are considered those aged 15–34, which is based on the definition used by the Government of Liberia and the African Union.

2.4. Women's Land Rights Conceptual Framework

The data from the research are presented using a Gender-Responsive Land Rights conceptual framework modified to fit the context of property rights of youth and women in rural Liberia. The Framework defines property rights as multi-dimensional, and states they should be effective, inclusive, and Gender Equitable. Effectiveness is defined as (1) the bundle of rights enjoyed by an individual or group is clearly defined in all its dimensions for all members of a community, including those who may have new rights to that land (Completeness); (2) the duration of the bundle of rights are clearly defined and enforced (Duration); and (3) the bundle of rights are clearly enforced in the face of contestation (Robustness). In addition, (4) are all these aspects of rights Inclusive and Gender Equitable?

The framework takes into consideration the institutions through which rights granted by customary and statutory law are mediated. The institutions are:

- Statutory and customary land-related laws, policies, regulations, conventions, and agreements that embody the rights determined and enforced by governments;
- Formal and informal institutions and actors who influence, decide, manage, or enforce land-related rights;
- Social norms that shape attitudes and beliefs on who should have land, for what purpose and through which means; and
- Individuals and communities whose land-related rights are protected, strengthened, limited or negated by the system.

For this analysis, the framework is operationalized in the form of questions that seek to assess (1) how rights are defined in the law, (2) the level of awareness of rights and attitudes toward rights for female and male youth, and (3) the extent to which different groups of youth are able to realize their rights in practice within their socio-political contexts. The main dimensions that are addressed in this analysis are completeness, robustness, gender equality and inclusiveness. The dimension of duration is addressed in the completeness dimension for this study and is therefore not used. The dimensions, their definitions, and the questions used to operationalize the framework are presented in Table 2 below:

Table 2. Operationalizing a gender-responsive land rights framework.

Land Rights Dimension	Definition	Questions Used to Analyze Findings
Completeness	Whether the bundle of rights enjoyed by an individual or group is clearly defined and enforced in all its dimensions for all members of a community, including those who may have new rights to that land	(1) Are the rights for female and male youth clearly defined under the LRA and related laws that govern land rights? (2) Do female and male youth and others have knowledge of youth land rights, and (3) Are female and male youth able to realize their rights?
Robustness	Whether rights can be enforced in the face of contestation	(1) Does the LRA clearly define enforcement of rights for female and male youth in the face of contestation, (2) Do female and male youth know how to enforce their rights in the face of contestation, and (3) Are female and male youth able to enforce their rights in the face of contestation.
Gender Equitable ——————— Inclusive	Whether the land rights system give all the ability to participate in decisions about the land	(1) Does the LRA clearly define land rights/land governance as inclusive and gender-equitable, (2) Do youth understand their rights as they relate to decision-making on land governance? (3) Are youth able to participate in land decisions at the community level?

3. Presenting Findings Using the Gender-Responsive Land Rights Framework

Using the framework, the quantitative baseline survey data are presented using descriptive statistics and are integrated into discussion of the qualitative findings and evidence from the literature reviews where relevant.

3.1. Is the Extent of the Bundle of Rights Clearly Defined and Enforced? (Completeness)

3.1.1. Are the Rights for Female and Male Youth Clearly Defined under the LRA and Related Laws That Govern Land Rights?

The LRA is a historic, comprehensive land law that provides new broad-based legal recognition of clearly defined bundle of rights to land for customary communities and individuals. Rights to access, control, and own land—collectively and individually—apply to female and male youth through robust equal protection provisions and inclusive community membership provisions yoked to recognized land rights.

The LRA incorporates fundamental principles of Liberia's Constitution and Land Rights Policy applicable to youth and their rights to land. Specifically, equal protection guarantees apply to all Liberians regardless of age and sex (art. 2). A core purpose of the LRA is to recognize and "ensure equal access and equal protection with respect to land ownership, use and management" (art. 3). This extends to giving equal legal protection to Customary Land ownership and Private Land ownership. Furthermore, the Customary Land rights bundle is clearly defined to include the right to exclude others, to possess and use land, to manage and improve land, and to transfer (art. 33). In doing so, the LRA recognizes and clarifies new rights to land for customary communities, including their members who are female and male youth.

Community membership is defined by the LRA in a way that includes all youth that meet the broad definition and requirements—regardless of age and gender. [4] Community membership broadly confers collective and individual rights to land. New generations born into the community also enjoy community membership and land rights (art. 34(5)). All community members meeting the LRA's definitions are part of the community and as such enjoy equal rights to use and manage their customary land, regardless of age or gender (art. 34(3)). All communities, acting collectively, control customary land through the power to make decisions about transfers of customary land to the government or non-community members (art. 36). Within a community's customary land, all community members, regardless of gender, have a right to exclusively possess a parcel of land for a residence (art. 39(3)). They also have the right to access land for agriculture (art. 40) and enjoy ownership and management of protected areas (art. 42(3)).

The LRA also recognizes private land ownership rights, held individual or jointly, enjoyed by all citizens, which includes female and male youth. Female and male youth also enjoy inheritance rights as sons and daughters including to parents in both statutory marriages under the Decedents Estates Law (§§ 3.2 & 3.4) and customary marriages under Equal Rights of Customary Marriage Law (§ 3.2). Finally, youth in the age range of 15 to 18 enjoy inheritance rights under the Children's Law (2011) (s. 17.1).

Despite the robust and inclusive provisions recognizing rights to (and rights within) customary land in the LRA and land-related framework, there are gaps that impact youth. For example, where patrilocal marriage is common combined with an increasing trend in marriage informality in Liberia [7], cohabiting youth, particularly female youth, moving into their partner's community, and residing there for under seven years, will either have less clear rights to customary land or no rights. In the first category, to be deemed a community member, the key to enjoying rights in customary land requires meeting the nebulous legal hurdle of qualifying as a presumptive spouse. [5] In the second category, youth from outside the community who are cohabiting with a community member but who do not meet the standard of presumptive spouse will not be considered community members, and as such will not enjoy rights to customary land under the LRA. Youth in these categories are land tenure insecure under the law.

3.1.2. Do Youth Have Knowledge of Their Bundle of Rights (i.e., Have They Been Clearly Defined for the Youth)? Are Youth Knowledgeable about Who Can Gain New Rights to Land and under What Circumstances? Is This the Case for All Youth?

Findings from the formative research highlights that youth in both Bong and Lofa counties know little about their land rights granted through statutory laws, including the Land Rights Act and the Community Rights Law. In general, youth are more knowledgeable about customary norms and laws regarding land rights. These are generally understood by youth as follows:

1. Males are the heads of households.
2. Elder males and landlords (male) have the most rights and privileges to land.
3. Land matters are for males, not females.
4. Elder males are the arbiters of land access and governance and need to be approached in the traditional way for land.
5. Youth must approach elders for land.
6. Youth can only access land within the community if he is a community member (i.e., his father is from the community).

---

[4] Under the LRA, a youth may legally qualify as a community member if he or she is a Liberian citizen and who fits in any one of the follow categories: (i) born in the community, (ii) parent(s) was born in the community, (iii) community resident of 7 years, or (iv) spouse of a resident community member (inclusive of statutory, customary, and presumptive spouses) (art. 2).

[5] "Persons who live together as husband and wife and hold themselves out as such are presumed to be married" (Civil Procedure Law, § 25.3(3)).

7. "Strangers" (i.e., parents were not born in the community) have more restricted access to land.

Some had heard about a few provisions of the Land Rights Act through radio programs. A few youth understood the difference between tribal certificates and deeds and recognized that those with deeds had stronger rights than those with tribal certificates. [6] Often, this understanding was the result of conflicts in which youth were involved, which resulted from people with deeds encroaching on land that had tribal certificates.

These findings were largely confirmed through the quantitative baseline survey. In the survey, respondents were asked if they had heard about the Land Rights Act, Marriage Law, the Constitution and the Land Authority. Responses indicate that only 55 percent of youth had even heard about the Land Rights Act, let alone understood any of its provisions. Almost twice as many males were aware of the laws, and more adults reported knowledge of the laws than youth, despite the higher education levels of youth. Figure 1 below highlights respondents' awareness about land laws and institutions by gender and age.

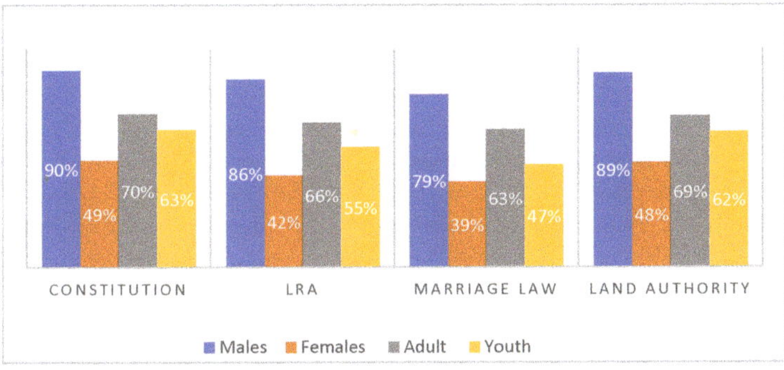

Figure 1. Community members' awareness of of land laws/institutions, by gender/age.

Knowledge Gaps on land rights emerged between women and men: women often answered they "didn't know" in response to questions on land rights; youth also often lacked knowledge on land rights (many of these were female youth), as exhibited in the Figure 2 below.

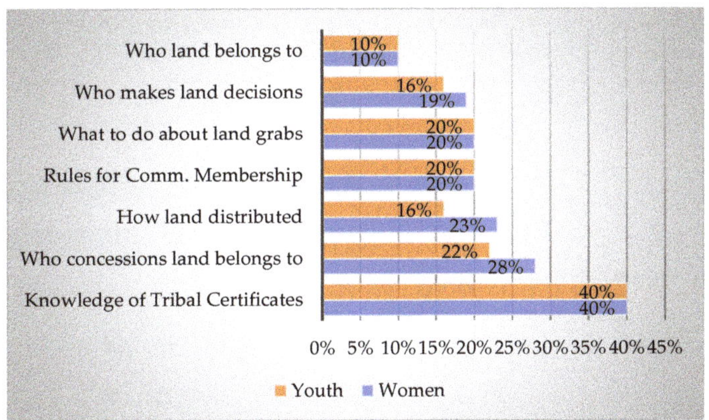

Figure 2. Youth and women who "Don't Know" about land rights under the Land Rights Act.

---

[6] Tribal Certificates are an initial step for registering land but until it is transformed into a deed, the land rights are only usufruct (use rights).

Views on women and youth land rights did not vary much by age, but they did by gender. Male respondents believed that youth have more land rights than did female respondents; for example, over 90% of men agreed to the statement in the survey questionnaire that "Youth who are community members have the right to make decisions about land" while only 58% of women agreed.

Interestingly, male participants also thought that women had stronger land rights than did female participants: almost 90% of men agreed to the statements that "Every woman has the right to participate in making decisions about land" and that "Daughters can inherit family land", while just over half of women agreed with these statements. In fact, men reported more positive attitudes and beliefs about the rights women should have than women did. Interestingly, nearly all of the respondents who said their communities made land decisions collectively reported that women had the right to participate in making decisions about land in their community.

In the formative research, a majority of youth understood that community membership is the prerequisite to accessing community land for farming or house plots, as well as for participation in land governance bodies. Youth understanding of community membership reflected current customary understandings that privileged those whose parents, or at the minimum whose fathers, are from the community. This understanding reflects customary norms and is different from the provisions of the LRA, which states that all (Liberian citizens) who have lived in a community for seven years, regardless of whether they were born outside the community, are community members. This provision puts "strangers" and their children, spouses (born in other communities), and wives in de-facto unions (without dowry) and polygamous marriages on equal footing with those who are traditionally considered community members.

The baseline survey results similarly highlight the disconnect between the legal definition of community membership in the LRA and the current norms around who is considered a community member. If one's parents were not considered official community members (i.e., the parents were "strangers") only 26% of the respondents said one would then be considered a community member, even if they were born in the community. Furthermore, only about half of respondents thought that living in the community for over seven years would make one a community member. In terms of who makes the rules for community membership, 60% of respondents said they are made by the traditional authorities; only 18% said they are made by the community as a whole. Figure 3, below, illustrates the beliefs around the various criteria for community membership. In the baseline survey, a high percentage of males (90%) reported that any resident, including women and youth, could plant life trees or cash crops, while fewer women (approx. 65%) felt this way. Similarly, more than half of males and adults felt that individuals of any age or gender could own land, but women and youth respondents were less sure of ownership rights, especially for themselves. This discrepancy between what men report and the actual practices on the ground is to be noted.

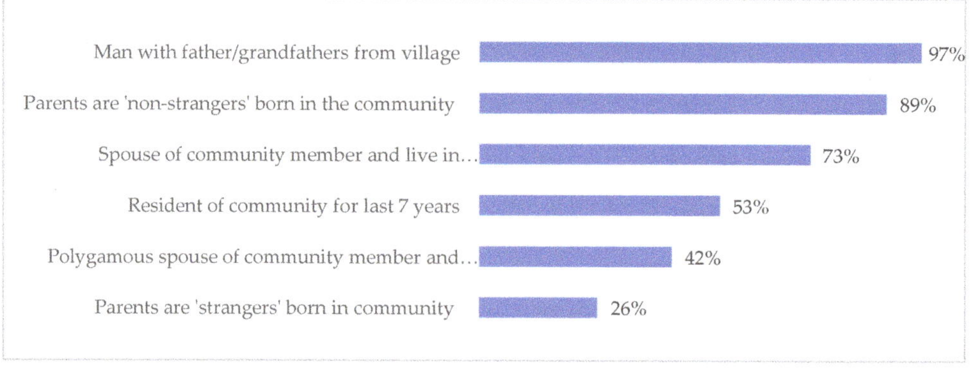

Figure 3. Community Members' beliefs on the criteria for community membership.

3.1.3. Are Youth Able to Enjoy the Full Extent of Their Bundle of Rights? Are Youth Able to Gain New Rights to Land If They Meet the Criteria for Eligibility? Is This the Case of All Youth?

For youth, family inheritance or allocation by customary authorities appear to be the most common means of accessing land for farming. In many clans, youth cannot inherit land or be allocated their own land to farm until they are considered "adults." Instead, they must work for their families and communities until this time, especially youth who are not born into families of the landowning lineages. Studies suggest that these practices are rooted in subsistence farming's reliance on surplus youth labor, especially in the northwest of Liberia, and the gerontocratic nature of many rural communities [8–10]. In addition, longer life expectancies of elders, a youth population bulge, and higher scarcity of land in some areas, has meant further delays and/or smaller land inheritances inheritances/allocations for youth. Delayed land rights prevent youth from planting "life trees," which are commercial tree crops like rubber or oil palm, through which one can assert a permanent claim to the land. These can only be planted on inherited land [11]. Smaller inheritances can mean parcels are too small for youth to earn a living. Increasingly, encroachment by outsiders on traditionally held lands has reduced the availability of land youth can receive, which is especially problematic in areas where valuable land is less plentiful [11].

The Eleven Clan Study found that in communities where land rights are vested in local authorities, the town or quarter chief would allocate newly married youth land for seasonal cropping [1]. These rights are typically for one or two agricultural seasons, with limitations on what crops can be grown. In communities where land rights are held by families, inheritance is more common. In some areas, informal rental markets provide an alternative for youth [1]. However, leases are often short-term with high rental fees, and cash cropping is not allowed.

Migrant youth faced extremely exploitive arrangements by customary authorities prior to the war, which fed youth into the militias [12] and the concern is that customary leadership might continue to exploit migrant youth who are considered strangers, particularly with regard to land access [12].

Some studies found youth were dissatisfied less in terms of access to land and more in terms of their lack of agency over the land (e.g., not being able to plant life trees). Male youth are increasingly challenging customary governance institutions by rejecting traditional practices, decisions made by customary authorities, and directives to contribute to town labor. Studies find that youth are increasingly showing interest in individual, documented land rights [9,13,14].

The primary avenue for youth access to land in Bong and Lofa counties is through their parents and traditional leaders. Most youth can typically access land only for short-term use (which allows farming of vegetables/grains) while long-term access to land for more profitable commercial crops or more permanent life trees is typically only granted to male youth from landholding families.

Male youth from landholding lineages have more secure access to land. They generally access land through their parents or family elders for farming or house plots. Even those who come from landholding families must wait to inherit land to enjoy their full spectrum of rights such as long-term access to plant life-trees. Male youth with older brothers may not inherit rights to land unless they are willing to work the land with them and are in good standing with their brothers. The traditional process of gaining access is for elders to recite the land history to youth which teaches the youth about their rights as well as the extent of their land boundaries and the nature of land-use. Male youth who are children of parents in polygamous marriages, may be denied rights even if they are from landholding lineages or if their parents own land, if they are not children of the formal wife.

Youth who are not from land-holding lineages may gain access by approaching elders in their community and sometimes from neighboring communities. Sometimes youth pay a nominal fee for use of the land or are expected to give small gifts to the elders who allocate land to them. Rights granted to these youth are largely for seasonal or short-term use for farming vegetables or grains and cannot be used for trees or cash crops because that would establish long-term rights to land. These youth may never gain long-term rights to land, unless they have the money to buy land.

Stranger youth face even more hurdles in accessing land. While they may be legally recognized as community members and hence have rights to customary land under the LRA, traditional norms tend to exclude these groups from accessing land. Depending on their situation they may be able to access land by finding a local elder to vouch for them ("stranger father") or leasing land. But their rights are less secure. While use rights to customary land are inherited, stranger youth mentioned that they were denied access to land even though their parents were allowed to use land. For example, some explained that they were told that land was not their parents in the first place, even if the parents had planted cash crops.

Some youth reported that those who lease land or are allocated land for short-term use have faced arbitrary evictions. Some youth also reported that they are not allocated land in time for planting or are given land that is not fertile. Sometimes, they are limited to how they can use land. For example, some youth mentioned that they are not allowed to plant certain crops or fish on the land.

Youth reported that land scarcity was one main reason that impacted their access. Increasing population density has increased demand for fertile land and house spots and made it less available for youth (mentioned in two communities in Lofa County, and two in Bong County). When land is scarce, stranger youth are more at a disadvantage. For example, stranger youth in Baila in Bong county, reported that they paid high rents for house plots because of land scarcity. Some youth mentioned that landlords sell-off land to outsiders, therefore there is not enough land for the youth. In some cases, most of the available land is taken for life trees, leaving little land for youth to cultivate vegetables.

3.1.4. Are Women Able to Enjoy the Full Extent of Their Bundle of Rights? Do Women Have Knowledge about Who Can Gain New Rights to Land and under What Circumstances? Is This the Case for All Women?

"Men have the right to land; women get permission from men because they (men) are the head and leaders of the family"—Male Focus Group Participant.

"Land business is a traditional thing so women cannot own land here. Besides, they do not even have the strength to fight or defend the land when there is conflict"—Male Focus Group Participant.

Most research respondents report that women's rights to land are derivative of men's rights. Single women gain access to land through their fathers, brothers, uncles, and other male relatives. Customary norms do not treat young single women as full members of the community because they are expected to eventually marry into a family from another community as is the practice in patrilocal systems. Married women gain rights through their husbands, and widows do not automatically acquire rights to their husband's land, rather they gain access through their deceased husband's brothers or other male relatives who control the land.

In Bong and Lofa counties, female youth are even more disadvantaged than male youth in access to land for agriculture or house spots. In one FGD, single female youth explained that they are rarely able to access land directly, and need to "beg" their brothers, fathers, uncles for land. Landholding families fear that control of their community land will pass to outsiders and therefore do not allow their daughters to inherit land.

Even when women do access land, their marital status determines the extent and duration of the bundle of rights they can enjoy. Single young women may access land for short-term access. For example, in Lofa County, Sucrumu village, single women from land holding families are actually able to access farming land easier than married women; however, these short-term rights are usually terminated when women get married. Long-term access for married women is mostly through male relatives. When women community members marry strangers, they may receive only short-term rights to land within the community, but their children may be recognized for long-term rights. The situation of women in de-facto (marriage without dowry) and polygamous marriages is not clear from this research.

In the survey, similar to the findings in the literature and the qualitative research, most respondents said women typically go through a husband or male relative to access farmland (only 25% said a woman could get land without such permission, and only 37% said female youth could independently access community farmland). De-facto unions appear to further disadvantage women's land access and use over more formal arrangements: while over 80% of respondents said women in customary and statutory marriages have the same rights to use customary land as their husbands, only 60% said women in de facto unions have the same rights. Respondents felt youth had rights to community land, but it seemed marriage may be a pre-requisite: while few respondents felt "land was only for adults," (15%), only 47% of respondents said unmarried youth are co-owners of community land, while nearly 60% said married youth are co-owners.

In the survey, almost all participants responded that if a woman's parents were not strangers (parents who were born in the community), she was also considered a community member. On the other hand, only 72% said being married to a community member made one a community member. Considering women often relocate from their natal area to their husband's community, this may indicate that women are not considered "members" of the communities they live in after marriage. Women in polygamous unions appeared to be further disadvantaged; only 42% of respondents said a woman in a polygamous union (even to a community member husband) would be considered a member. Interestingly, in these more ambiguous cases of community membership (e.g., stranger parents, polygamous unions), women and youth respondents were even less certain that one would qualify for membership than adults and men were. As mentioned earlier, land access is contingent upon community membership, therefore ambiguity about who is a community member impacts their rights to land.

In the formative research, some youth shared examples of women accessing land directly either because they are from landholding families and they get land from their fathers, or they have the funds to rent or buy land. For example, in Baila Village in Bong County, one woman who is a nurse bought her own land and has planted life trees such as rubber. In Kpaai, some families allow their girls to own land; the context under which this happens is unclear from the research. When women gain rights by renting land, access is usually clearly defined as short-term. Access for house spots are given as long-term if women have managed to access them (women need to approach traditional elders through male relatives for house spots).

However, most youth in the formative research highlight that traditional norms do not allow women to be involved in land matters. These youth believe that it is hard to change customary norms that limit women's access to land. In some cases, even if women are given short-term access to farmland, their rights are not respected, and others encroach on their lands. This was mentioned in an FGD in Yeala village in Lofa County. Furthermore, young women are sometimes not able to use the land they are given because they lack access to labor and other inputs. When the land they use is in the bush or far from the village settlements, women cannot go to the boundaries of the land (the bush) on their own because of safety reasons.

House Spots: Although community farmland is typically thought of as collectively owned, plots designated for homes (house spots) seemed to be considered individually owned; nearly all survey respondents (from both genders and age groups) said that community members were able to claim a house spot "forever," even women and youth—but women needed the permission and aid of their father or husband to claim a spot. However, almost all respondents said that women could inherit a plot, pass it on to their children, and that a widowed woman can stay on the spot. In terms of youth, nearly 70% said youth (male and female) can inherit family land, but in contrast, only 37% said that female youth can independently own land.

3.2. Are the Bundle of Rights Clearly Enforced in the Face of Contestation (Robustness)?

3.2.1. Does the LRA Clearly Define Enforcement of Rights in the Face of Contestation

The LRA contains high-level provisions that define the enforcement of land rights in the face of contestation. It obligates the Government of Liberia to protect and enforce all land rights and interests, which includes female and male youth rights to land (art. 10). An explicit objective of the LRA is to guarantee that all communities, families, and individual "enjoy secure land rights free of fear that their land will be taken from them, except in accordance with due process of law" (art. 3). Critically, Customary Land's existence and ownership is enforceable by operation of the LRA (art 32(2)), [7] and not affected because of lack of title or registration (art. 11(3)) or to lack or delay in completing the confirmatory survey of customary land boundaries across the country (art. 37(4)). These provisions give more legal force to the claims of individuals and communities, including the youth members, in the face of conflicting interest by government, companies, and private parties. Furthermore, LLA decisions (that lack objections and exceptions) can enjoy judicial enforcement, subsequent to filing a petition, and all LLA decisions are subject to judicial review through the Circuit Courts and appealable to the Supreme Court of Liberia (art. 37). The LRA also requires the government to provide adequate resources to implement the LRA's legal provisions (art. 36(13)).

The LRA also has high-level provisions to define enforcement of rights in the face of contestation at the community level. A community member's absolute right to his or her residential area is automatically transferred by operation of the LRA and requires the CLDMC to formalize these rights by issuing a deed, the absence of which does not defeat a community member's ownership rights (art. 49). The LRA prohibits a community from depriving a community member of his or her residential area. Community restrictions on the exercise of land rights by a community member, including youth members, are invalid if they violate the LRA or Constitution (art. 34(4)). [8] The community and its community members are jointly and severally responsible to protect the rights of private land owners located within customary land; this includes private land owners who are youth community members who have a right to own a residential parcel. Private landowners, including those who are non-community members, are also responsible to abide by community-adopted rules (art. 46(3)).

Finally, the LRA is a comprehensive land law that envisions more detailed enforcement provisions to come under implementing regulations. The LLA holds the authority to promulgate regulations necessary to effectively implement the LRA (art. 71). The LRA specifically requires the LLA to develop regulations to solve all customary land disputes between communities through customary law and alternative dispute resolution (ADR) mechanisms (art. 37(8)), including those available through customary ADR bodies. [9] With the LRA recently passed, the LLA is in the process of developing all necessary regulations. These regulations will determine how well and clearly female and male youth land rights will be enforced in the face of contestation. More clearly defined enforcement will hinge on how robustly LLA promulgated regulations respond to youth and gender considerations.

3.2.2. Do Youth Have Knowledge of Their Rights in the Face of Contestation or Changed Circumstances? Is This the Case for All Youth?

Youth were generally aware that they can approach elders or the courts to deal with land conflicts but were largely in favor of settling disputes outside of court because they reported that using courts are expensive and disputes may take many years to resolve. Some youth in the FGDs reported that disputes in their communities involving tribal certificates are settled by elders, while those involving deeds are generally settled in the courts.

Youth's faith in customary leaders to resolve conflicts was mixed. Some youth trusted the traditional authorities and felt that elders were open to their concerns and listened. However, some said that elders are biased toward landowning and wealthy families and disputes were usually settled in their favor. Some youth say that they do not have a voice with elders.

---

[7]   For example, as opposed to being enforceable only after the completion of an administrative process.

[8]   Although land can be taken if the community provides the community member with comparable land.

[9]   Under the LRA, alternative dispute resolution mechanism is defined as any process used to resolve disputes outside of court, and alternative dispute resolution body is defined as any entity, including a customary body, whose purpose is to resolve disputes outside of court (art. 2).

Many suggested that the best way to avoid land conflicts is by surveying and demarcation of land boundaries. For example, in two communities in Bong County, youth said that after land was surveyed conflicts came to an end. They also thought it was beneficial when fathers and elders show youth the land boundaries to help prevent encroachment and disputes. To better equip them to claim their rights, some youth expressed a desire to understand the provisions of the LRA; they also said they would like NGOs or the government to be involved.

While in some communities youth and elders have good relationships, many youth in the FGDs said that elders do not understand youth's needs and that elders need to be sensitized to the problems of youth and their need for land and also help build relationships between youth and elders. In a few FGDs, some youth thought that the government and NGOs should bring youth and elders together to train them and promote a dialogue, so that youth can better access land within the community. Some youth expressed the need to be more involved in land matters in order to increase their access to land.

In the survey, when asked what they can do about grabs of community land, the overwhelming majority of male survey respondents reported they could sue individuals (90%) and companies (83%), and half even said they could sue the government; for females, the percentages were much lower. However, 66% of respondents said the land that big companies are currently using for concessions belongs to the communities.

3.2.3. Are Youth Able to Enforce Their Rights in the Face of Contestation or Changed Circumstances? Is This the Case for All Youth?

There is little information in the literature about youth participation in customary dispute resolution mechanisms; this could be an area for further study. In general, there is a deep distrust of customary courts and their capacity to decide land issues in a fair manner [12]. The Eleven Clan Study found dissatisfaction with the customary system appears to be particularly prevalent among youth. In many clans, local authorities are viewed as biased, especially by the youth, who question the legitimacy of these authorities [1]. Youth did not typically hold significant positions of power in the process [1].

From the formative research we learned that contestation and conflict over land in Bong and Lofa Counties are characterized by: (1) Encroachment of boundaries between landowners and between communities, often exacerbated by lack of fertile land and unclear boundaries, which can inadvertently lead to encroachment as youth clear land in the bush for agriculture. Some research participants complained that tribal certificates do not protect their land from encroachment, (especially from those with deeds). They report that only deeds protect land and only those with deeded land can sell their land. However, most residents said they did not have deeds. In FGDs, many youth explained this was due to the difficulty and expense because land must be surveyed, and the cost is prohibitive; (2) intra-family disputes over land, including multiple claimants for family or tribal land and contestation between "legitimate" and "illegitimate" children for rights to their father's land; (3) disputes caused by people selling the same land to more than one buyer; and (4) disputes caused when people who have only short-term access to land to plant vegetables instead plant life trees to make a claim on the land; in some cases people even do this on land that is set aside for communal use; (5) ethnic conflicts were reported in two communities in Lofa county.

Most participants report that the most common way to solve land conflicts is to involve the elders who determine the rights based on their knowledge of the history of the land. In some cases, youth are also involved in settling community land disputes; this was mentioned in three villages in Lofa County and one in Bong County. There is a hierarchy of traditional elders that are involved in settling conflicts. First the local community elders and landlords are asked to arbitrate. If that is not fruitful, town and quarter chiefs are approached. Other institutions have also played a role in settling land conflicts. For example, in one community, the Liberia Refugee Repatriation and Resettlement Commission (LRRRC) was involved in settling land conflicts.

Both intra-family and community boundary disputes appeared to have a negative bearing on youth access to land in both counties. Some youth who participated in the formative research have experienced conflicts over boundaries fear violence and have no choice but to pursue other livelihoods or migrate to urban areas rather than fight for their rights. In two communities in Bong County, youth mentioned that conflict between neighboring towns or villages had even led to the loss of lives. Furthermore, youth say that they are the most affected by land conflicts because they do not gain rights to land being contested. In some cases, youth from families or communities who are in conflict cannot work together to use land.

3.2.4. Are Women Able to Enforce Their Rights in the Face of Contestation or Changed Circumstances? Is This the Case for All Women?

From the formative research, most youth reported that female youth are generally not involved in settling land conflicts. They explained that since men are mostly involved in clearing the bush, it is the men that encounter conflicts. However, in one village in Lofa county youth who participated in the formative research reported that elderly women leaders are allowed to help in dispute resolution. Some youth mentioned the need to organize themselves so that both male and female youth can be trained in conflict resolution.

3.3. Does the Land Rights System Give All the Ability to Participate in Decisions about the Land (Inclusive & Gender Equitable)?

3.3.1. Does the LRA Clearly Define Land Rights/Land Governance as Inclusive and Gender-Equitable?

The LRA formally devolves customary land governance to communities, including to their female and male youth members. Inclusive community membership provisions, combined with equal protection provisions, discussed above, ensure that the full community membership is included in community-level land governance bodies and holds customary land governance rights. The community membership, comprised of all community members acting collectively, is the highest decision-making body and is vested the authority to develop and manage customary land (art. 36). Female and male youth, as part of this collective, share this authority. Additionally, all Community Members, including female and male youth, are responsible for developing community by laws and a land-use plan and establishing Community Land Development and Management Committee (CLDMC) (art. 35).

The CLDMC is a community-level (executive) land body comprised of equal representation of youth, women, and men, who share decision-making authority (art. 36(6) & (7)). The CLDMC is accountable to the entire community membership, including its youth members (art. 36(4)). Even prior to the formal establishment of the CLDMC, youth and women along with elders, chiefs, and traditional leaders are responsible for developing and managing its customary land (art. 69).

While provisions in the LRA include female and male youth in community-level land bodies and grant them land governance rights, there are gaps. Whether female and male youth are able to exercise these governance rights, particularly when all community members are acting as a collectively, will depend on how well and inclusively LRA regulations operationalize these rights. Additionally, similarly to the issue raised above, cohabiting youth who move into their partner's community, and who reside there for under seven years, will either have less clear rights or no rights to participate in customary land governance.

3.3.2. Do Youth Understand Their Rights as They Relate to Decision-Making on Land Governance?

In general youth did not understand their rights to participate in decision-making on the land, but we did not learn anything specific to this particular aspect of youth rights. Youth however did indicate that they were unhappy to not be able to participate in land governance decisions. See following section.

3.3.3. Are Youth Able to Participate in Land Decisions at the Household and Community Level? Is This the Case of All Youth?

Youth participation in decision-making and land governance is variable. There are national, regional, and local youth organizations that advocate for land and resource rights on behalf of youth [1,13,15,16]. The Eleven Clan Study reported that in many clans, youth are well-respected and take part in land decision-making, the formulation of rules, and sometimes even in resolving disputes, though the level of involvement varied significantly [1]. However, in other clans, youth felt overlooked in decision-making about land issues, and were excluded from traditional governance structures. Some youth have expressed frustration about their lack of power in community land management processes [1,13,14] which has fueled a sense of exclusion and resentment [1,9].

Rural youth participation and control in local governance may have increased since the civil war. In areas that grant concessions, new governance structures have been established under the Community Rights Law for Forested Land [8] to negotiate with concessionaires. Even though some youth are represented (by requirement) on these governance structures, their participation may be weak [8,17]. In a recent study on women's participation in forest governance bodies conducted by Landesa in collaboration with the LLA and FCI, in four Community Forest sites in Bong, Grand Bassa and Margibi Counties, male and female youth seemed to be well represented on the forest governance bodies, especially in the Community Assembly. However, decision-making around dealings with companies or distribution of monetary benefits are often by elite members in the Community Forestry Management Body [18].

During the formative research, many research participants reported that male elders from landholding families have a firm grip on land related decision-making processes, while those that do not have land are excluded from governance decisions. In most cases, only select youth (usually youth leaders, or land-owing youth) are allowed to be part of land governance and that it is the elders' decision on whether youth are involved in land matters; youth cannot participate unless they are invited. Even if youth are involved, they felt that their voices do not carry as much weight, and that elders have final say on land governance decisions. They believe that elders fear of losing control of the land and therefore limit the scope for effective youth engagement in community land governance systems. Furthermore, they say that elders think that young people are not serious enough to be involved in governance. Most youth expressed a desire to be involved more fully in land governance decisions working in collaboration with the community elders. They felt that they would benefit from helping facilitate elders and youth to working together.

In many communities in Lofa, the relationship between youth and adults on land matters was mixed. In some communities, youth and adults were reportedly working together very closely when discussing or resolving key community land issues. However, this was questionable especially given that some youth were hesitant to discuss land issues in the absence of elders. For example, youth leaders in one town would only discuss land issues when a community elder could listen and respond to some of the questions. It was clear that some youth in this town could not discuss or engage in any meaningful land-related issues without prior approval from the elders. In another town, community elders viewed youth as a big problem and expressed that youth needed to be punished severely for being lazy and for concentrating on drugs and other community vices. A youth leader however strongly suggested that "youth were in charge" of the community and he praised the youth for being vigilant in the face of border threats posed by migrants from Guinea. In that same community, however, key informant interviews indicated that it was difficult for female youth to access land. In another community, youth-adult tension was very high and there was no indication of youth-adult partnerships or dialogue platforms on land matters. In many Lofa and Bong County towns, the existence of young quarter and towns chiefs depict encouraging signs regarding the future of youth engagement in community land governance processes in Liberia. The young quarter and town chiefs appeared to have a good working relationship with community elders and a good understanding of the key land issues affecting youth in their towns.

As illustrated in Figure 4 below, about half of the survey respondents reported that traditional authorities made decisions about land. Nonetheless, nearly a quarter said the members of the community made land governance decisions collectively—and the majority of these said that this collective decision-making was inclusive of all community members. None of the respondents reported that a statutory land management body made land decisions [10].

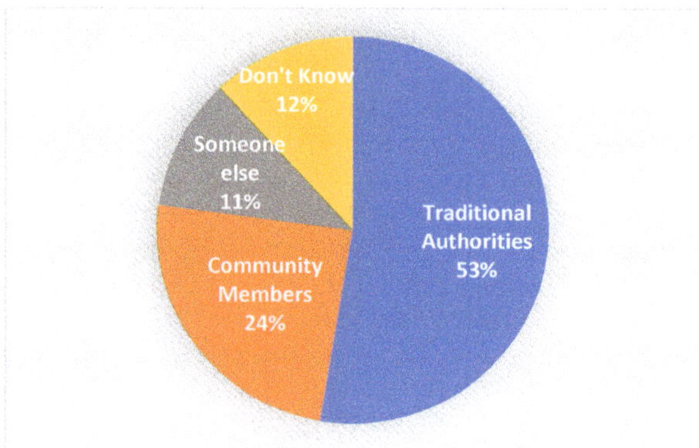

Figure 4. Community members' perception on who makes decisions about customary land in the community.

Because youth are often not landowners, and because they often lack intimate knowledge about land rights, community boundaries and the land governance systems, the research and literature found that they are generally excluded from land dispute resolution processes—including Alternative Dispute Resolution (ADR) systems, locally run systems set up by communities and/or CSOs to address unresolved disputes (typically after efforts by a traditional authority, or in lieu of them altogether). The formative research found that opportunities for community elders and youth to discuss and resolve land disputes are rare in both counties, which suggests that intergenerational tensions over land could increase in future, and force even more youth to abandon agriculture altogether.

3.3.4. Are Women Able to Participate in Land Decisions at the Household and Community Level? Is This the Case of All Women?

Female youth, in particular, are disempowered in terms of land governance, as the traditional view that land is the domain of men persists in rural Liberia. The Eleven Clan Study found that in some communities there are positions which have a strong influence over land that cannot be occupied by women. Dodd et al. [7] found that even where women held positions of authority, their power was often curtailed by traditional gender roles. A national report on youth engagement [19] found that young women are often the least engaged in their communities, hesitant to engage in politics, and overlooked by policymakers.

The formative research found that young women were excluded from land governance systems in deeply traditional communities, especially in parts of Lofa County, which could be a significant challenge for programs seeking to advance gender-equitable land rights for youth in these communities. When youth are involved in land governance, it is usually male youth that participate. There are some exceptions. For example, in one community in Bong County, youth in the FGD mentioned that young

---

[10] These responses are for a question asked about 'big' land decisions,' i.e., regarding land over 50 acres; responses were nearly identical on a question about land under 50 acres—The only exception being that when it came to decisions about small parcels, about 5% said the family/clan made land decisions.

women participate in land governance but are expected to act in a deferential way when they speak. In another in community, women were involved in a town hall meeting for the construction of a clinic.

Youth—female youth in particular—are often unaware of their legal rights to land, including the Land Rights Act (LRA). Unsurprisingly, youth in both Bong and Lofa Counties were more knowledgeable about the exclusionary, male- and elder-dominated customary norms and laws regarding land rights. These customary land rights practices revolve around adult males as the arbiters of land access and governance with limited opportunities for independent youth access to land.

In the formative research the general opinion of research participants is that women are not expected to be involved in land matters according to traditional norms. Women's voices are represented either through their male relatives or the chairladies. However, in the FGDs, many youth acknowledged that women should be empowered to participate in decision-making about land governance.

4. Discussion

4.1. Impacts of Land Tenure Security on Youth Livelihoods

Land tenure insecurity impacts rural youth's ability to sustain a livelihood based on agriculture. Our findings highlight that most youth can typically access land only for short-term use which allows them to cultivate vegetables and grains, while long-term access to land for more profitable commercial crops or more permanent life trees is typically only granted to male youth from landholding families.

Agriculture is the primary income activity for youth among the formative research participants in Bong and Lofa counties. Youth were involved in commercial cash crop farming (e.g., palm, rubber) as well as vegetable farming ("gardening"). In Lofa, youth often worked in *kuus* and were occasionally hired as daily contract workers. Youth's agricultural work was divided along gender lines, with men doing more of the physically demanding tasks such as clearing brush. Men also seemed to control the more profitable cash crop farming while women pursued gardening. Besides farming, male youth were involved in driving motorcycles for hire, especially in Lofa, and youth sometimes had small informal businesses; in Bong, women are involved in petty trading.

Youth in Bong and Lofa counties have few livelihood options and the majority wished to deepen their engagement in agriculture to better support themselves and their families. However, a majority of youth felt that lack of access to land or short-term access impacted their ability to farm and support themselves and their families; many end up working as agricultural laborers.

Gender was an important factor in determining access to land and land-based livelihood opportunities for rural youth. Young women's access to land for farming and housing was even more constricted than male youth's access. Findings from the formative research highlight that single women who are given access to land are usually not allowed long-term use and their land rights are often violated and less respected. Even when young women were able to access land, they faced hurdles in organizing labor and capital to use the land.

While land access posed as an impediment to earning a living in agriculture for youth, lack of support to buy inputs also emerged as a challenge to youth's pursuit of agricultural livelihoods. Poor transportation and access to markets, low crop prices, and poor weather also were challenges. The most often-cited suggestion for improving youth participation in agriculture was to support access to agricultural inputs, along with agricultural training and supporting kuus.

These findings are consistent with the literature that highlights that limited access to land impacts youths' ability to practice land-based livelihoods. Agricultural employment is informal and vulnerable to downturns, so most youth complement it with other activities, such as mining and plantation employment, transportation services, government work, and small businesses. Most depend heavily on forest products for consumption and sale [1]. One study found that most youth migrate because of the inequalities in accessibility to land for farming [20]. However, with nearly one-third of youth unemployed (2008 census), urban economic opportunities are also limited. Many of the most marginalized youth are not suited to work other than in the agrarian sector [9]. The lack of non-agrarian employment opportunities for youth, many of whom have had some forms of military experience from the war, is a source of social instability. Some engage in quasi-professional violence and illegal mining [21].

There is a widespread observation in the literature that Liberian youth are not interested in pursuing agriculture, for myriad reasons: they do not perceive that it offers a viable livelihood, the returns are too delayed as compared to daily work, and females may fear assault when farming far from home [11,22–24]. However, these perceptions are not backed by significant research, and some interventions and youth outreach officers report that youth are interested in agriculture, given the right conditions to earn a livelihood, a finding that is backed by our formative research findings.

4.2. Recommendation for Policy and Programming

The Land Rights Act presents a unique opportunity to recognize the rights of youth to access land and participate in land governance. However, closing the gap between statutory law that recognizes youth rights to land, and customary practices that are discriminatory toward youth, needs rural women and men to understand their rights under the LRA and buy in from the traditional gatekeepers who control access to land.

Our research in Bong, Lofa, and Rivercess counties provide useful policy and programming insights for promoting effective and gender-equitable youth land rights in rural Liberia. Landesa's ongoing LRSD interventions with CSO partners DEN-L and FCI are crafted to address some of the youth land access and governance challenges identified thorough our research, however, these are still at a relatively small scale, currently working directly in only 14 communities on the ground. To ensure impact at scale, policy and programming options might include:

- Land rights CSOs should co-develop and co-implement cross-sectoral youth livelihood and empowerment programs that seek to address not only youth land access and ownership challenges but also limited youth access to inputs, markets, poor transportation, low crop prices, and weather related challenges that discourage Liberian youth from pursuing a career in agriculture. The cross-sectoral youth and land interventions should also include nationwide radio and television programs showcasing successful youth farmers who can act as role models and drive home the idea of "agriculture as a business" and viable employment opportunity for Liberian youth.

- Supporting land rights education and awareness-raising to enable rural Liberian youth to better understand and defend their land rights as provided for in local customary and national statutory land rights frameworks. Community sensitization on the importance of youth land rights for sustainable rural youth livelihoods would be beneficial. The sensitization should especially target traditional gatekeepers such as customary leaders, landlords, elders, and local government officials, including town and quarter chiefs.

- Developing simplified language to popularize the provision of the Land Rights Act through online platforms such as WhatsApp and Facebook to increase the number of youth who are aware of the Land Rights Act and its provisions.

- Providing targeted and continuous youth land rights training at all levels of the traditional power structures and formal land administration systems to gradually break the deeply entrenched and oftentimes discriminatory land allocation and ownership systems that favor adult men and young men.

- Providing simplified land rights training and awareness raising Training of Trainers (ToTs) and community-based programs to grassroots-oriented youth associations, women's groups, and Community Based Organizations working with young farmers with particular attention to local organizations working with young women.

- Addressing ambiguities and questions around "who is a community member" by educating community elders, youth, and their representatives organizations about the community membership qualification provisions under the Land Rights Act to ensure that the community and youth's understanding of community membership evolves toward a more inclusive membership criteria that protects the land rights of youth "strangers" and their children, and spouses and wives in de-facto unions and polygamous marriages.

- Promoting adult-youth dialogue on land matters in order to improve opportunities for youth to access land. This could be done by strengthening existing community platforms for youth-adult dialogue where they already exist (e.g., Yeala in Lofa County), as well as developing and supporting new youth-adult land rights dialogue forums where these do not exist. These dialogue forums could enhance youth engagement in broader community land governance systems and boost opportunities for youth access to land.
- Promoting youth-adult dialogue on land matters by organizing community-, county-, and national-level multi-stakeholder youth land rights workshops bringing together youth leaders, customary authorities, government officials, private companies, and development partners to educate stakeholders about the importance of youth land rights, youth related provisions in the LRA and innovative approaches to foster youth land ownership, and effective engagement of young men and young women in land related decision-making processes.
- Developing youth-oriented land rights advocacy initiatives based on thorough youth segmentation and considerations for most disadvantaged groups of youth including strangers, unmarried youth, females, and those in de-facto unions and polygamous marriages.
- Developing and promoting youth sensitive land rental market policies to better protect the land rights of youth from unscrupulous landlords and traditional leaders especially in Bong County where there is an informal, unregulated land rental market.
- Encouraging the Government of Liberia to develop community-level social security schemes that can incentivize and quicken intergenerational land transfers.
- Land rights CSOs should work with customary authorities and government officials to prepare a database of vacant or underutilized land that can be reallocated to prospective youth farmers especially those residing in and around major urban centers like Gbarnga and Voinjama. The database can be shared with youth associations to help identify landless youth who are interested in farming and connecting the youth to landowners or allocation authorities.
- Training youth leaders in land dispute resolution mechanisms including Alternative Land Dispute Resolution to build the capacity of youth farmers to recognize and defend their land rights when they are infringed upon by customary and statutory authorities. Land rights CSOs should work with the Liberia Land Authority to ensure the development of regulations that stipulate quotas for youth engagement in customary land dispute resolution mechanisms including Alternative Dispute Resolution (ADR) approaches. The regulations should define clear procedures for the protection of male and female youth land rights.
- Engage youth and their representative organizations in community boundary harmonization, mapping and surveying processes to ensure that youth have a good understanding of the historical evolution of land issues in Liberian communities where boundary disputes and intergenerational tensions over land are common.
- Civil society organizations should work with customary authorities and community leaders across the country to ensure the participation of male and female youth in the collective community development of bylaws, land-use plans and Community Land Development and Management Committees (CLDMCs). The community bylaws should indicate collectively agreed quotas for male and female youth participation in land-use planning processes and representation in CLDMCs.
- Building the capacity of grassroots-oriented rural youth organizations by educating them about land rights and linking them up with customary and statutory authorities that mediate rural youth access to land and opportunities for land-based livelihoods.

- In Bong and Lofa counties, communities where there is increasing land scarcity especially close to major urban centers, identifying and promoting profitable small-scale, intensive farming activities that target landless unemployed, and connecting the youth with land owners and land allocation authorities to promote group leasing, contract farming, and access to urban markets to improve youth access to land and quick income generating opportunities.

5. Conclusions

Our paper documented rural youth livelihood opportunities and land access and governance constraints for diverse groups of young men and young women using evidence from select communities in Bong, Lofa and Rivercess counties of Liberia. The paper highlighted that rural youth engage in a wide of range of livelihood activities of which agriculture is an important activity. For youth who want to engage in farming, it emerged that secure and long-term access to land is a major challenge especially for young women and youth who are considered strangers in the community. In all three counties where we conducted research, the land access challenges faced by youth are compounded by the existence of a customary land rights system dominated by adult males who often exercise a firm grip on land-related decision-making and land access opportunities. While most youth have a good understanding of the exclusionary customary land rights system, they do not seem to have a full understanding of their own land rights within the system and there are limited opportunities for youth to defend their land rights in non-confrontational ways. We noted how the lack of youth-adult forums to discuss land issues along with few opportunities for youth to participate in community land governance further constrain opportunities for youth access to land. In some communities, the continued exclusion of young men and young women in community land governance systems increases intergenerational tensions over land and further limits opportunities for youth to engage in farming.

Policy and programming options must address the individual and community level barriers that impede rural youth's access to land. To address individual-level youth barriers, land rights education, and awareness raising is critical. To improve youth access to land in deeply traditional settings such as Lofa County where land is communally owned, community sensitization targeting traditional leaders and landlords is fundamental. Youth land rights programs should be designed with particular attention given to gender, identity, and age dimensions that determine youth access to land. The success of these programs will largely depend on the extent to which they can help bridge the gap between statutory provisions and customary norms and practices that continue to discriminate against young men and young women in most parts of rural Liberia.

Author Contributions: Conceptualization, all authors; methodology, E.L., E.U.; T.M., M.-L.D., P.D. and T.H.; formal analysis, E.L., T.H., M.-L.D., T.M.; writing—original draft preparation, E.L., M.-L.D., T.M.; writing—review and editing, E.L., E.U.; visualization, T.H.; supervision, E.L.; project administration, E.L.; E.U.; funding acquisition, Landesa. All authors have read and agreed to the published version of the manuscript.

Funding: This research was funded by Kings Philanthropies through Landesa's Land Rights and Sustainable Development (LRSD) Project. The APC was funded by Landesa.

Acknowledgments: We would like to acknowledge the organizations and individuals that contributed to the research presented in this paper: The community members who participated in our research in Bong, Lofa, and Rivercess counties and the local and regional government officials and traditional leaders who gave us their time for Key Informant Interviews and supported our research in other ways. Our CSO partner DEN-L who conducted the formative research on youth and land. Special thanks to Peter Dolo for his leadership in the formative research and Dorothy Toomnan for conducting research in Bong County. DEN-L also facilitated our baseline survey data collection activities in their intervention communities. Our CSO partner FCI who facilitated our baseline survey data collection activities in their intervention communities. Research contractor Ecomsult, who conducted our Baseline Survey in Bong, Lofa and Rivercess counties. Special thanks to Jeremiah Collins and his team for doing an excellent job with data collection under challenging field conditions.

Conflicts of Interest: The authors declare no conflict of interest. The funders had no role in the design of the study; in the collection, analyses, or interpretation of data; in the writing of the manuscript, or in the decision to publish the results.

References

1. Namubiru-Mwaura, E.; Knox, A.; Hughes, A. Liberia Land Policy and Institutional Support (LPIS) Project. Customary Land Tenure in Liberia: Findings and Implications Drawn from 11 Case Studies. Report Prepared for USAID, under the Auspices of the Land Commission of the Republic of Liberia and coordinated by Landesa and Tetra Tech ARD. Available online: https://www.land-links.org/wp-content/uploads/2016/09/USAID_Land_Tenure_Liberia_LPIS_Synthesis_Report.pdf (accessed on 29 June 2020).
2. United States Agency for International Development Office of Food for Peace (USAIDFFP). Country-Specific Information: Liberia Fiscal Year 2016 Development Food Assistance Program: Liberia Community Development Project. Available online: https://www.usaid.gov/sites/default/files/documents/1866/2016%20Final%20Liberia%20CSI.pdf (accessed on 29 June 2020).
3. CIA The World Factbook: Liberia. Central Intelligence Agency, Central Intelligence Agency, 1 Feb. 2018, CIA World Factbook. 2018. Available online: www.cia.gov/library/publications/the-world-factbook/geos/li.html (accessed on 29 June 2020).
4. UNDESA World Population Prospects, Revision. Available online: https://population.un.org/wpp/tr (accessed on 29 June 2020).
5. TRC Final Report of the Truth and Reconciliation Commission of Liberia (TRC) Volume I: Findings and Determinations. Available online: https://reliefweb.int/sites/reliefweb.int/files/resources/3B6FC3916E4E18C6492575EF00259DB6-Full_Report_1.pdf (accessed on 29 June 2020).
6. Bruce A Strategy for Further Reform of Liberia's Law on Land. Land Governance Support Activity USAID/Tetratech. Available online: https://www.land-links.org/wp-content/uploads/2017/05/USAID_Land_Tenure_LGSA_Report_Reform_Strategy_Liberia_Law_Land.pdf (accessed on 29 June 2020).
7. Dodd, M.-L.; Duncan, J.; Uvuza, J.; Nagbe, I.; Neal, V.D.; Cummings, L. Women's Land Rights in Liberia in Law, Practice, and Future Reforms: LGSA Women's Land Rights Study. Technical report, USAID. Available online: https://www.land-links.org/wp-content/uploads/2018/03/USAID_Land_Tenure_LGSA_WLR_Study_Mar-16-2018.pdf (accessed on 29 June 2020).
8. De Wit, P.; Stevens, C. 100 Years of Community Land Rights in Liberia: Lessons Learned for the Future. [Presentation]. Paper Presented at the World Bank Conference on Land and Poverty. Washington, D.C. Available online: http://documents.worldbank.org/curated/pt/896191468762635261/pdf/317730PAPER0SDP770conflict0WP0211web.pdf (accessed on 29 June 2020).
9. Richards, P. To fight or to farm? Agrarian dimensions of the Mano River conflicts (Liberia and Sierra Leone). *Afr. Aff.* **2005**, *105*, 571–590. [CrossRef]
10. Utas, Mats. Building a Future? The Reintegration and Remarginalization of Youth in Liberia. In *No Peace No War: An Anthropology of Contemporary Armed Conflicts*; Richards, P., Ed.; Ohio University Press: Athens, Greece, 2005; Available online: https://books.google.com/books?id=3U80lcHx94oC&lpg=PA1&ots=oEztzoCaYN&lr&pg=PA242#v=onepage&q&f=false (accessed on 29 June 2020).
11. Scarborough, G. Growing a Future: Liberian Youth Reflect on Agriculture Livelihoods. USAID, Mercy Corps. Available online: https://www.mercycorps.org/sites/default/files/Liberian-Youth-Reflect-on-Agriculture-Livelihoods-Mercy-Corps-2017_0.pdf (accessed on 29 June 2020).
12. Brottem, L.; Unruh, J. Territorial tensions: Rainforest conservation, postconflict recovery, and land tenure in Liberia. *Ann. Assoc. Am. Geogr.* **2009**, *99*, 995–1002. [CrossRef]
13. Bloh, O.; Yarsiah, J.; Otto, J. Individual Land Ownership Versus Collective Land Ownership. Current Practices, Opportunities, Challenges & Options. Report Prepared for the Sustainable Development Institute (SDI). Available online: https://www.sdiliberia.org/sites/default/files/publications/SDI%20Report%20Individual%20Land%20Ownership%20Versus%20Collective%20Land%20Ownership.compressed.pdf (accessed on 29 June 2020).
14. Knight, R.; Kpanan'Ayoung Siakor, S.; Kaba, A. *Protecting Community Lands and Resources: Evidence from Liberia*; Report prepared by Namati, Sustainable Development Institute, and IDLO; Available online: https://namati.org/resources/protecting-community-lands-and-resources-evidence-from-liberia/ (accessed on 29 June 2020).
15. Liberia News Agency [AllAfrica]. All Africa Bong Youth Wants Govt Intervene in Fuamah Land Conflict. Available online: allafrica.com (accessed on 29 June 2020).

16. Daffah, M. Liberia: Youths seek global intervention for resolution of Nimba land dispute. In *BBC Monitoring International Reports*; Text of Report Published by Liberian Non-State Station Star Radio Website; 29 August 2010.

17. Urey, E.K. *Political Ecology of Land and Agriculture Concessions in Liberia*; The University of Wisconsin-Madison: Madison, WI, USA, 2018.

18. Louis, E.; Teage, C.; Dodd, M.; Urey, E.; McClung, M. *A Report on High Level Findings from Research on Women's Participation in Forest Governance Bodies in Nimba, Grand Bassa and Margibi Counties*; Landesa: Seattle, CA, USA, July 2020.

19. Brownlee, T.; Krawczyk, L.; Krumrei, K.; McCachren, C.; Raval, N.; Visser, C. Youth to Youth: Measuring Youth Engagement Liberia. Report Prepared by Search for Common Ground, Liberia's Ministry of Youth & Sports, and American University. Available online: https://www.sfcg.org/wp-content/uploads/2014/09/Youth_Engagement_Report_Full.pdf (accessed on 29 June 2020).

20. Williams, C.A.; Ocha, G.B. Land ownership for youth in agriculture: A case study of Bolivia, Brazil and Liberia. [Presentation]. Presented at the World Bank Land and Poverty Conference, Washington, DC, USA, 23–27 March 2015.

21. Blattman, C.; Annan, J. The consequences of child soldiering. *Rev. Econ. Stat.* **2010**, *92*, 882–898. Available online: https://www.chrisblattman.com/documents/research/2010.Consequences.RESTAT.pdf (accessed on 29 June 2020). [CrossRef]

22. Macaulay, F. The Pivot to Yes: Positive Youth Development and Our Agriculture Program in Liberia. [Blogpost]. Making Cents International. Available online: http://www.youthpower.org/resources/making-cents-blog-pivot-yes-positive-youth-development-and-our-agriculture-program-liberia (accessed on 29 June 2020).

23. Making Cents. Our Work Projects. Liberia Agriculture, Upgrading Nutrition, and Child Health (LAUNCH). Client: ACDI/VOCA (USAID). Available online: http://www.makingcents.com/liberialaunch (accessed on 29 June 2020).

24. USAID. Liberian Youth and Agriculture: Building a Food Secure Future in Liberia through Youth Entrepreneurship. [Power Point Presentation]. 2016 Global Youth Economic Opportunities Summit. Available online: https://youtheconomicopportunities.org/sites/default/files/uploads/resource/Changing%20Youth%E2%80%99s%20Perceptions%20of%20Agriculture%20Experiences%20from%20Liberia.pdf (accessed on 29 June 2020).

 *land*

Article

Benefits and Constraints of the Agricultural Land Acquisition for Urbanization for Household Gender Equality in Affected Rural Communes: A Case Study in Huong Thuy Town, Thua Thien Hue Province, Vietnam

Nhung Pham Thi [1,2,*], Martin Kappas [1] and Daniel Wyss [1]

[1] Division of Cartography, GIS and Remote Sensing, Faculty of Geoscience and Geography, Georg-August University Goettingen, 37077 Goettingen, Germany; mkappas@gwdg.de (M.K.); daniel.wyss@uni-goettingen.de (D.W.)
[2] Faculty of Rural Development, University of Agriculture and Forestry, Hue University, Hue 53000, Vietnam
* Correspondence: nhungphamthihuaf@gmail.com or nhung.pham-ti@geo.uni-goettingen.de; Tel.: +(84)-944-495-372

Received: 16 June 2020; Accepted: 23 July 2020; Published: 28 July 2020

Abstract: The Vietnamese Government has implemented agricultural land acquisition for urbanization (ALAFU) since 2010 which has caused a high level of social-economic transition in the country. In this paper, we applied the gender and development approach to discover how ALAFU has influenced the household gender equality in affected areas in Thua Thien Hue province, Vietnam. The data for this paper was mainly collected from two household group surveys, four group discussions, and six key informant interviews. Group 1 covers 50 affected households whose agricultural land was acquired for urbanization, while Group 2 consists of 50 households whose agricultural land was not taken away. The findings reveal that ALAFU has led to reduced access to agricultural land for group 1, but has contributed to an increase of economic status for women in both groups by creating non-farming job opportunities with a good income. However, most of their new jobs are still informal, contain potential risks, and the unpaid care work burden is heavy. Moreover, although the rate of women participating in household decision making has increased, the quality of participation is limited. Their participation in social activities and vocational training courses has improved insignificantly. Therefore, if the Government continues to promote ALAFU, they should take structural gender inequalities into account to achieve their sustainable development goals.

Keywords: agricultural land acquisition; urbanization; household gender equality; unpaid care work; women's economic position; women's participation in household decision making

1. Introduction

Agricultural land acquisition for urbanization (ALAFU) is the popular and cheap option of the urbanization process of developing countries [1–4], but it has disordered the life and livelihoods of people in the affected rural areas because they have strongly depended on agricultural land, and have low education and working skills causing low adaptive capacity during social-economic change [5–7]. Indeed, some research has shown negative impacts associated with ALAFU, such as decreasing farming employment, increasing poverty and environmental degradation, and rural-urban migration, which reduces arable land, degrades soil quality, raises the price of food commodities, and increases competition between the agricultural and residential uses of natural resources [7–12]. However, new opportunities have also appeared including high access to developed extension services, good educational systems, and medicinal health care, a high demand for agricultural products and diversified non-farming employment [7,13,14].

Gender equality is one of the important components of sustainable development [15,16]. It relates closely to agricultural land accessibility [17] and urbanization [18,19]. In fact, the Food and Agriculture Organization (FAO) has stated that unequal access to land resources between men and women, which is common in rural areas, has hindered progress toward rural gender equality [20]. Additionally, according to Frank Ellis (2000), access to land is the most important asset of rural people. In many cases, it is a valuable unique asset for their livelihoods [21]. Hence, losing access to land has a direct impact on their livelihoods, causing a change of gender-based income and labor division in the family. In addition to the loss of access to agricultural land, urbanization has gradually merged the affected rural communities into urban areas [22]. One study revealed that urbanization has worsened the female dual-burden of paid and unpaid work, and that the sharing of unpaid work within the family has increased for males [18]. A second study via United Nation Women (UNW) concludes that urbanization does not yet advance gender equality [16]. Both papers do not consider gender equality in affected households (GEIAH). Another recent study shows that ALAFU has contributed to improving the socio-economic status of women in affected households, [23] but this is not enough evidence to conclude that GEIAH has generally improved. Unfortunately, the above-mentioned studies have not adequately considered GEIAH, therefore research on the benefits and constraints of ALAFU for GEIAH could significantly contribute to sustainable development in both urban and rural areas.

Over time, the social-economic transition has gradually improved gender equality in Vietnam's rural households. However, there are still inequalities between male and female members in families (especially between wife and husband) in terms of economic status, participation in household decision making and social activities, unpaid care work burden, as well as the access to social services [15,24,25]. These terms could be deeply affected by ALAFU.

Therefore, the objectives of this paper are to discover the changes in GEIAH through comparison of the above aspects between wives and husbands before and after ALAFU in Huong Thuy Town, Thua Thien Hue province, Vietnam, where ALAFU has been implemented over recent years [26]. Moreover, to understand whether such changes are associated with agricultural land acquisition (ALA) or urbanization, we compared two household groups in the affected areas. Group 1 consisted of 50 affected households whose agricultural land was acquired for urbanization, while group 2 consisted of 50 households affected through urbanization but with remaining agricultural land. We assumed that ALA could decrease the GEIAH, but then new opportunities associated with urbanization could also improve GEIAH.

This paper shows the gender impact of ALAFU that policymakers and planners, local authorities, women's unions, NGOs, and investors who acquired agricultural land for their business need to consider in planning, decision-making, and policies related to land management and gender development as well as rural and urban development. The structure of the paper is as follows: Section 2 introduces land acquisition policy and practice in Vietnam; Section 3 describes the conceptual framework of the study; Section 4 is about the research methods and study site; Section 5 contains the results; followed by the discussion and policy implications in Sections 6 and 7.

2. Agricultural Land Acquisition for Urbanization in Vietnam

According to the land law in 2013, land resources belong to the Vietnamese people and the Vietnamese state as the representative owner. The state issues land use rights for people through land use certificates. People can sell, transfer, inherit, and offer their land use right to others, but they have to return their land use rights when the State requests support for the social-economic development of the country. So ALAFU is a process whereby the State withdraws the agricultural land use rights from farmers in order to extend the urban area, develop infrastructure and industrial zones. The State compensates the loss of land use rights through financial and other support measures based on regulations of the land law [27]. Thus, ALAFU consists of three main components: the ALA (1), compensation for losing farmland use rights and support to recover livelihoods (2), and urbanization (3). The first is the ALA component, which states that farmers have to return their agricultural land rights to the State. In the land use plans for each province, the Department of Natural Resources and Environment (DONRE) and the Land Fund Development Center consider and propose ALA plans based on the social-economic development strategy of the provinces and then submit the plans to the Ministry of Natural Resources and Environment (MONRE). Farmers do not have a voice in this step and they only receive ALA announcements from local authorities before ALA implementation. The second is the compensation component, which consists of compensation for losing land use rights and support in allocating alternative jobs or recovering livelihoods. In the Land Law, the State regulates compensations for the affected farmers, either financially or through provision of new land use rights in new areas depending on the land funds of each affected area. If by cash, the State decides on the compensation for each type of acquired land without the participation of affected farmers, and then totals the compensation based on the area of acquired land. If by land use right, the land area for which the affected people receive compensation must be equal or bigger than the acquired land area. The support includes vocational training courses and rice for 12 months (12 kg of rice per month per person). The support can also be converted into cash depending on the decision of the affected people. The last component is urbanization, which leads to the growth of cities due to industrialization and modernization. Urbanization in Vietnam always goes along with infrastructure and industrial development, creating non-farming employment, developing technology, and social equality [28].

In practice, to support the rapid urbanization since the 2000s, the State has acquired over 10 million hectares of agricultural land, most of which with fertile soil [29–32]. ALA projects have selected monetary compensation methods, and many studies have indicated that the compensative land price is inadequate compared to the value of the acquired land, and that compensation is often delayed for a long time [30,32,33]. Nevertheless, the cash amount is still significant enough to recover livelihoods, and is used for many purposes such as repairing or building houses, buying new equipment (e.g., motorbikes, washing machines, fridges, kitchen equipment, mobile phones), investing in children's education, investing in a new job, paying for health care, and adding to savings in the bank [7]. The vocational training courses offered do not coincide with the labor market demand and the affected people, and have not improved working capacities, especially of the middle-aged population group who are facing unemployment or must accept temporary jobs [34]. Considering urbanization, the Ministry of Construction (MOC) states that there was a total of 833 urban areas across the country with a current urbanization percentage of around 38.5% in 2018. The figures are expected to increase to 50% by 2025. Urbanization has created non-farming jobs for millions of rural laborers, contributing to socio-economic development and the infrastructure in both rural and urban areas [2,13]. However, the rate of urbanization causes many problems such as inadequate development between social infrastructure and technology, and it increases the poor–rich disparity [13] which could restrict the livelihoods of affected households, especially of women through their low adaptive capacity.

In the 1993 land law, all rural people were issued an agricultural land use right with an equal land area based on the land fund in each commune. From 1993 until the present, the Government has not re-allocated the agricultural land use rights of farmers, meaning that all rural people who were born after 1993 were not issued agricultural land use rights, and, in the case of people who died after 1993, their land belongs to their family after their death. Moreover, education and industry have been strongly promoted across the country since the 2000s, and, as the income from agriculture is low, many rural young people born after the 1980s have tried to get higher education to escape agricultural employment, and no longer depend on agricultural land. These people offer or sell their agricultural land use rights to others who still practice agriculture [23].

3. Conceptual Framework

Nowadays, gender researchers and development projects can apply different approaches to understand gender equality including Women in Development (WID), Women and Development

(WAD), and Gender and Development (GAD). Of these, the Gender and Development approach of 1980 does not only focus on women like WID and WAD, but on both women and men through assessment of their relations and interactions with society and family. This approach defines specific characteristics of women and men to help understand the similarities and differences in respect to economic development and how they respond and adapt to changes in socio-economic conditions, and it helps to identify the roles and responsibilities of women and men [35]. Therefore, the approach taken in this study elucidates the changes of GEIAH.

FAO has defined gender as: "Gender is the relations between men and women, both perceptual and material. Gender is not determined biologically as a result of the sexual characteristics of either women or men, but is seen socially. It is a central organizing principle of societies, and often governs the processes of production and reproduction, consumption and distribution" [36]. According to Frank Ellis (2000), gender is the socially determined division of roles, responsibilities, and power between men and women. These socially constructed roles are usually unequal in terms of power and decision making, control over assets and events, and freedom of action and ownership of resources, among other things [21]. Gender equality is when men and women have the same rights and opportunities in all sectors of society and their different behaviors, needs, and aspirations are equally valued [37]. Based on these definitions, gender equality on a household level refers to equal responsibilities, roles, and power division in all household activities.

In practice, gender equality in Vietnam's rural areas has been considered by the socio-economic national strategy and rural development programs in the past decades. However, the position of women is still low. In rural areas, 65% of women work in agriculture and 45% work as self-employed without social security benefits and access to insurance. They still take strong responsibilities in the unpaid care work [38]. The participation of women in decision making has improved but is still limited and only a few rural women can access vocational training courses, extension services, finance, technology, markets and trader networks, as reflected by 70.9% of rural women laborers being unable to access vocational training [15,24,25,38]. Many aspects still need to be improved, including the economic position, the participation in household decision-making and social activities, the unpaid care work burden, and the access to the social services of women. The gender preconceptions such as "men in public life, women stay at home", "men build the house, women make the home" are still deeply rooted in rural households [15,24,25].

At present, most of the heads of farming households (husband or wife) in rural areas in Vietnam are from the middle-aged group and their children, who have not been issued agricultural land use rights, do not want to go into agriculture. In these households, the wife often stays at home, does agricultural work, and takes care of the family. The husband often takes a non-farming job or does both farm and non-farming jobs to secure the family's income. [39]. Therefore, ALAFU could strongly impact husband and wife's roles in farming households and cause a restriction in agriculture [31,32,39] in which women participate more than men [5,38]. ALA has also created non-farming opportunities through urbanization with the result that the affected people can abstain from traditional cultivation work and take other employment opportunities with higher incomes in the non-farming sector if their working capacity is good, which lead to improved status [23,40]. On the other hand, it can also increase unemployment and increase the dependency of women on men [41]. Additionally, ALA increases the family food supply burden that is often the responsibility of women [12,38]. Furthermore, ALA contributes to rural-urban migration [29,35], leading to the feminization of agriculture [38] and increasing the responsibilities of women in unpaid care work such as caring for children and elders, washing, cleaning, and cooking. Compensation and support measures could help the affected people overcome and adapt to the new situation after ALA.

In this study, we have selected the aspects that we assume might clearly be impacted by ALAFU, consisting of economic status, participation in decision-making and social activities, the burden of unpaid care work, and access to the social services for both wife and husband in the two affected household groups before and after ALAFU. Their economic status is measured by the type of employment, income, and working days; the participation in decision making and social activities is measured by the participation percentage of wife and husband; the burden of unpaid care work is measured by the time allocation and the division of responsibility between wife and husband; and access to social services is measured by the accessibility to vocational training courses, credit services, and health care.

Based on these definitions and the situation in Vietnam, we have built this research framework, which shows the relations between ALAFU and GEIAH (Figure 1).

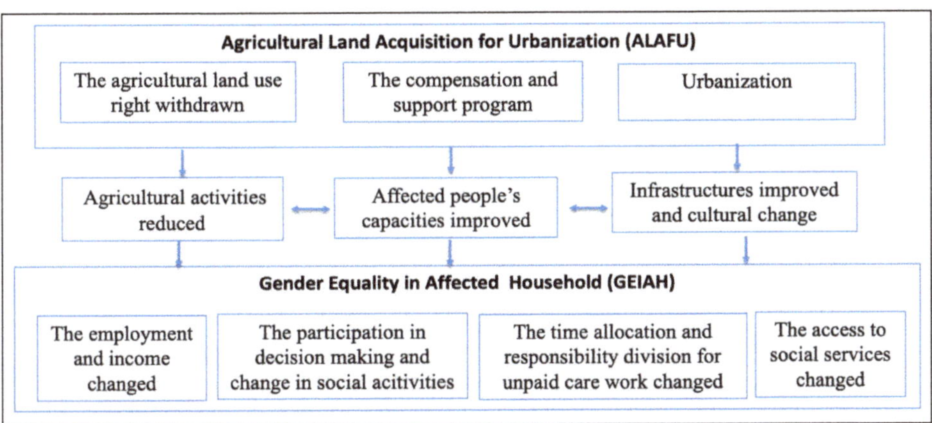

Figure 1. The linkage between Agriculturad Land Acquisition for Urbanization and Gender Equality in Affected Household.

4. Study sites and Methods

4.1. Study Sites

Thua Thien Hue is a province of central Vietnam. According to the Department of Natural Resources and Environment (DoNRE) of Thua Thien Hue province, rice cultivation plays the main role in agriculture, but it has to face difficulties because most agricultural land areas have bad soil quality, and weather conditions are becoming more and more severe [42]. According to "Resolution No 72/NQ-CP of the Government of Vietnam", DoNRE of the province has converted 7083 ha of agricultural land to non-agricultural land during the period 2010–2015. In 2014, Government plans were that this province should become a centrally controlled city in the future, and they approved the agricultural land conversion of 19,000 ha during the period 2016–2020 to extend Hue to be five times larger in the near future [26,43]. In 2019, the rate of urbanization of the province reached 52.7% and could reach 60–65% in 2020. The average income per person in the province was 3,136,000 Vnd/month in 2018 [44].

Huong Thuy Town, adjoining the south of Hue city, has traffic and transport advantages including road, railway, and airway to connect with large cities such as Da Nang, Ha Noi, and Sai Gon. The total land area is 45,466 ha, in which agricultural land occupies 80.9%, non-agricultural land occupies 18.3%, and unused land makes up 0.8%. According to reports from the Huong Thuy Town People Committee (HTTPC), 3527.8 ha of the agricultural land area was converted to non-agricultural land during 2005–2018, occupying 6.3% of the total agricultural land area of the town [31]. Since 2010, Huong Thuy became one of three satellite cities of Hue city and undertakes the main functions of the industrial and tourist development of the province. According to decisions No 368/QĐ-UBND and No 123/QĐ-UBND of the province, Huong Thuy Town has converted 2000 ha of agricultural land in order to speed urbanization in the period 2019–2020. The average income per person of the province was 4,333,000 Vnd/month in 2018 and the rate of urbanization reached almost 70% in 2019 [44].

To suit the research objectives, we selected two communes consisting of Thuy Van and Thuy Thanh in Huong Thuy Town to be our study sites. These communes, located closely to Hue city and with more than 200 ha of previously used agricultural land, have been converted to non-agricultural land such as road and residential land. On average, each household lost about 1500 m^2 of agricultural land, so that only 400 m^2 remains per household [33]. In the near future, these communes will be merged with Hue city. Most households whose land was acquired were not entirely dependent on agriculture before ALA. The Figure 2 shows mapping of study sites.

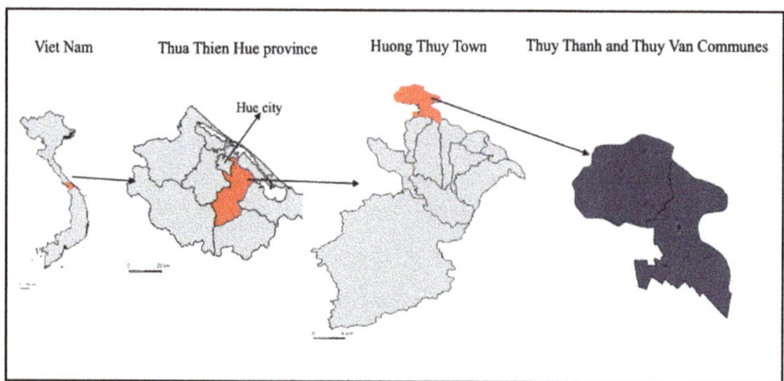

Figure 2. Mapping of study sites. (Source: Global Administrative Areas (GADM), 2018).

4.2. Methods

Important information for this study does not only build on statistical data but also on the thoughts, feelings, perspectives, stories, explanations, and expectations of interviewees, local people, or information suppliers, in order to better understand the progress of change in respect to GEIAH. Therefore, it applies a mixed methods approach combining quantitative and qualitative procedures. Quantitative data are collected through household surveys, whereas group discussions, key informant interviews, in-depth interviews, and observations provide the qualitative data.

To get both qualitative and quantitative data, we used the rapid rural appraisal tools (RRA). RRA supports sufficient communication and interaction with communities and provides tools to collect diversified data [45]. In this study, we used RRA tools consisting of semi-structured questionnaires, checklists, village walking, and observations to collect primary data during two time periods (before ALA in 2012 and the surveyed time in 2017).

To meet the research objectives, two groups from the Commune People Committee of Thuy Thanh and Thuy Van were targeted: A list of households whose acquired agricultural land exceeded 50% of their former agricultural land area (group 1), and a list of households whose agricultural land is still entirely owned (group 2). Then, we selected a new list of 50 households in group 1 and 50 households in group 2. As described in the conceptual framework, all 100 surveyed households were couple families consisting of husband and wife in order to reveal gender role changes between the couples. Lively group discussions during household surveys and in completing the study produced various views, thoughts, and stories, providing diversified information for ALAFU.

To collect primary data, 100 households were surveyed using a semi-structured questionnaire. This survey collected statistical data in respect to employment, income, working days, participation in household decision-making, unpaid care work division, and the access to social services of both wife and husband before ALAFU (2012) and after (2017), as well as reasons for the change in these terms. All the data from the household surveys were analyzed statistically, giving the average value of each index in percentage. To make sure we met both husband and wife, the heads of villages helped us to contact and make appointments with households from the list. In a second step, we held discussions with four groups: a wives' group, a husband's group, a mixed wives and husbands version of group 1, and a mixed wives and husbands version of group (2), with each group containing eight to ten people. In these group discussions, we showed the statistical data from the household surveys to receive their confirmation or feedback, then asked participants to list reasons for the change for each gender equality dimension to rank the importance level of each reason by using a couple comparison method.

Using this ranking scale helped us get more information from explanation, perception, and the feelings of the affected people whose voice is important for this study. Based on the result of ranking, we know which reasons are important and how these reasons related to ALAFU. Thirdly, we conducted in-depth interviews with six local key informants, including the head of a women's union and head of the people's committee at two communes. Most open questions focused on their views about GEIAH and its change over time, the change of cultural issues and social norms in respect to gender issues, the benefits and constraints from ALAFU for improving the GEIAH, and their ideas about improving the GEIAH. Information from these interviews offered us alternative views about ALAFU to cross-check data, and information from the affected households, to avoid biased conclusions. Fourthly, we visited the villages multiple times, and attended village meetings and social events on the study site to observe the participation of men and women in order to better understand the situation of the GEIAH.

During primary data collection, we also collected available secondary data from reports of the Commune People's Committee, the Women's Union, and the Town People's Committee. We also analyzed a broad variety of other sources (e.g., scientific papers and daily magazines) related to our research topic.

5. Results

5.1. Characteristics of the Surveyed Households

As explained in Section 3, we selected 100 surveyed households based on the household lists supplied by the local authority. Some characteristics of the surveyed households are shown in Table 1. To see the different impacts of ALAFU between both groups, we compared characteristics of two surveyed household groups before ALAFU. Numbers are similar for the two surveyed household groups in respect to the age of household leader, household income, total agricultural land area of household, as well as income contribution from agricultural activities to the family's total income. The data also show that both household groups did not depend totally on agriculture before ALAFU. The income contribution from agriculture is just around 45% of the total household income. This could reduce the livelihood shock for them after ALAFU. However, the average age of the household leader in both groups is around 48 years, limiting flexibility to adapt to major career changes or to improve the household gender equality (HGE).

Table 1. Characteristics of the surveyed households.

Characteristics	Unit	Group 1 (N = 50)	Group 2 (N = 50)
The average age of household leader before ALAFU	Year old	48.8	47.2
Income/person/year before ALAFU *	Million Vnd	19.7	19.3
Total agricultural land area of household before ALAFU	m^2	2050	2100
Rate of income from agriculture of household before ALAFU	%	45.6	44.5

* Note: 1 USD ≈ 20,000 Vnd. (Source: Household survey, 2018–2019).

5.2. Changing Employment and Income

One of the most important factors that influence the HGE is the employment and income of wife and husband. Improving the income of women has been highlighted in many development initiatives to achieve gender equality [46]. Basically, the rural women's economic position is lower than men because most of them participate in agricultural activities with low income. As a result, their power and position in the family are often lower than their husbands [47]. The question of ALAFU is how the employment and income of wife and husband in the affected households change and whether this change could improve GEIAH. We have listed our results in Table 2.

The change in employment and income between the wives and husbands in each group are not similar. In both groups, the income activities of the wives changed more clearly than the activities of the husband. Many wives changed from being farmers to being small businesswomen and hired laborers after ALAFU. In group 1, the percentage of wife farmers decreased from 36% to 12%, while the percentage of the wives being small businesswomen and hired laborers increased from 10% to 32% and from 12% to 20%, respectively. In group 2, the percentage of wife farmers decreased from 36% to 10%, most of them having changed to hired laborers or holding dual jobs (farmer and hired laborer). However, the income activities of the husband in both groups insignificantly changed. The percentage of the husbands in group 1 who are farmers slightly decreased from 30% to 22%, but dual job holders (both farmer and hired laborer) increased from 18% to 24%. It is interesting that while the employment of the wife in group 2 changed a lot, this did not happen with the husband's employment, even though their main job also changed. For example, in group 2, 36% of husbands are farmers, but 77.8% of them changed from cultivating rice and breeding to floriculture, which developed into a high-income activity since 2010. Another example is that 32% of husbands are hired laborers but 37.5% of them have worked around their commune instead of going far away from home (as before ALAFU).

Table 2. Changes in employment and income of wife and husband (Percentage of household).

| Employment Type | Groups 1 (N = 50) | | | | Group 2 (N = 50) | | | |
| | before ALAFU (2012) | | after ALAFU (2017) | | before ALAFU (2012) | | after ALAFU (2017) | |
	Wife	Husband	Wife	Husband	Wife	Husband	Wife	Husband
Civil servant	2	8	2	8	0	2	0	2
Small businessman	10	6	32	6	12	4	12	4
Worker	16	8	12	8	16	8	20	8
Hired laborer	12	30	20	32	10	32	20	32
Farmer	36	30	12	22	36	36	10	36
Dual jobs (both farmer and small businessman)	12	0	10	0	14	0	14	0
Dual jobs (both farmer and hired laborer)	12	18	12	24	12	18	22	18
The average number of working day/year (day)	226	270	270	290	220	260	264	300
The average income/day (1000 Vnd) *	101	181	170	220	100	183	170	220
Contribution to total income of household (Percentage)	28.9	62	41.5	49	27.8	60.1	34.5	50.7

* Note: 1 USD ≈ 20,000 Vnd, 22,000 Vnd before ALAFU and 2017, respectively (Source: statistical data from household survey, 2018–2019).

Besides the change in employment, the average number of working days and the income per day of both husband and wife in both groups also increased after ALAFU. Of these, the working days per year of the wives in group 1 and group 2 rose from 226.2 to 315.7 days and from 220 to 310 days, respectively. Their income per day rose around 86% in group 1 and 70% in group 2 compared to before ALAFU. As a result, their contribution to the total income of the household increased from 28.9% to 41.5% in group 1 and from 27.8% to 34.5% in group 2. For the husbands, there was not much change in their employment, but the average number of working days and income per day also increased. However, their income contribution declined to around 14% in group 1 and 9.4% in group 2 compared to before ALAFU. The data also show that the difference in income between wife and husband still exists, but significantly narrowed after ALAFU. This reveals that the economic status of the wives in both groups improved significantly compared to that of their husbands, especially for the wives in group 1. We investigated the reasons for the changes, which is reflected in Table 3.

Table 3. Reasons for changing employment and income (Percentage of household).

Reasons Lead to Changing Economic Status		Groups 1 (N = 50)			Group 2 (N = 50)		
		Wife	Husband	Important Ranking *	Wife	Husband	Important Ranking *
Changing employment	Losing agricultural land (ALA)	50	20	2	30	10	2
	Getting cash compensation (ALA)	30	40	3	80	80	1
	Improving infrastructure and access information (urbanization)	40	30	4	56	52	3
	Increasing non-farming job opportunities (urbanization)	60	60	1	20	20	3
	Others	20	10	4	16	14	4
Changing income per day	Losing agricultural land (ALA)	20	10	3	0	0	
	Increasing non-farming job opportunities	80	70	1	80	80	2
	Others	30	20	2	20	20	1
Changing working day per year	Losing agricultural land (ALA)	20	20	2	0	0	
	Improving infrastructure access information (urbanization)	80	80	1	80	80	1
	Increasing non-farming job opportunities (urbanization)	30	30	3	20	20	2
	Others	20	10	4	20	18	3

* Note: Result from group discussion. (Source: Household survey and group discussion, 2018–2019).

We divided the reasons for change into three groups: (1) a reason associated with ALA (losing agricultural land, compensation, supporting vocational training courses); (2) a reason associated with urbanization (good infrastructure, new companies/factories opening, rise in number of residents, increasing access information, and modern culture); (3) other reasons (e.g., age, health, working skills). The interviewee could give more than one answer if their change was motivated by multiple reasons. The surveyed data in Table 3 show that the main reasons for both husband and wife changing their employment in both group 1 and group 2 are urbanization and ALA, of which urbanization is the most common and important. In the group discussion, the wives in group 1 stated, *"We have more advantages than the wives in group 2 because we are not concerned about agricultural land anymore, and we have received the cash compensation for losing the land. Therefore, we could get alternative jobs more easily than them"*. It is surprising that 30% of the wives and 10% of the husbands in group 2 also answer that the ALA has caused their employment change. In addition, 70% of group 2 have the same explanation for their employment change such as, *"Although we still have our agricultural land, our neighbors have changed their jobs because of ALA. Their new jobs seem to be not only better than agricultural work but also suitable to our working capacity. The benefit from agricultural land contributes insignificantly to our family's income, moreover, we believe that our agricultural land might be acquired in the near future as well. Therefore, ALA is one of the reasons for our employment change"*. Most surveyed households agreed with the statement from one key person in their communes, *"Urbanization has created many kinds of jobs that are not only suitable to our capacity and near our villages and homes, but also provide a good income. Besides that, an improving infrastructure offers us more conveniences to do with work."*

5.3. Changing Allocation of Time and Responsibility Division for Unpaid Care Work

Unpaid care work (UCW) mainly includes caring for family members (especially children, the sick, and elderly), buying and making food, washing, and cleaning [48]. Such work is very important to maintain family life but is not counted in the paid work system. Unfortunately, social preconceptions often reckon that women can do this work better than men, leading to gender inequality and hindering job mobility for women [48]. In this study, we assume that ALAFU could improve this issue. To understand the situation, we investigated the allocation of time and the division of responsibility for UCW between husband and wife for both groups through household surveys and group discussions. The results show the difference in the allocation of time between wives and husbands, and the change of time allocation for each of them after ALAFU. The total time allocated to UCW for both wives and husbands in both groups was shorter than before ALAFU. The total average of reduced time for the wife and husband is 2 h and 0.6 h, respectively. The wife still has to spend over five hours per day for UCW, while the husband spends just over one hour. This difference reveals that inequality still exists after ALAFU.

In terms of time allocation, the division of responsibility for husband and wife is shown in Tables 4 and 5. It has improved a little but the wives in almost all the surveyed households in both groups still take the main responsibilities of UCW. Most of them have to take the main daily repeated UCW including buying food and clothes, preparing food, cleaning, washing, and caring for children, the sick, and the elderly. Meanwhile, most of the husbands take responsibility for building and repairing the house, which is not often required.

Table 4. Changing allocation of time for unpaid care work (Hours/day).

Allocation of Time	Group 1 (N = 50)				Group 2 (N = 50)			
	before ALAFU (2012)		after ALAFU (2017)		before ALAFU (2012)		after ALAFU (2017)	
	Wife	Husband	Wife	Husband	Wife	Husband	Wife	Husband
Preparing food	2.5	0	2.2	0.1	2.6	0	2.3	0.2
Cleaning and washing	2.5	0.7	1.8	0.5	2.5	0.7	2	0.5
Caring for family members	2.2	1.4	1.2	1	2.2	1.3	1	1
Total	7.2	2.1	5.2	1.6	7.3	2	5.3	1.5

(Source: Household survey, 2018–2019).

Table 5. Changing responsibility division in unpaid care work (Percentage of household).

Who Take the Main Responsibility in	Group 1 (N = 50)						Group 2 (N = 50)					
	before ALAFU (2012)			after ALAFU (2017)			before ALAFU (2012)			after ALAFU (2017)		
	W	H	B	W	H	B	W	H	B	W	H	B
Buying food and clothes	100	0	0	100	0	0	100	0	0	100	0	0
Preparing food	100	0	0	90	0	10	100	0	0	70	0	30
Cleaning and washing	100	0	0	96	0	4	100	0	0	80	0	20
Caring for children, sick and the elderly	68	0	32	50	4	46	70	10	20	50	20	30
Repairing the house and household equipment	0	80	20	0	70	30	0	86	14	0	50	50

Note: W = wife, H = husband, B = both wife and husband. (Source: Household survey, 2018–2019).

As with the changes in other areas, we investigated the reasons for changes in the allocation of time and responsibilities of UCW. The results from the two group discussions are similar (Table 6). A reduction of the women's time for UCW associated with improved household equipment and infrastructure, support from children, and changing employment can be seen. Of these, improving household equipment and infrastructure are ranked the most important. This is also confirmed in discussions with the husbands' group. One woman explained in our in-depth interview *"We have used a part of our cash compensation or income to buy new household equipment such as a washing machine, fridge and other things. They save time for us. Buying food and shopping is more convenient and takes less time because many food stores, street markets and supermarkets have appeared around our house. Moreover, our children have grown up we don't have to take care of them much, they can even share the housework."*

Table 6. The reasons for changing allocation of time and dividing responsibilities in unpaid care work.

The Change of Participation in	Reasons for Change	Important Ranking
Changing allocation of time	Changing employment	3
	Improving household equipment and infrastructure	1
	Support from children	2
Changing responsibility division	Changing employment	1
	Changing attitude	2

(Source: Group discussions, 2018).

Considering the change in dividing responsibility occurring in a few surveyed households, participants said the wife is absent all day so the husband has to do some of the work. In discussions with the husband group, a male participant said, *"In a few households, due to the change in the wife's employment, she is absent all day because of a new small business. She tries to do almost every UCW in the early morning or late afternoon. Therefore, the husband does some of the simple work that she does not have time for, such as making simple food for lunch. Most of us have changed our minds and are more open, but honestly, we cannot do these jobs as well as they do. So the women still undertake almost every UCW in a family."* And a female participant said, *"Although we have spent more time and labor-power for new paid work after ALAFU, all our family members have got used to the previous division of responsibility and don't want to make chaos. So, we women try to undertake the UCW as usual"*.

5.4. Changing Participation in Household Decision Making and Social Activities

The participation in household decision making or social activities of women and men is one of the dimensions that reflect the HGE issue. It normally relates to the economic position, culture, and ideology [46]. One study has shown that women's participation in household decision-making (WPHDM) positively relates to their income and occupation [49]. In this study, we assume that ALAFU leads to rising income and social change, and both of them promote the WPHDM and social activities as well. The change in the participation of both wives and husbands in the two groups is shown in Table 7.

The data shows significant improvement in the WPHDM in both groups after ALAFU and their limited participation before ALAFU. The men's participation in decision-making related to children and the elderly also increased. At present, around 50% to 80% of the surveyed households have made decisions through discussion and the agreement of both husband and wife. In other surveyed households, the decisions were mainly made by the husband. Such improvement is almost equal between the two groups. In general, the participation of the wives is still lower than that of the husbands. In particular, the women's participation strongly increased in decision-making relating to family livelihoods and family finances. The decisions relating to housing and property are still often made by the men, although the women's participation in this also increased slightly compared to before ALAFU. As for participation in social activities, such as commune/village meetings, husband and wife's family events, neighbors, and relatives (e.g., weddings, funerals, anniversaries), the women's participation improved a little, but the men often remain the family representative here. This situation has similarly occurred in both groups. In addition, while the data were being collected, we casually visited two events at the study sites (a social event and a village meeting) and noted that over 80% of the participants were men.

Table 7. Changing the participation in decision-making and social activities (Percentage of household).

| Who Have Participated in | Group 1 (N = 50) | | | | | | Group 2 (N = 50) | | | | | |
| | before ALAFU (2012) | | | after ALAFU (2017) | | | before ALAFU (2012) | | | after ALAFU (2017) | | |
	W *	H *	B *	W *	H *	B *	W *	H *	B *	W *	H *	B *
Decision-Making relating to employment	4	60	36	4	34	62	6	54	40	4	34	62
Decision-Making related to caring for children and the elderly	52	0	48	20	0	80	50	0	50	38	2	60
Decision-Making related to family finance	0	70	30	0	28	72	0	50	50	0	28	72
Decision-Making related to housing and properties	0	80	20	0	56	44	0	76	34	0	54	46
Hamlet or village meetings	8	92	0	20	80	0	10	90	0	20	76	4
Social activities in the commune	6	84	10	20	56	24	8	80	10	24	52	24
Family events	0	60	40	8	40	52	0	92	8	8	82	10

* Note: W = wife, H = husband, B = both wife and husband. (Source: Household survey, 2018–2019).

We showed the above results in the group discussions to receive their confirmation and investigate the reasons for the changes. We asked people in group discussions to give reasons and then rank them according to importance. For the wives, the information in Table 8 showed three reasons for change. Those are changing employment, improved knowledge and attitude, and encouragement from their husbands. Of these, changing employment was the most important. They stated, *"Most of us have changed our job due to ALAFU and we now work independently with our husbands, therefore, we have to make decisions connected with our jobs through discussions with our husbands to get the best options. Moreover, our income and social knowledge are also making us more confident in contributing ideas that our husbands could use to make the final decision. The participation is both our right and the way we share responsibilities with our husbands. We don't want to make decisions alone."* The husband's group explanation reads, *"Urbanization has changed our lives—not only our income activities but also our attitudes to HGE. We have discussed issues with our wives to reach agreement. Making household decisions together is the way we respect each other and share responsibility as well. Our family is happier since most decisions are made by both of us".* However, in terms of the women's participation in social activities, participants explained that the social prejudice that "men go in public, women stay at home" still exists despite social changes due to urbanization. Both husband and wife argue that participation in such social activities are responsibilities rather than interests, so the husband should take them on. The wives only participate in these events when their husbands are busy or the event requests the participation of both husband and wife.

Table 8. Reasons for the change in participation in household decision-making and social activities.

The Change of Participation in	Wife Group		Husband Group	
	Reason	Important Ranking	Reason	Important Ranking
Household decision-making	Changing employment	1	Changing employment	1
	Changing knowledge/attitude	2	Changing attitude	2
	Encouragement from husband	3		
Social activities	The husband is busy	1	Changing employment	1
	Changing attitude	2	The wife's participation	2

(Source: Group discussions, 2018).

5.5. Changing Access to Social Services

Access to social services such as financial credit, training courses (including vocational and educational training), and the health care service can improve working capacity and health for both women and men, leading to positive changes in their attitudes, their working capacity, and their health. At present, the opportunities to access these services are increasing. According to our household survey, 100% of interviewees answered that they can access social services more easily now than before ALAFU. However, just 50% of wives and 30% of husbands have access to or have participated in these services. Others do not want to access them because they have good economic conditions and health, or they still hesitate. Of these, most of those accessing the financial service to get a small loan from banks are wives in group 2 and most of the participants in money-saving groups are wives in group 1. In terms of health care service, the access of wives in both group 1 and group 2 increased similarly. For both these services, women have taken part more than men.

The leader of a women's union said *"Most women can take a small loan from the Vietnam Bank For Social Policies Or Vietnam Bank for Agriculture and Rural Development with low interest because of the national policy for rural women since the 2000s. However, the percentage of women who could pay a loan on time increased after ALAFU because of their good income. Moreover, having a good income, many women have joined the money-saving group. Therefore, they can save money for themselves and use it in urgent cases. ALAFU is one of the most important reasons leading to this change. The percentage of women who have participated in the periodical health check is higher. In general, the accessibility to social services for people (women included) is better."* However, our household survey shows that neither wives nor husbands have participated in vocational training courses after ALAFU. A leader of a farmer's union said. *"At present, the vocational training courses are always available and priority is given to households affected by ALAFU. However, most of the husbands and wives in the farming households are middle aged and they prefer to base their work on their experience and learning from others in communes rather than on learning from training courses. Moreover, vocational training courses seem not to suit their needs. Therefore, most of them have not participated in any vocational training courses after ALAFU, although many of them (farmers) have changed employment. The households affected by from losing land have accepted cash instead of participating in training courses supported by the ALAFU project. Yearly, only young people have participated in the vocational training courses."*

6. Discussion

Previous studies have indicated that losing agricultural land has forced affected women to participate in non-farming jobs and has improved their economic position [23,50]. This study also supports the evidence that decreasing access to agricultural land due to urbanization has improved women's economic position in the affected households. Indeed, the factors of the ALAFU including the ALA, the cash compensation for acquired land, the improved infrastructure, and the high demand for non-farming jobs due to urbanization have incited women to move from the agricultural sector to the non-farming sector, from cultivation to breeding, from single jobs to dual jobs. As a result, their income and economic position in the family has improved significantly. It means that ALAFU has offered opportunities instead of difficulties in improving the affected women's economic position. This positive impact of ALAFU coincides with progress in socio-economic development in many rural areas in Vietnam. Since the 2000s, many rural areas have become peri-urban, and non-farming jobs have been attracting rural people, while the income from cultivating agricultural land is very low [23,51]. This has led to a male laborers' migration from rural to urban areas, and the feminization of agriculture [38]. Then, only a small group of rural women convinced their families to move on from the non-farming sector. These households allowed other people to cultivate their agricultural land for free, or even left the agricultural land uncultivated to focus totally on non-farming jobs [52]. One study has concluded that access to agricultural land is one of the causes that has limited the non-farming job opportunities of most rural women [23]. This explains why, after ALAFU, the women's economic position improved in group 1 (whose agricultural land was acquired) and is better than group 2 (whose agricultural land was kept).

However, other previous studies have revealed that the rural people affected by ALA, especially the affected rural women, have to face unemployment, and depend more on their husbands or male members in the family after ALA [7,53,54]. Increasing access to agricultural land for women has been recognized as a way of narrowing the gender gap, and this view has been applied in many development projects [17]. It has been urgently promoted in many developing countries where the percentage of rural women participating in agriculture is still higher than men [41]. Although our research findings are different, they do not disclaim rural women's efforts to increase their access to the land. However, it supports the evidence that, in order to improve the rural women's economic position, changing from traditional cultivation on agricultural land to non-farming employment could be also one of the better options, but it should be a long-term strategy based on the actual rate of socio-economic development. The affected people (women included) in our study have some prerequisite conditions for successfully changing employment. The ALA has occurred near their communes since the 2000s. Meanwhile, the benefit from agricultural land is low (rice land) and their agricultural area is small, so they have had time to prepare and improve their capacity, and have gradually switched their income activities to the non-farming sector.

The evidence shows that before ALAFU, their family's income due to agricultural activities was just around 40% of the total family's income, and not only most men, but also some women started take non-farming work as a full-time or part-time job. Moreover, the focus of ALA in our study site is on the Hue city expansion, where the infrastructure in and around their communes has improved. Thus, the travelling distances from their communes to Hue city, other cities, and industrial zones are shorter and more convenient, making non-farming jobs available in or nearby their communes. So, the ALA seems to be not only the last step to encourage affected women to escape agriculture, but also increases their finances through the cash compensation for losing the land use right, although the compensation is still inadequate. Consequently, ALAFU has positively impacted women's economic position. However, in the cases of remote rural areas where urbanization and industrialization are still far away, the livelihoods of rural people (women included) are still based totally on agricultural land, and the benefit from agricultural land is high. Thus, increasing access to agricultural land for women still needs to be promoted as, if ALA occurs, it will bring financial hardship and unemployment.

This study reveals the interesting point that 10% of wives in both groups still continue to work in agriculture with their husbands, and their income has improved because most of them have changed from rice to flower cultivation. The new agricultural strategy, although requiring more skill, has succeeded because the women have carefully learned the cultivation technique from each other, and their output is meeting the high demand from urban residents. This reveals that increasing access to agricultural land for women should go together with improving the women's work skills and finding solutions to increase the benefits from the land. Otherwise, access to land is not very significant for enhancing household gender equality.

Improving the economic position and employment of women in the context of urbanization could improve other dimensions of household gender equality. A higher income has also positively affected the access to financial services by women because they are then able to repay a loan on time or even make small savings. This impact is very significant for the progress of GEIAH. Some studies have concluded that occupation and income positively affect WPHDM [49]. Moreover, traditional gender ideology, which could be counteracted by urbanization, also strongly controls the WPHDM [46]. This means that when both women's employment and gender ideology improve, the WPHDM also increases. Such impacts also show in our research. Although most of the interviewees believe that this change is related to ALAFU, it has also been identified in many studies as the indispensable result of developmental progress in Vietnam [55–57]. Moreover, though the income contribution of the wives in group 2 is lower than in group 1, the WPHDM improvement in both groups is similar. This indicates that women's income does not seem to be an important factor influencing the WPHDM, which was found in some previous studies [48,58]. In addition, although women have made household decisions together with their husbands, most men still make the final decisions on most household issues, especially those related to housing and valuable property. This situation is very common in Vietnam in both urban and rural areas [57].

The positive effect of ALAFU does not occur within other areas of GEIAH as our hypothesis shows. Following an improvement in income and the WPHDM, participation in social activities and the UPW burden of women remain almost unchanged. Women in both groups still take the main responsibility for the UPW and their time allocated to the UPW is still more than twice that of the men. This shows that the gender prejudice in the division of UPW has not yet been broken despite the modern lives resulting from urbanization, rising income and employment, and escaping agriculture. Women still continue to undertake the responsibility of supplying the family with food, caring for family members, cleaning, and washing. They still do not want or have an opportunity to participate in community activities which could make them more confident generally in their lives. The gender ideologies reflected in the sayings "men in public life, women at home", "men build the house, women make the home", and "men only do the big things, women should do the small things" are rooted deeply in rural people's thinking.

Many studies have also mentioned this as a challenge to GEIAH progress [48,57]. Although women are busier with their new paid work, they still take responsibility for the UPW as before ALAFU. Taking on both paid and unpaid work could put women in an overworked situation and lead to stress or vulnerability [58], even though their access to health care has increased. Unfortunately, ALAFU has led to such negative impacts on HGE, but the support policy for the affected households of ALAFU projects has failed to integrate any gender solutions so far. Moreover, increasing participation in vocational training courses for women, which could increase their work skills, did not occur as was expected through the support policy, which developed the ALA projects to support affected people (women included) through courses supporting employment recovery. Unfortunately, they have converted these courses to cash, and financially support affected people, and this way seems to be the best for both sides at present. However, because of lack of sufficient training, most of the new jobs with higher income that affected people have taken are temporary jobs. These jobs contain many potential risks as they do not come with a work contract or unemployment insurance [59]. Therefore, the change in women's employment may be better in terms of income but could not be sustainable unless their capacity is also improved through training courses.

7. Conclusions and Recommendation

This paper studies the impacts of ALAFU on GEIAH by comparing gender issues between women and men in group 1 and group 2 before and after ALAFU. Besides reducing access to agricultural land

and the existing unreasonable issues of compensation and support, ALAFU projects have improved the infrastructure, created non-farming job opportunities and have begun to change the gender ideology. In this context, women's economic position in affected communes has improved and their employment has switched to non-agricultural activities, which has created a higher income for them. As a result, their income contribution and their savings have increased. The WPHDM has also increased, but its quality is still limited. The improved income does not have much influence. Other dimensions such as the division of UCW, access to training courses, and participation in social activities have insignificantly improved. As a result, the ALAFU has contributed to the improvement of GEIAH, although women still have to face the potential risks of temporary jobs without security, and continue to undertake the UCW burden.

Based on these findings, this paper recommends the ALAFU be a long-term plan prepared carefully not only to further economic development but also to create more gender equality in the affected communes. Besides improving the cash compensation mechanism, as was recommended in previous studies, projects to improve the affected people's working skills before they change to the non-farming sector should be implemented through vocational training courses or career consultation. To attract the participation of affected people (women included), the topic, content, and method should be based on their needs, on consultations with employers and the labor market. The support policy of ALAFU projects should integrate gender issues by estimating the gender impacts based on the gender situation in each affected area, in order to offer training courses to change gender preconceptions or to help affected households understand more about gender impacts and share UCW with women. This study also recommends agricultural land usage conversion if benefits of agricultural land access are small and without significance to the livelihoods of rural people. This should be considered as a solution to improve GEIAH. The conversion needs to be carefully prepared, should be based on the real situation in each rural area, and could include the transition from traditional agriculture (rice cultivation) to modern agriculture (such as organic cultivation or high-tech agriculture), or from agricultural land to industrial or urban land.

Author Contributions: Conceptualization, N.P.T., M.K. and D.W.; methodology, N.P.T.; software, N.P.T.; validation, N.P.T.; formal analysis, N.P.T.; investigation, N.P.T.; resources, M.K. and D.W.; data curation, N.P.T.; writing—original draft preparation, N.P.T.; writing—review and editing, N.P.T., M.K. and D.W.; visualization, M.K. and D.W.; supervision, M.K. and D.W.; project administration, M.K. and D.W.; funding acquisition, M.K. and D.W. All authors have read and agreed to the published version of the manuscript.

Funding: This research was funded by Ministry of education and Training of Vietnam and DAAD scholarship via providing the PhD scholarship for the first author from 2017–2020 at the Institute of Geography, Cartography, GIS and Remote Sensing Department, University of Goettingen, Germany.

Conflicts of Interest: The authors declare no conflict of interest.

References

1. Schneider, A. Monitoring land cover change in urban and peri-urban areas using dense time stacks of Landsat satellite data and a data mining approach. *Remote Sens. Environ.* **2012**, *124*, 689–704. [CrossRef]
2. Kelly, P.F. Everyday Urbanization: The Social Dynamics of Development in Manila's Extended Metropolitan Region. *Int. J. Urban Reg. Res.* **1999**, *23*, 283–303. [CrossRef]
3. Kontgis, C.; Schneider, A.; Fox, J.; Saksena, S.; Spencer, J.H.; Castrence, M. Monitoring peri-urbanization in the greater Ho Chi Minh City metropolitan area. *Appl. Geogr.* **2014**, *53*, 377–388. [CrossRef]
4. Azadi, H.; Ho, P.; Hasfiati, L. Agricultural land conversion drivers: A comparison between less developed, developing and developed countries. *Land Degrad. Dev.* **2011**, *22*, 596–604. [CrossRef]
5. ILO. *Research Report on Rural Labour and Employment in Viet Nam.*; International Labour Organization: Hanoi, Vietnam, 2011; ISBN 978-92-2-125715-8.
6. Thinh, H.B. *Rural Employment and Life: Challenges to Gender Roles in Vietnam's Agriculture at Present*; Pathways out Poverty: Washington, DC, USA, 2009.

7. Nguyen, T.H.T.; Tran, V.T.; Bui, Q.T.; Man, Q.H.; de Vries Walter, T. Socio-economic effects of agricultural land conversion for urban development: Case study of Hanoi, Vietnam. *Land Use Policy* **2016**, *54*, 583–592. [CrossRef]

8. United Nations Human Settlements Programme. *State of the World's Cities 2010/2011: Cities for All: Bridging the Urban Divide*; UN-HABITAT: Nairobi, Kenya, 2010.

9. Bah, M.; Cissé, S.; Diyamett, B.; Diallo, G.; Lerise, F.; Okali, D.; Okpara, E.; Olawoye, J.; Tacoli, C. Changing rural–urban linkages in Mali, Nigeria and Tanzania. *Environ. Urban.* **2003**, *15*, 13–24. [CrossRef]

10. Gossop, C. Low carbon cities: An introduction to the special issue. *Cities* **2011**, *28*, 495–497. [CrossRef]

11. Mandere, N.; Barry, N.; Stefan, A. Peri-urban development, livelihood change and household income: A case study of peri-urban Nyahururu, Kenya. *J. Agric. Ext. Rural Dev.* **2010**, *2*, 73–83.

12. Thanh, H.; Anh, T.; Quang, L.; Giang, D.; Phuong, D. International Institute for Environment and Development. 2013. Available online: www.jstor.org/stable/resrep01291 (accessed on 24 March 2020).

13. MOC. *Báo cáo đánh giá quá trình đô thị hóa ở Việt Nam giai đoạn 2011–2020 và mục tiêu, nhiệm vụ của giai đoạn 2021–2030, kế hoạch 5 năm 2021–2025 (Assessment of Urbanization Process in Vietnam Peroid 2011–2020, Target and Mission of Peroid 2021–2030 and Plan for Period 2021–2025, 2021–2025)*; Ministry of Construction: Ha Noi, Vietnam, 2019.

14. Sati, V.P.; Deng, W.; Lu, Y.; Zhang, S.; Wan, J.; Song, X. Urbanization and Its Impact on Rural Livelihoods: A Study of Xichang City Administration, Sichuan Province, China. *Chin. J. Urban Environ. Stud.* **2017**, *5*, 1750028. [CrossRef]

15. UN. Gender Equality and Sustainable Urbanisation 2010. Available online: https://www.un.org/womenwatch/feature/urban/factsheet.html (accessed on 20 January 2020).

16. UN Women. Gender Equality and the New Urban Agenda. Available online: https://www.unwomen.org/en/digital-library/publications/2016/10/gender-equality-and-the-new-urban-agenda (accessed on 13 December 2018).

17. FAO. *Gender and Access to Land*; FAO Land Tenure Studies; Food and Agricultural Organization: Rome, Italy, 2002; ISBN 978-92-5-104847-4.

18. De Bruin, A.; Liu, N. The urbanization-household gender inequality nexus: Evidence from time allocation in China. *China Econ. Rev.* **2019**, 101301. [CrossRef]

19. Tacoli, C. The Benefits and Constraints of Urbanization for Gender Equality 2013. Available online: https://pubs.iied.org/10629IIED/ (accessed on 15 January 2020).

20. FAO. *FAO Policy on Gender Equality: Attaining Food Security Goals in Agriculture and Rural Development*; Food and Agricultural Organization: Rome, Italy, 2013.

21. Ellis, F. *Rural Livelihoods and Diversity in Developing Countries*; Oxford University Press: Oxford, UK, 2000; ISBN 978-0-19-829695-9.

22. Fan, P.; Ouyang, Z.; Nguyen, D.D.; Nguyen, T.T.H.; Park, H.; Chen, J. Urbanization, economic development, environmental and social changes in transitional economies: Vietnam after Doimoi. *Landsc. Urban Plan.* **2019**, *187*, 145–155. [CrossRef]

23. Nhung, P.T.; Kappas, M.; Faust, H. Improving the Socioeconomic Status of Rural Women Associated with Agricultural Land Acquisition: A Case Study in Huong Thuy Town, Thua Thien Hue Province, Vietnam. *Land* **2019**, *8*, 151. [CrossRef]

24. Newman, C. *Gender Inequality and the Empowerment of Women in Rural Viet Nam*; WIDER: Helsinki, Finland, 2015.

25. ILO. Rural Women at Work: Bridging the Gaps. Available online: https://www.ilo.org/wcmsp5/groups/public/---ed_protect/---protrav/--ilo_aids/documents/publication/wcms_619691.pdf (accessed on 20 March 2019).

26. Van, D. *Huế mở Rộng Thành Phố Gấp 5 lần, người Dân Đồng Thuận*; 2020; Available online: https://baotainguyenmoitruong.vn/hue-mo-rong-thanh-pho-gap-5-lan-nguoi-dan-dong-thuan-298006.html (accessed on 20 January 2020).

27. Cira, D.; Arish, D.; Kilroy, A.; Lozano, N.; Wang, H.G. *Vietnam Urbanization Review*; 2012; Available online: https://www.researchgate.net/publication/272088348_Vietnam_Urbanization_Review (accessed on 15 December 2019). [CrossRef]

28. VOV5. Vietnam Seeks Solutions for Problems Caused by Rapid Urbanization. Available online: https://english.vietnamnet.vn/fms/society/192232/vietnam-seeks-solutions-for-problems-caused-by-rapid-urbanization.html (accessed on 1 March 2020).

29. Vietnam Government. *Báo Cáo Quốc Gia Về Kế Hoạch Sử Dụng Đất Đai Đến Năm 2020 Và Kế Hoạch Sử Dụng Đất Giai Đoạn 2016-2020 (National Report Land Use Planning up to 2020 and Planing Land Use for Peroid 2016–2020)*; Vietnam Government: Ha Noi, Vietnam, 2016.

30. Quang Tuyen, T. Livelihood Strategies for Coping with Land Loss among Households in Vietnam's Sub-Urban Areas. *Asian Soc. Sci.* **2013**, *9*, 33. [CrossRef]

31. Phuc, N.Q.; van Westen, A.; Zoomers, A. Compulsory land acquisition for urban expansion: Livelihood reconstruction after land loss in Hue's peri-urban areas, Central Vietnam. *Int. Dev. Plan. Rev.* **2017**, *39*, 99–121. [CrossRef]

32. Tuyen, T.Q.; Lim, S.; Cameron, M.P.; Huong, V.V. Farmland loss and livelihood outcomes: A microeconometric analysis of household surveys in Vietnam. *J. Asia Pac. Econ.* **2014**, *19*, 423–444. [CrossRef]

33. WB. *Sửa đổi luật đất đai để thúc đẩy phát triển bền vững ở Việt Nam*; World Bank: Ha Noi, Vietnam, 2012.

34. Tuyen, T.Q. Laborer Status and the Effect of Vocational Training on Employment and Income of Rural Laborers in Kien Giang Province, Vietnam. *Glob. J. Hunmsn-Soc. Sci.* **2015**, *15*, 1–4.

35. Razavi, S.; Miller, C. *From WID to GAD: Conceptual Shifts in the Women and Development Discourse*; 1995; Available online: http://www.unrisd.org/80256B3C005BCCF9/(httpPublications)/D9C3FCA78D3DB32E80256B67005B6AB5 (accessed on 15 November 2019).

36. FAO Gender. The Key to Sustainability and Food Security. 1997. Available online: http://www.fao.org/english/newsroom/highlights/1997/introG-e.htm (accessed on 4 September 2019).

37. McGinn, K.L.; Oh, E. Gender, social class, and women's employment. *Curr. Opin. Psychol.* **2017**, *18*, 84–88. [CrossRef]

38. UN Women. *Policy Brief and Recommendations on Rural Women in Viet Nam Prepared for World Food Day in 2014 Family Farming: Feeding the World, Caring for the Earth*; UN Women: Ha Noi, Vietnam, 2014.

39. Thinh, B.H. *Công Nghiệp Hoá Và Những Biến Đổi Đời Sống Gia Đình Nông Thôn Việt Nam (Nghiên cứu trường hợp xã Ái Quốc, Nam Sách—Hải Dương)*; 2009. Available online: http://www.hids.hochiminhcity.gov.vn/c/document_library/get_file?uuid=89aa0397-14f9-4593-bfab-b057646e842f&groupId=13025 (accessed on 10 September 2019).

40. Pham, H.T.; Tuan, B.A.; Le Thanh, D. Is Nonfarm Diversification a Way Out of Poverty for Rural Households? Evidence from Vietnam in 1993-2006. *SSRN Electron. J.* **2010**. [CrossRef]

41. Rao, N. *Women's Access to Land: An Asian Perspective*; UN Women: Accra, Ghana, 2011; p. 20.

42. DoNRE. *Báo cáo Kiểm kê đất đai Năm 2010 tinh Thừa Thiên Huế*; DoNRE: Hue, Vietnam, 2011.

43. Que, S. *Thừa Thiên Huế Vững Bước lên độ thị Loại 1*; 2014; Available online: https://baodautu.vn/thua-thien-hue-vung-buoc-len-do-thi-loai-1-d15086.html (accessed on 10 February 2019).

44. Thua Thien Hue province Báo cáo kinh tế xã hội tinh Thừa Thiên Huế (The Anual Report of the Social-Economic Situation of Thưa Thien Hue Province). Available online: https://thuathienhue.gov.vn/vi-vn/Thong-tin-kinh-te-xa-hoi/cid/BC720E2C-8122-493C-8850-531999ECDD6F (accessed on 20 March 2020).

45. Crawford, I.M. *Marketing Research and Information Systems*; Marketing and Agribusiness Texts; Food and Agriculture Organization of the United Nations: Rome, Italy, 1997; ISBN 978-92-5-103905-2.

46. Bradshaw, S. Women's decision-making in rural and urban households in Nicaragua: The influence of income and ideology. *Environ. Urban.* **2013**, *25*, 81–94. [CrossRef]

47. Sweetman, C. *Gender, Development, and Citizenship*; Oxfam: Nairobi, Kenya, 2004; ISBN 978-0-85598-736-7. Available online: https://policy-practice.oxfam.org.uk/publications/gender-development-and-citizenship-121133 (accessed on 15 October 2019).

48. UNWomen. *Tài Liệu thảo Luận chính Sách công việc chăm sóc Không Lương: Những vấn đề đặt ra và gợi ý Chính Sách cho Việt Nam*; UNWomen: Ha Noi, Vietnam, 2016.

49. Roy, P.; Haque, S.; Jannat, A.; Ali, M.; Khan, M.S. Contribution of women to household income and decision making in some selected areas of Mymensingh in Bangladesh. *Progress. Agric.* **2017**, *28*, 120. [CrossRef]

50. Das, B.K.; Guha, N. How do Women Respond in the Context of Acquisition of Agricultural Land? A Micro Level Study in Semi-urban South Bengal, India. *Indian J. Hum. Dev.* **2016**, *10*, 253–269. [CrossRef]

51. Tuyen, T.Q. *A Review on the Link Between Nonfarm Activities, Land and Rural Livelihoods in Vietnam and Developing Countries*; 2012; p. 20. Available online: https://mpra.ub.uni-muenchen.de/55850/ (accessed on 20 March 2019).

52. Liem, P.S. *Nông Dân bỏ Ruộng: Thách Thức lớn đối với Người làm Chính Sách (Famer Leaving Agricultural Land: Big Challenge for Policy Maker)*; Nhà Xuất Bản Nông Nghiệp: Hanoi, Vietnam, 2015; pp. 254–259.

53. Thao, N.T.B. *Acquisition of Agricultural Land for Urban Development in Peri-Urban Areas of Vietnam: Perspectives of Institutional Ambiguity, Livelihood Unsustainability and Local Land Grabbing*; Okayama University: Okayama, Japan, 2015.

54. Oxfam. *Vấn đề sử dụng đất và thay đổi quyền sử dụng đất ở miền trung Việt Nam*; Oxfarm: Ha Noi, Vietnam, 2012.

55. Loi, V.M.; Quang, T.T.; Thai, N.H.; Tram, N.K.B.; Thach, D.V.H. Phân công lao động và quyền quyết định trong gia đình nông thôn đồng bằng Bắc Bộ. *Tạp Chí Nghiên Cứu Gia Đình Và Giới* **2013**, *1*. Available online: http://ifgs.vass.gov.vn/journal/Tap-chi-Nghien-cuu-Gia-dinh-va-Gioi--ban-tieng-Viet-/So-1---2013 (accessed on 20 March 2020).

56. Van, L.N. Một khía cạnh về mối quan hệ vợ chồng qua các cuộc điều tra xã hội học gần đây ở Việt Nam", Tạp chí Nghiên cứu Gia đình và giới. *Tạp Chí Nghiên Cứu Gia Đình Và Giới* **2012**, *2*, 16.

57. Khuat, T.H.; Van Anh, N.T.; Thao, N.T.P.; Quynh, N.H.N.; Diep, T.T.N. *Nguyen Ngoc Huong Social Determinants of Gender Inequality in Vietnam*; 2016; Available online: https://www.researchgate.net/publication/309487636_Social_Determinants_of_Gender_Inequality_in_Vietnam (accessed on 25 March 2020). [CrossRef]

58. Tacoli, C.; International Institute for Environment and Development; Human Settlements Programme; United Nations Population Fund; Population and Development Branch. *Urbanization, Gender and Urban Poverty: Paid Work and Unpaid Carework in the City*; 2012; ISBN 978-1-84369-848-7. Available online: https://pubs.iied.org/10614IIED/ (accessed on 25 March 2020).

59. ILO. *The 2016 Report on Informal Employment in Viet Nam*; Hong Duc Publishing House: Ha Noi, Vietnam, 2018; ISBN 978-604-89-2814-8.

 *land*

Communication

Absent Voices: Women and Youth in Communal Land Governance. Reflections on Methods and Process from Exploratory Research in West and East Africa

Stefanie Lemke * and Priscilla Claeys

Centre for Agroecology, Water and Resilience, Coventry University, Ryton Gardens, Wolston Lane, Ryton-on-Dunsmore, Warwickshire CV8 3LG, UK; priscilla.claeys@coventry.ac.uk
* Correspondence: stefanie.lemke@coventry.ac.uk

Received: 30 June 2020; Accepted: 7 August 2020; Published: 10 August 2020

Abstract: An increasing number of African States are recognizing customary land tenure. Yet, there is a lack of research on how community rights are recognized in legal and policy frameworks, how they are implemented in practice, and how to include marginalized groups. In 2018–2019, we engaged in collaborative exploratory research on governing natural resources for food sovereignty with social movement networks, human rights lawyers and academics in West and East Africa. In this article, we reflect on the process and methods applied to identify research gaps and partners (i.e., two field visits and regional participatory workshops in Mali and Uganda), with a view to share lessons learned. In current debates on the recognition and protection of collective rights to land and resources, we found there is a need for more clarity and documentation, with customary land being privatized and norms rapidly changing. Further, the voices of women and youth are lacking in communal land governance. This process led to collaborative research with peasant and pastoralist organizations in Kenya, Tanzania, Mali and Guinea, with the aim to achieve greater self-determination and participation of women and youth in communal land governance, through capacity building, participatory research, horizontal dialogues and action for social change.

Keywords: gender; women and youth; communal land governance; right to land; collective rights; participatory action research; transdisciplinary approach; COVID-19; West and East Africa; constituencies

---

1. Introduction

After a long and intense process of negotiation involving representatives from peasant organizations, pastoralists, fisherfolk, agricultural workers and Indigenous Peoples from all over the world, the United Nations (UN) Declaration on the Rights of Peasants and other people working in rural areas (UNDROP) was adopted by the UN General Assembly in December 2018 [1]. This new international legal instrument represents a key milestone in the struggle for food sovereignty and people's control over land and natural resources. The Declaration recognizes, for the first time, the human rights to land, seeds, biodiversity and food sovereignty. In the Declaration, these rights are recognized not only as individual, but also as collective rights.

The topic of collective human rights was extremely controversial during the negotiations. Discussions around the right to land, as observed by Priscilla Claeys who conducted participant observation during the negotiation process at the UN Human Rights Council between 2012 and 2018 [2], revealed conflicted views and visions about the management and governance of land, but also the meanings, values and relations between rural people and their land. Representatives of rural constituencies insisted that a great proportion of the world's land and natural resources continues to be managed collectively. They called on alternatives to private property, and denounced land grabbing. They also stated that rural communities, especially women, are often dependent on commons, such as pastoral land, forests or rangelands for their livelihoods, as has been shown by other authors [3,4].

These claims were met with resistance by several states [2]. Some diplomats argued that human rights are inherently individual rights, while others stated that nothing justifies granting special rights to peasants as a group. It was also stated that the only collective right recognized in international human rights law was the right to self-determination and that collective rights could not be recognized for any other group than Indigenous Peoples. Peasant and other rural organizations responded that they exist as a group, because they endure systemic and systematic discriminations as a group, and that without their rights to land, seeds, biodiversity and food sovereignty, they would not be able to continue feeding their communities and the world at large. Their voice was heard, and, to a large extent, the Declaration meets their demands and expectations, including that of being recognized as a collective, political subject and rights-holder [2].

Having long worked on topics relating to food sovereignty, the right to food, food governance, agrarian movements, nutrition and gender dynamics, we were excited by this development and eager to support the implementation of the Declaration [5–10]. We wanted to better understand, and document, how collective rights are experienced and implemented in practice, and how the rights of communities that depend on land, seeds and natural resources for their livelihoods could be better protected. We were particularly interested in exploring these issues in Africa, considering the prevalence of communal rights regimes in the continent. Customary tenure is estimated to be the dominant form of tenure over 78 per cent of Africa [11,12]. We felt that lessons could be drawn from efforts to codify and protect communal land rights in the face of growing appropriation of land and nature, that could inform struggles elsewhere.

As scholar-activists, we wanted to co-develop the research agenda with peasant organizations and other rural constituencies, and the research process and outcomes to be useful for local actors in their ongoing struggles to advance the right to land and other natural resources on the continent. Yet, the constraints associated with most research grants are such that developing a research agenda from the bottom-up is rarely possible. Based on our experience of many years in research funders usually require pre-determined project goals, assumed outcomes and impact measures at the stage of applying for funding. This limits the possibility of engaging in truly participatory and transdisciplinary approaches that would enable all partners to co-develop research priorities and research processes from the initial conception phase. To be able to do this requires establishing trustful relationships among research partners, respecting each other's worldviews, as well as overcoming unequal power relations between 'conventional sciences' and alternative knowledge systems [13,14].

With this goal in mind, we approached the 11th Hour Project, which we knew was sympathetic to the food sovereignty movement in Africa, and counted several food sovereignty actors in its list of grantees. We pitched our idea and approach, and were successful in obtaining seed money to initiate an exploratory research process on Governing Natural Resources for Food Sovereignty in West and East Africa. We were able to draw on a long and rich tradition of Participatory Action Research (PAR), which basically involves researchers and local actors working together to understand a problematic situation and change it for the better, following an iterative cycle of research, action and reflection [15–17].

In this article, we report on the methods, process and outcomes of this project, which we conducted from February 2018 to July 2019. Section Two provides detailed insights into the research process and participatory methodology used. Section Three presents key outcomes emerging from this process, and how it helped us collectively identify an important research gap around gender and communal land. Section Four briefly outlines our new research project on women's communal land rights.

2. Research Process and Methods

In this section, we reflect on the successive steps that led us to adopt our current new research focus on women's communal land rights. We place emphasis on the process and the methodologies we used, with a view to sharing lessons learned, as well as some critical reflections moving forward.

2.1. Establishing Networks in West and East Africa

Right from the start, we envisioned some kind of comparative research involving organizations from both West and East Africa. Despite differences regarding colonial histories, legal and policy frameworks, and the extent to which customary rights are recognized in law, the issues affecting customary land tenure regimes are shared across the two regions. In addition, many representatives from peasant and other rural organizations know each other from attending organizational events, be it international fora or UN processes (such as the UN Committee on World Food Security, CFS; the Food and Agriculture Organization, FAO; the International Fund for Agricultural Development, IFAD; or the International Convention on Biodiversity, CBD). We further felt it would be valuable to enable mutual learning between the two regions. Yet, communications between food sovereignty/agrarian networks are limited, partly due to language barriers.

We conducted an initial desk-review to select potential case studies—cases of legal mobilizations to protect/defend collective/communal land rights—but found it difficult to identify relevant experiences as there is little documentation readily available on the internet.

Through key contacts in both regions, obtained via our professional and academic networks (including the Civil Society and Indigenous Peoples' Mechanism (CSM) for relations with the CFS, the human rights network FIAN International, the Coalition of European Lobbies for Eastern African Pastoralism (CELEP) and academics in the region), we liaised with a variety of actors. Our approach was to organize two workshops, one in each region, to identify potential partners and research priorities.

We co-organized the first workshop in West Africa with the Institute for Research and Promotion of Alternatives in Development (*Institut de recherche et promotion des alternatives en développment (IRPAD)*, a progressive think-tank and action research center. We selected IRPAD for our longstanding collaboration with its Director Mamadou Goïta who is connected to global and regional actors and processes and was involved as expert panelist in the process of negotiation of the UNDROP. In East Africa, our local partner organization was the Alliance for Food Sovereignty in Africa (AFSA). We selected AFSA for its reputation and ability to bridge with various networks at both global and regional level. We had already established a connection with its Director Million Bellay, who was also involved as expert panelist in the process of negotiation of the UNDROP.

A field trip to Mali in March 2018 enabled us to meet with the IRPAD team and plan for the workshop that we co-hosted with them in June of the same year. We discussed the methodology, which questions to discuss collectively and were able to rely on Mamadou Goïta's extensive network to identify potential participants. In addition, we were able to connect with the National Federation of Peasant Organizations (*Confédération nationale des organisations paysannes, CNOP*), and the Women's Coalition for Food Sovereignty (*Coalition des Femmes pour la souveraineté alimentaire, COFERSA*). We were privileged to attend a two-day forum in Segou with more than 400 farmers and communities from across Mali affected by landlessness and appropriation of land, organized by the grassroots Malian Convergence against Land Grabbing (*Convergence malienne contre les accaparemments de terre, CMAT*). Our trip to Mali proved extremely useful to understand the issues underlying the protection of community rights in the country. It also enabled us to interview people who had been driving the process of drafting and adoption of the new Agrarian Land Law, which recognizes the possibility for community rights over land to be managed at the village level [18].

A second field trip to Uganda in January 2019 enabled us to meet with AFSA and plan for the workshop that we co-hosted in March of the same year. We discussed workshop objectives with the AFSA team and shared insights from the first workshop we had co-organized the year before. In addition to co-designing the workshop with AFSA, we were able to connect with a number of agrarian organizations, NGOs and academics working on land, including the Eastern and Southern Africa Small Scale Farmers' Forum ESAFF – UGANDA, Katosi Women Development Trust (KWDT), progressive lawyers from the University of Makarere and the Advocates for Natural Resources & Development ANARDE, and FIAN Uganda.

2.2. Participatory Workshops in Mali and Uganda

Participants at the two workshops represented a diversity of voices from the food sovereignty movement in the two regions, while opening opportunities for dialogue between actors who may be in different networks and do not often speak to each other. We took great care in elaborating the methodology for the two workshops. The first step was elaborating the participants' lists, in partnership with IRPAD for West Africa and AFSA for East Africa. For the first workshop, in Bamako, we paid attention to gender as well as regional balance, and invited organizations from Benin, Burkina Faso, Guinea, Mali, Niger and Senegal. We also made sure to invite a mix of representatives from peasant organizations, pastoralists/herders, NGOs and lawyers, as well as academics. For the second workshop, in Entebbe, we similarly applied a regional quota to have a good balance of participants from Ethiopia, Kenya, Tanzania and Uganda, respectively. We ensured gender parity with a view to guarantee the strong participation of women. In addition, we ensured a diversity of constituencies, representing small scale farmers' organizations, pastoralists, fishers and Indigenous Peoples, as well as some NGOs, feminist/women's rights' organizations, academics and progressive human rights lawyers.

Efforts to ensure a diversity of constituencies and address power dynamics among participants were inspired by research conducted by Priscilla Claeys and Jessica Duncan [19] on the mechanisms used by actors in the Global Food Sovereignty Movement to facilitate convergence and unity in diversity. Actors in this movement have long tried to prioritize the voice of peoples' organizations over that of NGOs, to enable affected constituencies to directly participate in food governance and not let others speak on their behalf [20]. Complex mechanisms balancing regions, gender, generations and constituencies are used in the Civil Society Mechanism to the UN Committee on World Food Security, for example, to enable participants to speak with one voice while protecting/fostering separate identities (for a historical overview of how constituencies developed within the Global Food Sovereignty movement and how they are applied in the CSM see [19,21]. Critically reflecting on our efforts to be as inclusive as possible, we are aware of the fact that we mostly engaged with members of organizations and social movements who are in a more privileged position. In other words, the most affected did not attend our workshops. This is partly due to the fact that we identified participants through international networks (and accessing these networks requires considerable social and political capital, including mastering several languages). This might result in missing the voices of those who are most marginalized. In follow-up research, an important task in collaboration with our partner organizations will be to ensure that they reach out to those who are usually excluded so as to fully take into account the various power relations and potential conflicts between various users of land and other natural resources (as addressed in both workshops by Axis 3 and Axis 4).

We applied participatory and horizontal methodologies to ensure the full and active participation from all participants, such as the World Café and the Fishbowl (see Figures 1 and 2).

(a)   (b)

Figure 1. World Café, Workshop Bamako, Mali, June 2018. (**a**) Discussions at the three tables; (**b**) summarizing key points emerging from discussions for the next group of particpants. Photos: Ann-Christin Weiler (**a**) and Stefanie Lemke (**b**).

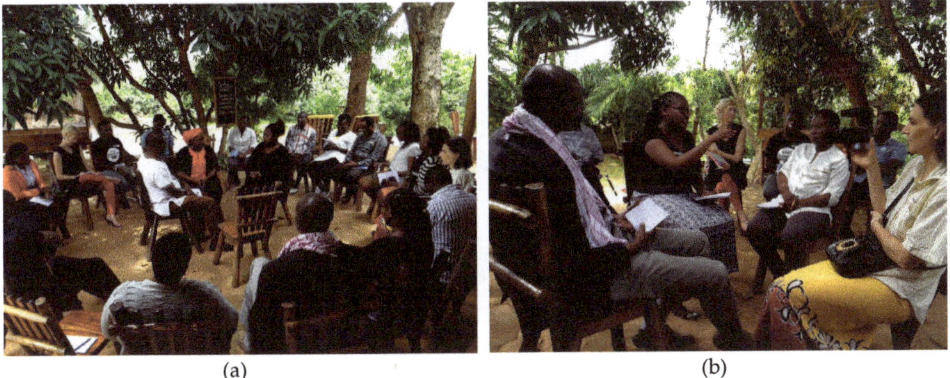

(a)   (b)

Figure 2. Fishbowl, Workshop Entebbe, Uganda, March 2019. (**a**) Arrangement of the outer and inner circle, the 'fishbowl', one chair remains empty; (**b**) filming discussions in the 'fishbowl'. Photos: Stefanie Lemke.

In the World Café, participants are divided into three tables. Each table is assigned a facilitator. The facilitator introduces the question and initiates the discussion. Everyone at the table exchanges ideas for about 20 min. Then, participants move on to the next table, except for the facilitator. He/she shares with the second group the ideas developed by the first group. Rather than starting from scratch, the second group thus builds on the ideas that were collected. After 20 min, the three groups move again to the next table. In the end, all participants go from table to table, listen to the facilitator's report and add information or ask questions [22].

In the fishbowl, four to five chairs are arranged in an inner circle, the 'fishbowl'. The remaining chairs are arranged in a circle outside the fishbowl. Participants are selected to fill the fishbowl; the rest of the group sit on the chairs outside the fishbowl. One chair is left empty. The moderator introduces the topic and the participants start discussing the topic. The audience outside the fishbowl listens in on the discussion. Any member of the audience can, at any time, occupy the empty chair and join the fishbowl. When this happens, one person in the fishbowl must voluntarily leave and free a chair. The discussion continues with participants entering and leaving the fishbowl [23].

We approached participants to ask if they would agree to take turns in facilitating sessions. This role involved organizing the discussion, managing speaking times, ensuring everyone can speak, taking notes on flipcharts and sharing outcomes with the larger group. Most participants got to facilitate a session, which allowed us to share tasks and also created a sense of collective ownership over the process and its outcomes.

The methodology was very much appreciated by participants and enabled in-depth discussions and interactions. Our perception was that the strong presence of women enabled them to make their voices heard. We observed mild tensions initially when starting discussions around women's rights and gender. For example, several participants made jokes or comments about the great number of women present. However, these tensions quickly dissolved, thanks to the participatory format that encouraged everyone to engage. Participants shared how they had made new contacts and networks; how lively and relaxed the atmosphere was; how they connected with others at a deep and personal level; how they enjoyed the open, efficient and involving methodology that makes it impossible NOT to participate; and how several of them were aiming at replicating the fishbowl and world café methodologies in their own work.

Our only regret was to have failed to achieve adequate representation of the youth in both workshops. This was due to the fact that we had not considered this enough at the invitation stage, and also because it is difficult for young people to achieve leadership positions in organizations, and to be nominated to represent their organization at workshops or events. Based on this experience, we concluded that we need to better include the youth in future, both in the process and in our research. This was stated also by our research partners in follow-up conversations.

Through our initial discussions with Mamadou Goïta, we had identified five main axes to guide our discussions. We used the same five axes in both workshops:

- Axis 1: a collective mapping exercise to identify ongoing struggles for collective rights to land, seeds and other natural resources;
- Axis 2: a collective mapping of the legal and policy frameworks that currently recognize collective/community rights to land, seeds and natural resources;
- Axis 3: a discussion of relations of power within community rights systems, and how to address the marginalization and the needs of specific groups. We paid specific attention here to women and youth, but also to other groups that we collectively identified, such as fishers or pastoralists;
- Axis 4: a discussion of the conflicts that may arise between various users of land and natural resources, and effective mechanisms to address these;
- Axis 5: future collaboration and processes to collectively identify gaps in research and action that could support movement struggles.

In a final feedback round participants commented on how they perceived the workshop process, methods and outcomes. Participants highlighted how much they had learned from each other, for example on pastoralism, seeds, fishing or struggles against mining; how useful they found it to reflect on the legal aspects of community land issues; how the information they got at the workshop could not have been found anywhere else, as this information is not available in books or on the internet; how they discovered opportunities for collaboration at regional level; and how the voices of pastoralists, fishers and women are still struggling to emerge in food sovereignty discussions and that this workshop was a great opportunity to engage on these issues.

2.3. Desktop Study

In parallel, we commissioned a desktop study of legal and policy frameworks in both regions, entitled "Governance of land and natural resources in Africa. Overview of legal and policy frameworks on collective/customary land tenure; marginalization; and conflicts", conducted by our colleague, Stefania Errico [24]. The objective of this study was to undertake:

- a preliminary analysis of the extent to which collective rights to land, seeds and natural resources (including customary rights and the commons) are recognized and protected in legal and policy frameworks in Western and Eastern Africa; and,

- a first identification of the key factors that may have an impact on the realization of collective rights to resources in these regions and consequently on the potential contribution of collective rights to equitable and resilient food systems.

The desktop study followed the logic of the five axes developed for the two workshops, and focused on the following three main sections: 1. mapping of legal and policy frameworks (Axis 2); 2. power relations and the rights of specific groups (women, youth, marginalized groups) (Axis 3); and 3. the rights of different rural constituencies and potential conflicts between different users of land and natural resources (Axis 4). The study provided more analysis and background on some of the key issues that had emerged in the workshops, and helped to identify gaps in research based on a review of the literature.

2.4. Video Documentation

We documented both workshops through video filming, photographs, audio recording and extensive notes. We were assisted in filming by Marine Lefebvre from SOS Faim Luxembourg and two Masters students based at the University of Hohenheim, Germany. Back in the UK, we created a series of short movies, in French and in English, and a full movie comprising all short movies. We had no prior experience with this and really enjoyed the creative process that entails drafting a script, and watching and selecting hours of footage. We quickly realized we would need to complement the footage we had with some graphic animation. Further, we were very lucky to have Stéphane Parmentier compose some beautiful music for us, mixing traditional African and electronic tones. We also benefited from exchanges with our colleagues at the Centre for Agroecology, Water and Resilience, Coventry University (CAWR) who gave us very useful feedback through the editing process.

In the full movie (see link in Supplementary Materials), the introductory section presents the objectives of the project and situates it within the broader context of struggles around natural resources and why community rights matter, making the link with the UN Declaration on the Rights of Peasants and other people working in rural areas. Our intention with the movies was to illustrate both workshop process and interactions as well as outcomes. The methodologies applied to facilitate horizontal discussions on sensitive topics are documented in detail, due to the feedback received by workshop participants who highlighted that our approach was innovative and that they were hoping to use this type of approach and some of the methods applied in their own work. We decided to mix footage from Mali and Uganda as a way to highlight commonalities, but also differences between the two regions, and to enable a sharing of experiences and learning from each other.

3. Results

3.1. Key Outcomes of Exploratory Research

The two workshops in Mali and Uganda and the supporting desktop study "Governance of land and natural resources in Africa. Overview of legal and policy frameworks on collective/customary land tenure; marginalization; and conflicts" [24] served to collectively identify key achievements and challenges with regard to governing natural resources in the two broader regions, as well as identifying key areas for future research.

Among the key achievements in protecting collective rights to resources are:

- growing recognition of community land rights in the law and in policy agendas;
- social movement mobilizations in defense of communal land rights (mainly in West Africa);
- multi-actor platforms for dialogue with governments;
- progressive human rights lawyers working with peasant and pastoralist organizations (mainly in East Africa).

Among the key challenges in protecting collective rights to resources are:

- declining natural resources and biodiversity, climate change, and growing conflicts over land;
- commercialization (large-scale land deals), privatization and selling of land, elite capture;
- focus on individual titling to protect tenure security, which may result in women losing land;
- power dynamics and marginalization, with women and the youth being largely excluded from communal land governance;
- lack of legal protection of communal land rights and tensions between customary and national law. For example, while gender equality is part of many constitutions, this is not implemented in practice.

These findings were supported by the desktop study [24], showing that the main obstacles to the respect, protection and realization of collective/customary land rights include biased approaches against certain forms of land use, contradictory laws and policies, the absence of mechanisms and capacity to ensure the actual implementation of relevant pieces of legislation recognizing customary rights and the weakening of traditional authorities, in a context marked by increasing pressure on customary lands. As the desktop study further revealed, the commodification and individualization of land rights have occurred within customary systems, as a result of increasing land value and market integration, and have had a repercussion, among others, on communities' cohesion and their capacity to speak with one voice on land matters and defend their rights 'collectively'. Tenure insecurity and the growing marginalization of certain groups remain crucial sources of conflicts throughout the continent, whether they occur between local resource users or with external actors [12,25–28].

The key areas identified by participants for future research and much needed documentation, in partnership with peasant, fisherfolk, pastoralist and Indigenous Peoples' organizations, include:

- ongoing struggles by peasant, fisherfolk, pastoralist and Indigenous Peoples' organizations to defend these rights;
- how community/customary rights over land and natural resources operate in practice and how they have evolved over time;
- the extent to which community rights are protected in the law and the impact of various institutional setups on their enjoyment;
- local and traditional knowledges for communal natural resource management;
- local strategies for climate change adaptation and resilience;
- obstacles and opportunities for the participation of women, youth and other marginalized groups in communal resource management and policy-making in general.

The desktop study [24] equally emphasized the need for more research on how customary regimes are changing and evolving at a rapid rate. In the context of current debates on the recognition and protection of customary/collective rights to land and resources, there is a need for more clarity and documentation of customary practices. Particular attention needs to be paid to marginalized groups, including women, youth and Indigenous Peoples. With regard to the youth, there are growing efforts by youth organizations in relation to access to land and other natural resources, which seek to address the lack of perspectives and future livelihoods in rural areas. More research is also needed on gendered use of common resources within collective tenure and the norms of their allocation and use.

3.2. Reflecting on Key Research Outcomes and Process, Co-Designing Follow-Up Research

Reflecting on the main learning points from this process during the fall of 2019, a major takeaway for us was the lack of a women's voice on communal land governance. From what was shared with us by workshop participants, it was clear that land issues continue to be seen as the domain of men, with women and the youth being largely excluded from land governance decisions. The absence of a women's voice in land governance was of course nothing new. What we found most intriguing however was the "return" of customary regimes on the agenda of key actors in the food sovereignty movement, such as the Malian national peasant organization CNOP and the Alliance for Food

Sovereignty in Africa (AFSA). Their insistence on promoting and defending communal land ownership, in lieu of private and absolute property, echoed efforts by agrarian movements involved in the negotiations of the UNDROP to achieve recognition of the right to land as an individual and collective right. Just as their international allies were making a strong case for collective rights at the Human Rights Council, the actors we talked to at our workshops were advocating for the legal recognition of communal land rights (CLR), because they see this as key to protecting communities against land grabbing. Yet, we also found that actors in these movements do not all equally emphasize the importance of ensuring inclusivity in communal land governance and the participation of all community members in decisions concerning the land. Even in these progressive food sovereignty movements, women's rights can be marginalized, as we witnessed when engaging with actors who had advocated for the new Agrarian Land Law in Mali. One of them explained to us that their first priority is to secure communal land rights, and that the position and 'participation' of women in this process will be dealt with later. Again, this resonated with debates held during the negotiations of the UNDROP. While women's rights to land are recognized in the Declaration, there are important limitations in the text concerning their rights to inherit land [29], and it will be particularly important to monitor the implementation of the UNDROP on this point. The participation of the youth in communal land governance is similarly lacking although there is great awareness of the need to do more in this area.

Between November 2019 and February 2020, we conducted a series of follow-up interviews with representatives from selected partner organizations, with the aim to co-develop a more focused research theme. The limited decision-making power of women and the youth within communal land governance was confirmed during interviews, but what emerged strongly was the need to do more action research on the lacking intersection between gender and communal land rights, in partnership with peasant and pastoralist organizations. Turning to the literature, we found there was very little to build on, especially considering rapidly changing gender dynamics and customary regimes. While some of these processes might be documented by organizations working on these issues, this type of information is often not easily available.

In the next section, we shortly present the new research project which was approved by the 11th Hour Project in June 2020 and will take place over the next 18 months, with a focus on women's communal land rights.

4. New Research Project on Women's Communal Land Rights (WCLR)

4.1. Project Rationale

The gender-communal land rights nexus deserves more attention for four reasons. First, efforts by development actors have focused on enabling women to acquire or work on the land individually. This is however particularly problematic in Africa, where most of the land is under customary tenure, and where access to collectively held land is essential to women's livelihoods. Second, in most customary land systems in Africa, women are still not recognized independent rights to land and they rarely participate in decisions about communal land governance (CLG). In practice, they gain access to them through their relationship with a male relative (usually the husband or father) [25–27]. Women are rarely compensated for their losses in case of land dispossession, and may lose access if the relationship no longer exists (in case of divorce or death of the husband). In addition, customary institutions are rapidly changing, and the traditional norms that used to ensure that women would have access to land are no longer enforced or in place. Third, communal land rights are increasingly being turned into individual plots that are sold to investors, triggering intra-household competition between men and women—and between generations—over productive resources. In many places, women are losing access to land as a result of the family property being privatized by senior male members, a phenomenon known as 'family land grabbing' [30]. Fourth, efforts to provide secure land tenure for communities through the formalization of communal land ownership, while valuable and important, often have negative outcomes for women, as their interests are not properly considered in the implementation of state programs to strengthen collective tenure [28].

4.2. Project Partners, Objectives and Research Priorities

We selected four partners for this project, all of them peasant (women) or pastoralist organizations. It was important to us to prioritize grassroots organizations over NGOs. We decided to focus on two countries in West Africa: Mali and Guinea, and two countries in East Africa: Kenya and Tanzania. Reflecting on the size and format of this new project, we started with a small group to be able to build this process slowly, with the aim to expand and include other partners at a later stage.

The main project partners are:

- Kenyan Peasants League (KPL), a peasant organization in Kenya established in 2016 and member of La Via Campesina, which promotes peasant farming and agroecology;
- Pastoral Women's Council (PWC), a pastoralist women's organization created by Maasai women in 1998, Tanzania, to defend the rights of pastoralist women and girls;
- CNOP-G, the national confederation of peasant organizations in Guinea (a member of ROPPA and COPAGEN), set up in 2000 to defend the interests of peasants and which represents 700,000 family farms, with a membership of 52 per cent women;
- COFERSA, the national coalition of women for food sovereignty, a network of 45 women's cooperatives gathering more than 4000 women across Mali, established in 2009.

The overarching objective of this project is to draw lessons from and scale up efforts to advance WCLR (in East and West Africa). The project will rely on four main areas of work (Figure 3):

1. *Capacity-building*: support efforts at becoming more gender-sensitive and gender-transformative; co-develop capacity to conduct participatory action research (PAR);
2. *Participatory Research*: document and draw lessons from efforts to advance women's communal land rights (WCLR) in different spaces (household/village level, local government, traditional authorities, district level);
3. *Facilitate dialogues:* between women and men, across generations, at household (HH) level, among women, among men; identify, create and multiply tools and processes that work to facilitate inclusive and empowering dialogues; create social cohesion;
4. *Action for social change*: support actions that advance and scale-up WCLR, in line with the individual needs and priorities of the partner organizations.

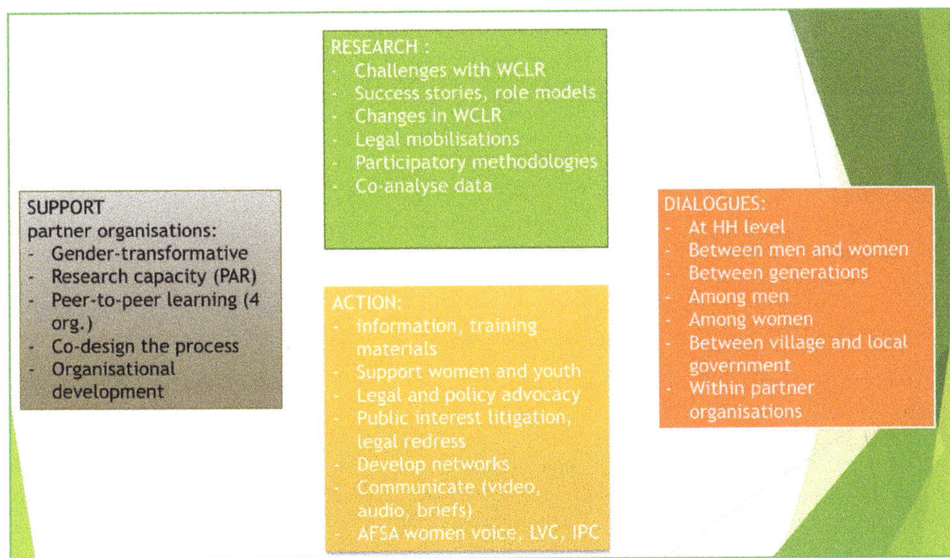

Figure 3. Overview of project goals and the four main areas of engagement outlining project activities.

The four African-based partners have each selected specific focus areas for research. These focus areas have been identified through dialogue to ensure that they align with organizational needs, and address gaps in data and documentation. Research outcomes will be used to support advocacy objectives or be fed into training materials. Each organization will implement a range of activities in relation to the above project goals (see Figure 4).

	KPL	PWC	CNOP-G	COFERSA
Country	Kenya	Tanzania	Guinee	Mali
Region	East-Africa	East-Africa	West-Africa	West-Africa
Key focus	Widows (and marginalized women & youth)family land grabbing	Women's participation in governance	Access to land for women and youth	Access to land for women and youth, family land grabbing
Key research activities	Mapping obstacles and barriers; mapping initiatives and success stories; peer-to-peer learning; inventory of customary practices/norms			
Key activities	Legal and policy advocacy, legal redress for widows, PIL	Joint action plan, advocacy around land, support capacity building	Training, advocacy around land	Training, awareness raising, support capacity building

Figure 4. Overview of regions, project foci and key activities as elaborated by the four organizations.

The total budget will be shared equally between the five organizations (CAWR and four partners). The Centre for Agroecology, Water and Resilience, Coventry University (CAWR), will be coordinating and supporting project design, implementation and dissemination. There will further be an advisory board to provide feedback and reflections, where several members of the larger network that was established during the previous research will participate. Our aim is to keep close links with all partners, and to expand the current research to include other partner organizations. The research process, methodology and outcomes will be shared more widely, for example via an online platform.

The anticipated impacts of this project include: reduced resistance of household members towards women's participation and greater self-determination; reduced resistance of community members towards including women and youth in community land governance; shared understanding at community level of the advantages of broader inclusivity in community land decisions; greater meaningful and active participation of women and young women and men in community land governance; greater visibility of issues affecting women's communal land rights; prioritization of WCLR in agenda-setting by several organizations and networks in the food sovereignty and women's rights and feminist movements.

4.3. Adapting Our PAR Methodology to the COVID-19 Pandemic

The final stages leading to the approval of this project took place in the midst of the COVID-19 pandemic. This forced us to reflect further on how we would conduct this research in times of uncertainty and possibly the inability to travel. We decided the pandemic would be an opportunity for us to step back and share even more responsibilities with our partners. The project was always designed as participatory and collaborative. One of the key objectives of this new project is to support our partners with building and reinforcing their own capacity for research, notably through trainings on how to apply PAR methods within their organizations. We will engage in regular dialogue via online platforms, collectively shape the research processes and actions, while giving partner organizations the leadership and responsibility to undertake most of the activities. We are able to do so thanks to the trust we have established over the past two years. By facilitating regular meetings with our project partners for the duration of the project we will ensure clear communication, which will allow for adapting safety measures and project activities if required. As with the preceding project, we will carefully and in a transparent manner document the research and action processes in collaboration with our partners, including the challenges and the opportunities that might arise to build the internal capacity of our local partner organizations in conducting participatory action research.

In addition, all activities conducted during this project will apply safety and hygienic measures as relevant based on the local context, such as: providing and wearing face masks, keeping 2 m distance, providing soap and water for washing hands, providing sanitizers, limiting the size of groups to maximum 15 people, facilitating meetings outside as much as possible, and avoiding spreading the virus from urban to rural areas. We will further closely monitor and adhere to the respective national guidelines regarding safety measures in the five participating countries.

Finally, we will travel less and better. Even before COVID-19, we had decided to limit our flights and CO_2 emissions to a minimum, by reducing the number of trips and staying for extended periods of time when we travel. This is in compliance with our internal policy to engage in responsible travelling. If possible, our plan is to do some targeted but extended fieldwork, over a period of four to six weeks, in the summer of 2021.

Supplementary Materials: Video S1: Governing natural resources for food sovereignty in Africa, available at https://tinyurl.com/y46p7xsv.

Author Contributions: This discussion article and the underlying research and methodological approach were jointly conceptualized, written, reviewed and edited by S.L. and P.C. They are therefore equal co-authors. The same applies to project administration and funding acquisition. All authors have read and agreed to the published version of the manuscript.

Funding: This research was funded by the 11th Hour Project, The Schmidt Family Foundation. Funding for the follow-up research on Women's Communal Land Rights is also received from the 11th Hour Project.

Acknowledgments: We thank Stefania Errico for assisting with compiling background literature on the topic and for writing the consultancy report in the first phase of this project. We thank IRPAD and AFSA for co-hosting the two workshops in Mali and Uganda; Marine Lefebvre from SOS Faim Luxembourg, and Master students Ann-Christin Weiler and Alya Belkhodja for video filming during the two workshops; Ben Cook for post-editing, Larry Campbell for graphic design, Geetha Neelakatan, Moussa Sidibe for voice-overs in English and French; Susie Maugham,

Sue Chambers, Geetha Neelakatan and Josh Elliot for support with administrative issues at Coventry University, and our colleagues at the Centre for Agroecology, Water and Resilience at Coventry University for providing valuable feedback on a draft version of the video. We greatly appreciate the financial support provided by the 11th Hour Project, The Schmidt Family Foundation, to undertake this research. We would like to especially acknowledge our friend, the late Stéphane Parmentier, who so sadly and unexpectedly passed away in June 2020. He composed the music for the video. His music page can be found here: https://soundcloud.com/user-34584257.

Conflicts of Interest: The authors declare no conflict of interest. The funders had no role in the design of the study; in the collection, analyses, or interpretation of data; in the writing of the manuscript, or in the decision to publish the results.

References

1. United Nations Declaration on the Rights of Peasants and Other People Working in Rural Areas. Resolution Adopted by the General Assembly on 17 December 2018, A/RES/73/165 21 January 2019. Available online: https://digitallibrary.un.org/record/1661560/files/A_RES_73_165-EN.pdf (accessed on 29 June 2020).
2. Claeys, P.; Edelman, M. The United Nations Declaration on the rights of peasants and other people working in rural areas. *J. Peasant Stud.* **2019**. [CrossRef]
3. Doss, C.; Summerfield, G.; Tsikata, D. Land, gender and food security. *Fem. Econ.* **2014**, *20*, 1–23. [CrossRef]
4. Tsikata, D.; Yaro, J.A. When a good business model is not enough: Land transactions and gendered livelihood prospects in rural Ghana. *Fem. Econ.* **2014**, *20*, 202–226. [CrossRef]
5. Claeys, P. The creation of new rights by the food sovereignty movement: The challenge of institutionalizing subversion. *Sociology* **2012**, *46*, 844–860. [CrossRef]
6. Claeys, P. *Human Rights and the Food Sovereignty Movement: Reclaiming Control*; Routledge: London, UK, 2015.
7. Claeys, P. The rise of new rights for peasants. From reliance on NGO intermediaries to direct representation. *Transnatl. Leg. Theory* **2019**, *9*, 286–399.
8. Lemke, S.; Bellows, A.C. Bridging nutrition and agriculture. Local food-livelihood systems and food governance including a gender perspective. *J. Technol. Assess. Theory Pract. Spec. Issue Feed. World Chall. Oppor.* **2011**, *20*, 52–60. Available online: https://tatup.de/index.php/tatup/article/view/771 (accessed on 8 August 2020).
9. Lemke, S.; Bellows, A.C. Sustainable food systems, gender, and participation: Foregrounding women in the context of the right to adequate food and nutrition. In *Gender, Nutrition, and the Human Right to Adequate Food: Toward an Inclusive Framework*; Bellows, A.C., Valente, F.L.S., Lemke, S., de Lara, M.D.N.B., Eds.; Routledge: London, UK, 2016; pp. 254–340.
10. Lemke, S.; Delormier, T. Indigenous Peoples' food systems, nutrition and gender: Conceptual and methodological considerations. *Matern. Child Nutr.* **2017**, *13*, e12499. [CrossRef] [PubMed]
11. Alden Wily, L. Collective land ownership in the 21st century: Overview of global trends. *Land* **2018**, *7*, 68. [CrossRef]
12. Boone, C. Legal empowerment of the poor through property rights reform: Tensions and trade-offs of land registration and titling in Sub-Saharan Africa. *J. Dev. Stud.* **2018**, *55*, 384–400. [CrossRef]
13. Lemke, S.; Waters-Bayer, A. *A Report on the International Colloquium of the High-Level Panel of Experts (HLPE) and the University of Hohenheim: Food Security and Nutrition in the Context of the 2030 Development Agenda: Science and Knowledge for Action*; University of Hohenheim: Stuttgart-Hohenheim, Germany, 27 September 2016. Available online: https://gfe.uni-hohenheim.de/international-colloquium (accessed on 8 August 2020).
14. Pingault, N.; Caron, P.; Kolmans, A.; Lemke, S.; Kalafatic, C.; Zikeli, S.; Waters-Bayer, A.; Callenius, C.; Qin, Y. Moving beyond the opposition of diverse knowledge systems for food security and nutrition. *J. Integr. Agric.* **2020**, *19*, 291–293. [CrossRef]
15. Chambers, R. *Rural Development: Putting the Last First*; Longman: London, UK, 1983.
16. Pretty, J.; Guijt, I.; Thompson, J.; Scoones, I. *Participatory Learning and Action: A Trainer's Guide*; IIED: London, UK, 1995.
17. Reason, P.; Bradbury, H. (Eds.) *The Sage Handbook of Action Research: Participative Inquiry and Practice*; Sage: Thousand Oaks, CA, USA, 2008.
18. Coulibaly, M.; Claeys, P.; Berson, A. The right to seeds and legal mobilisation for the protection of peasant seed systems in Mali. *J. Hum. Rights Prac.* **2020**, under review.

19. Claeys, P.; Duncan, J. Do we need to categorize it? Reflections on constituencies and quotas as tools for negotiating difference in the global food sovereignty convergence space. *J. Peasant Stud.* **2018**. [CrossRef]
20. McKeon, N.; Kalafatic, C. *Strengthening Dialogue: UN Experience with Small Farmer Organizations and Indigenous Peoples*; United Nations Non-Governmental Liaison Service: Geneva, Switzerland, 2009; 49p.
21. Claeys, P.; Duncan, J. Food sovereignty and convergence spaces. *Political Geogr.* **2019**. [CrossRef]
22. World Café Method. Available online: http://www.theworldcafe.com/key-concepts-resources/world-cafe-method/ (accessed on 30 June 2020).
23. Fishbowl. Available online: https://www.unicef.org/knowledge-exchange/files/Fishbowl_production.pdf (accessed on 30 June 2020).
24. Errico, S. *Governance of Land and Natural Resources in Africa. Overview of Legal and Policy Frameworks on Collective/Customary Land Tenure; Marginalisation; and Conflicts*; Unpublished Report; Coventry University: Coventry, UK, 2019.
25. Krantz, L. *Securing Customary Land Rights in Sub-Saharan Africa*; Working Papers in Human Geography; Department of Economy and Society, Göteborg University: Göteborg, Sweden, 2015; Volume 1.
26. Scalise, E. Indigenous women's land rights: Case studies from Africa. In *State of the World's Minorities and Indigenous Peoples*; Minority Rights Group (MRG): London, UK, 2012.
27. Food and Agriculture Organization of the United Nations (FAO). *The State of Food and Agriculture 2010–2011*; Food and Agriculture Organization of the United Nations (FAO): Rome, Italy. Available online: http://www.fao.org/3/a-i2050e.pdf (accessed on 30 June 2020).
28. Giovarelli, R.; Richardson, A.; Scalise, E. Gender and Collectively Held Land: Good Practices and Lessons Learned from Six Global Case Studies. Resource Equity and Landesa. 2016. Available online: https://www.landesa.org/resources/synthesis-report-gender-collectively-held-land/ (accessed on 30 June 2020).
29. Bourke Martignoni, J.; Claeys, P. Without feminism there is no food sovereignty? Negotiating gender equality in the United Nations Declaration on the rights of peasants and other people working in rural areas. In Proceedings of the European Conference on Politics & Gender (ECPG), Amsterdam, The Netherlands, 3–6 July 2019.
30. International Institute for Environment and Development (IIED). *Innovation in Securing Land Rights in Africa: Lessons from Experience*; IIED: London, UK, 2006.

 *land*

Article

Smallholder Agricultural Investment and Productivity under Contract Farming and Customary Tenure System: A Malawian Perspective

Emmanuel Olatunbosun Benjamin

Department of Agricultural Production and Resource Economics, Technical University of Munich (TUM), 85354 Freising, Germany; emmanuel.benjamin@tum.de; Tel.: +49-8161-71-2772

Received: 16 June 2020; Accepted: 17 August 2020; Published: 18 August 2020

Abstract: Land tenure security, especially customary residence systems, is found to influence the agricultural investment decision-making and productivity of smallholder farmers across sub-Saharan Africa. However, as country-specific customary residence systems and farming models evolve over time, their impact on food security and livelihood remains unclear. This study investigates the impact of customary residence systems on both agricultural investment (in tea shrubs and agroforestry) and productivity among contracted smallholder tea outgrowers in Southern Malawi. A survey of 228 farmers was conducted in 2018, and a linear probability and ordinary least squared (OLS) models were used for the analysis. The results suggest that matrilocal residence practices positively influence agricultural investment. The study concluded that despite the dominance of matrilineal-matrilocal systems in Southern Malawi, there is a need for policy to address gender gaps in the region because women are still vulnerable and insecure even in these assumed women-friendly customary systems. It is recommended that future research explores other prevailing tenure security systems.

Keywords: land tenure security; contract farming; tea; agroforestry; investment; productivity; Southern Malawi; sub-Saharan Africa

---

1. Introduction

The 51.3 million smallholder farms (i.e., farms with less than 2 hectares) in sub-Saharan Africa (SSA) are the primary source of livelihood and labor for a large proportion of the population [1]. Thus, land is important for agricultural development in SSA. Any effort to alleviate poverty in sub-Saharan Africa has to pay attention to land and agricultural issues. Land tenure security implies access to use and ownership of land [2]. Land tenure security can be either statutory or customary in nature. Land tenure security enhances the utility that farmers derive from their farmland [3]. For instance, farmers in developing countries that are land secure may enter into new contracts and/or markets [3]. Diverse types of contract farming arrangements have been observed across SSA over the last couple of decades. Therefore, it is plausible to assume that smallholder farmers with adequate land tenure security and access to markets will invest in their land to increase productivity and alleviate poverty. However, studies on the relationship between land tenure security and agricultural investment across sub-Saharan Africa have produced ambiguous results [4]. The differences in these results are often due to limited detailed empirical analysis and vague assumptions on land tenure security [4,5]. Given the fact that some of the existing studies do not reflect the reality in the respective countries by neglecting interregional differences and customary tenure arrangements (see [4]), there is a possibility of misplaced and biased policies that may hinder economic development. Therefore, it is important to conduct a country- or region-specific empirical analysis of the relationship between land tenure security and agricultural investment and productivity to have an adequate understanding of reality on the ground [4–6]. For instance, Malawi loses approximately 11 million USD per annum in agricultural output due to insecure access, use, and ownership of land [7]. This implies that insecure land rights and tenure can disincentivize farmers from an investment that is vital for agricultural productivity despite contract farming opportunities. Furthermore, the government of Malawi recently enacted a land reform law that allows customary land to be registered. The rationale behind this reform was to address tenure insecurity as well as monitor and track the level and incidence of tenure insecurity and relevant conflicts. According to Lovo [8], Lunduka [9], and Place and Otsuka [10], tenure (in)security in Malawi falls under the customary residence system of a landholder. These customary residence systems determine the rights to use, access, and ownership of land for roughly 69% of the Malawian population [8]. The objective of this study is to investigate the effect of diverse customary residence systems on contracted tea farmers' agricultural investments in tea and trees (henceforth agroforestry) in Southern Malawi. Furthermore, it is important to analyze which form of customary residence system, among others, increases agricultural productivity.

Malawi has a matrilineal and patrilineal customary residence system based on locational characteristics. The former is widely practiced in the southern part of the country and the latter in the north. In Southern Malawi, where farmers primarily cultivate tea for revenue generation through some form of contract farming, there is a need for them to understand the drivers of investment in their agribusiness to ensure a sustained livelihood and thus alleviate rural poverty. Moreover, population growth is on the increase and is making agricultural land scarce in Southern Malawi, which means it is important to concentrate efforts on increasing productivity [11]. Tea cultivation in Southern Malawi is characterized by the coexistence of large and small farms [11,12], which may influence the productivity of smallholders due to likely spillover effects. Therefore, this study also investigates diverse factors that drive productivity among the smallholder farmers cultivating tea in Southern Malawi.

As mentioned earlier, scholars have produced ambiguous empirical results about the relationship between the customary residence system and investments in land in Malawi. This is mostly because the exact pattern and measurement of tenure security and its effects on agricultural investments under contracted farmers remain unclear. These systems have continuously adapted to contexts from a historical perspective, i.e., appropriation of land during colonization and more recently the fragmentation of land. Whether the customary residence practices affect tenure security and how that relates to investment in land is work in progress that requires a comprehensive approach.

The Food and Agriculture Organization of the United Nations (FAO) [13] argued that as functioning markets develop and as land becomes scarce due to population growth in rural sub-Saharan Africa, land transfers will increase from land-abundant to land-poor households. However, land scarcity will also keep land resources and management within a certain lineage, and they will only be transferred within the community or family members based on customary residence systems in parts of SSA, e.g., Malawi and Burkina Faso [14,15]. This customary residence system may be inefficient and may constrain economic development, in part due to the inability to attract capital investment (e.g., access to credit) [16,17]. Interestingly, limited empirical evidence exists that shows agricultural development (e.g., investments and adoption of innovation) to be constrained by customary residence systems in Africa [18,19]. For example, Austin [20] and Berry [21] found that a growing number of farmers in Ghana and Nigeria, respectively, had invested in cocoa under customary land tenure despite not having a formal land title. Ultimately, customary land tenure systems have been able to avoid imposing unsuitable private land title systems, and these should be considered and investigated from a social, historical, and economic perspective [22,23]. Place and Otsuka [10] and Lunduka [9] argued that agricultural investments are weaker in matrilineal-matrilocal residence systems (i.e., husband moves to the wife's village) compared to patrilineal systems, while other authors in [8] have found no evidence.

The hypothesis in this study is that certain types of customary residence systems of contracted smallholder farmers influence investment in tea and agroforestry. Furthermore, specific customary residence systems, among other factors, may influence the productivity of tea farms in the study area. In 2018, 228 contract farmers were surveyed in the Mulanje District, in the southern region of Malawi.

The socioeconomic characteristics of the tea farmers and other variables, such as customary residence systems, land tenure, farm characteristics, farmer block, asset, housing facility, transportation, tea expenditure and yield, number of trees, and access to credit, were collected.

The effect of tenure security on investments and productivity was estimated from the cross-sectional data set with linear probability as well as ordinary least squared (OLS) models [24]. The former provides unbiased parameter estimates with minimum variance. The results of this study suggest that for Southern Malawi, matrilocal residence practices influence investment in tea cultivation compared to patrilineal residency. Furthermore, the proximity of smallholder farmers to large commercial tea estates has a positive influence on the farmer's productivity [25].

This study is structured as follows: Section 2 describes the material and methods, including the conceptual framework with a focus on the customary residence systems and investments as well as contract farming in the study areas of Southern Malawi. This section also provides an overview of the methodology. In Section 3, the study results are presented, followed by a discussion of the results and limitations in Section 4. In Section 5, the conclusion of the study is drawn.

2. Materials and Methods

There is a need for a broader approach to the adaptation and evolution of the relationship between land tenure security and agricultural investment and productivity due to both the complex nature of the former as well as multiple-actor involvement. Thus, the proposed conceptual framework is a system of concepts and assumptions supported by earlier studies [5]. This study also relies on the work of Arnot et al. [26], Miles et al. [27], Place [5], and Ghebru et al. [28] on tenure security and investment, thereby providing a comprehensive overview. Tenure security exists if farmers expect that the evolution of institutional rules will not decrease the expected utility of their assets. For instance, a landholder is considered tenure secure if their perception of losing their land in the future is low, i.e., nondecreasing expected utility. The expected utility derived from the land by the asset holder is denoted as (Equation (1)):

$$E\left(U^R\right) = f(R_0, r(R_t), C_0, c(C_t)) \tag{1}$$

where $E\left(U^R\right)$ is the expected utility derived from land resource R. R_0 is the status quo of resource benefits, e.g., soil quality, and $r(R_t)$ is its future state of benefits, e.g., level of income. C_0 is the status quo of institutional rules, e.g., taxes and permits, while $c(C_t)$ is their future state of institutional rule. Both the future states of resource benefits and institutional rules can be denoted respectively as (Equations (2) and (3)):

$$r(R_t) = \alpha x_t + \varepsilon_{rt} \tag{2}$$

$$c(C_t) = \alpha z_t + \varepsilon_{ct} \tag{3}$$

where αx_t denotes the predicted outcomes of the future state of resource benefits contingent on investment, e.g., in conservation agriculture, while αz_t is the predicted future state of institutional rule, e.g., lobbying; ε_{rt} and ε_{ct} are respectively the unpredicted components of the future state of resource benefits and institutional rules due to weather, natural catastrophes, war, corrupt practices, etc. For tenure security to prevail in Africa, given the fact that the future state of resource benefits evolves faster over time compared to institutional rules, which are coordinated by social institutions, the emphasis should therefore be on $c(C_t)$. Thus, a landholder assured of the future state of institutional rules has a high level of tenure security independent of the initial property right substance. The expected utility of resources under tenure (in)security can be denoted as (Equations (4) and (5)):

$$E\left(U^R \ Secure\right) \rightarrow f\frac{\partial E\left(U^R\right)}{\partial c(.)} \geq 0 \tag{4}$$

$$E\left(U^R \; Insecure\right) \rightarrow f\frac{\partial E\left(U^R\right)}{\partial c(.)} < 0 \tag{5}$$

This implies that farmers feels tenure secure if they expect that developments in institutional rules will increase their expected utility. Accordingly, this study adopts this into the conceptual framework by incorporating the development of social norms and rules as a tenure security component. This study presents a comprehensive link between land tenure security and investment incentives based on a certain distinct line of reasoning. First, it is based on the aforementioned argument that future institutional rules change will not adversely affect farmers' expected utility and the benefits derived from investing in the land. Second, it involves the use of land as collateral in order to facilitate access to credit to economically empower landholders. Finally, this type of tenure security can promote the development of land markets, due to transferability. This would incentivize landholders to invest in land, ultimately increasing its market value. The relationship between tenure security and investment has several caveats driven by the individualization of land rights, land commodification, and demographic changes.

In analyzing the relationship between tenure security and investment, it is important to further expand the scope of variables. For instance, other social norms and contract farming are variables that could influence the relationship. Social norms are embedded in customary residence systems, particularly family lineage, and provide different land tenure rights for men and women. However, in parts of SSA, women's land claims are often lower to those of other male members of the family, irrespective of their status [5]. Certain customary residence norms may result in securing land tenure for all. Contract farming for smallholders has the potential to address the failure of functioning output and input markets in rural areas [27]. Access to functioning markets has a higher economic benefit to farmers that have secure tenure [28]. There are different types of contract farming between retailers and farmers, retailers and cooperatives that act on behalf of their members, as well as large estate farms and neighboring smaller farms. In essence, contract farming is established as the so-called nucleus estate schemes, i.e., large agribusinesses and/or farms with high productivity enter into contractual agreements with surrounding smallholders [29]. The farmers deliver products of a specified quantity and quality to buyers at a negotiated price. To increase and improve the supply chain, the buyer often provides extension services such as agricultural practice training as well as production-inputs, loans, logistics, etc. [30]. There are limited studies on the relationship between contract farming and tenure security because contract farming itself does not affect tenure security directly but indirectly through other factors, e.g., investment, productivity. For instance, only farmers confident in their ability to produce food products that meet the quantity and quality requirements will enter into contractual agreements. Grosh [31] found that buyers would rather contract limited smallholders with higher productivity. This could also imply that contracted smallholders may be better off financially and be able to purchase land resulting in a concentration of holdings rather than fragmentation. This would allow for higher investment in agricultural input and productivity on the path of smallholders. Given the higher investment in agricultural input and productivity, due to land tenure security, as well as the corresponding demand, there is economic growth in the agrarian community. Figure 1 illustrates the broader perspective of the aforementioned relationship between tenure, investment, and contract farming.

Malawi is a land-locked country in Eastern Africa that borders Tanzania, Mozambique, and Zambia. Malawi has a total land area of 118,484 square kilometers (km^2) of which 79.4% is arable land [32]. Malawi's main economic driver is agriculture and has a population of 17.5 million, which is expected to double by 2038 [33]. More than 80% of the Malawian population relies on agriculture for income generation, with tobacco, sugar, and tea being the largest contributors to the economy alongside foreign aid [34]. Malawi has one of the highest population growth rates as well as the one of the lowest Human Development Index (HDI) ranks in Africa [35,36]. Thus, the country is experiencing land scarcity as well as poverty, which is due to low productivity in the agriculture sector [33]. In Malawi, government and traditional rule coexist, forming a dual system of governance [37]. Malawi legally recognizes three land tenure systems: customary, estate, and public. The customary land tenure system is the predominant system [32]. In Malawi, a variety of customary tenure systems coexist as various ethnic groups have different land rights norms and practices [23]. For instance, in the north of the country, there are the Ngoni and Tumbuka ethnic groups that predominantly practice the patrilineal systems while in the south are the Chewa, Yao, and Lomwe that practice the matrilineal system. Across the ethnic groups of Malawi, smallholders mostly transfer land via inheritance or gifting.

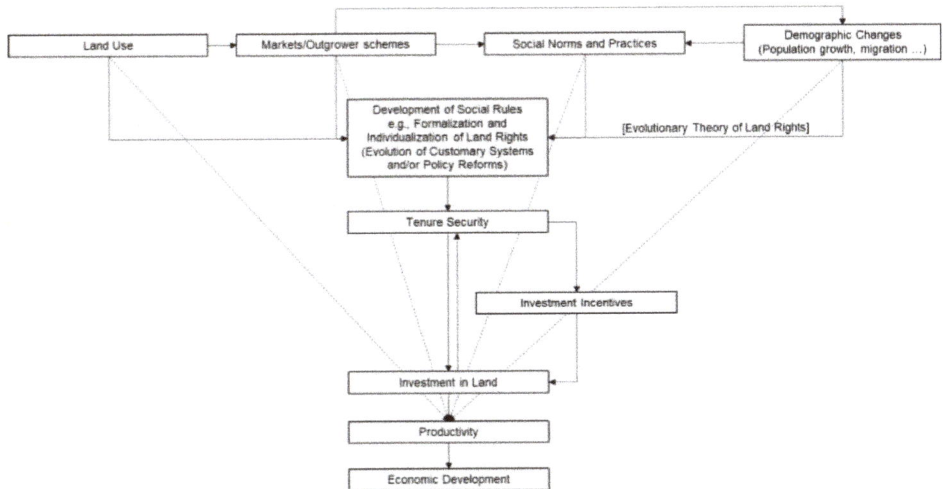

Figure 1. Conceptual framework of tenure security and investments in land in their broader context. Adapted from Feder [29], Place [5] and Ghebru et al. [28].

Land can be inherited along either the male or the female lineage, depending on the regionally dominant system. In the matrilineal systems, which is common in the southern region of Malawi, land is passed along the female line [38]. In the northern region, the patrilineal system is practiced, where land is passed from father to sons [23]. In both matrilineal and patrilineal systems, descendants also receive land as a gift in case of special events such as marriage, birth, or maturity of the heir [14]. The inheritance system is also linked to the residence and marriage, which strongly influences the land rights of the husband or the wife [38]. Couples in Malawi reside either matrilocally, patrilocally, or neolocally after marriage. In matrilocal systems, the husband moves to the wife's location, while in patrilocal systems, the wife moves to the husband's location. Neolocal residence occurs when couples move to a place where they have no family ties. In matrilineal-matrilocal societies, the husband is expected to leave the wife's village and return the land back to her lineage if she dies or they divorce [8]. The man is an outsider in his wife's village, often not respected by his wife's male relatives [39,40]. Moreover, the brother of the wife and his descendants have first claim to the resources. Thus, the matrilineal-matrilocal system creates tenure insecurity in that men outside the family lineage can be deprived of the land they cultivate. This may create a situation where married male smallholders living in a matrilineal-matrilocal system are disincentivized from investing in agriculture. Place and Otsuka [10] ranked the Malawian male tenure security according to the customary inheritance and residence system, which is a kind of assurance measurement (see Table 1).

This customary-based tenure security illustrates the complexity of tenure security and its measurement [3]. Furthermore, social norms and practices, markets, land use, and demographic changes have evolved over time, influencing customary inheritance and residence norms. For instance, land reforms across SSA over the years are recognizing customary rule; this would affect the dynamics of customary inheritance and residence norms.

Table 1. Customary-based tenure security ranking of male land use decision maker.

Inheritance-Residence	Description	Tenure Security
Matrilineal-matrilocal	A husband moves to the wife's village; the land belongs to the family of the wife.	Low
Matrilineal-patrilocal	A wife moves to a husband's village; the land belongs to the family of the wife	
Matrilineal-neolocal	Wife and husband move to a village, not of their origin; the land belongs to the community of that village.	
Patrilineal-matrilocal	A husband moves to a wife's village; the land belongs to the family of the husband	
Patrilineal-patrilocal	A wife moves to a husband's village; the land belongs to the family of the husband	High

Source: Place and Otsuka [10].

Tea is one of the major smallholder cash crops cultivated in the southern districts of Malawi. Tea is mostly grown in the Thyolo and Mulanje districts, as the area provides adequate agroclimate conditions [41]. The orange boundaries in Figure 2 represent the agricultural estate boundaries where most of the tea in the study area around Blantyre are produced. According to Chirwa and Kydd [42], after the introduction of tea contract schemes in 1966, contracted smallholders' cultivation areas have grown from 30.8 to 2,900 hectares (ha) of customary land by 2004, producing 810 kilograms per hectare of tealeaf. Even though contracted smallholders produced less than 7% of all Malawian tea in 2013, the future growth of Malawi's tea sector depends on these farmers [10]. The main tea production comes from the estates because contracted smallholders have low tea bush density on their land (6000–8000 plants/ha) in comparison to estates (15,000 plants/ha) [10]. Tea can provide income for up to 100 years with minimal risk of total crop failure, but it requires a high initial investment in time and money and does not provide direct income in the first few years [10,41]. Another necessary long-term investment that supports tea production is tree planting. To increase the productivity of tea plants, trees are planted to shade the tea plants and plucked tealeaves [42,43]. In the existing literature on tenure security and investment, trees are commonly used as a long-term investment as they can be security- and productivity-enhancing [23,44,45]. The investment in trees increases tenure security as well as establishes, to some extent, access and ownership. However, the security-enhancing effect here is trivial given that tenure security in Malawi is based on inheritance and residence norms because they determine access and rights, regardless of the previous investments [46].

This study samples 228 smallholder tea outgrowers in the Mulanje District using a simple random sampling technique of tea associations and cooperatives. These farmers, along with their associations and cooperatives, as well as the cultivation area provide a representative sample for Southern Malawians. The study area is one of the largest tea-growing areas in Malawi with over 5000 smallholder farmers in different farm associations and cooperatives. The particular cooperative under consideration negotiates contract conditions and tealeaf prices for the farmers with the tea processing and growing companies. These companies pay farmers on a monthly basis per kilogram of tealeaf provided and are often owners of tea plantations in the area. A number of tea estates also support the farmers with extension services and tea seedling nurseries. Furthermore, farmers are divided into farmer blocks to increase efficiency. Every farmer block has at least one tealeaf collection point where farmers deliver their produce. At the collection points, tea processing and growing companies weigh, record, and transport tealeaves to the processing facility. Most cooperative farmers grow tea on customary land while processing and growing companies grow tea on freehold land (see Figure 3 for an overview).

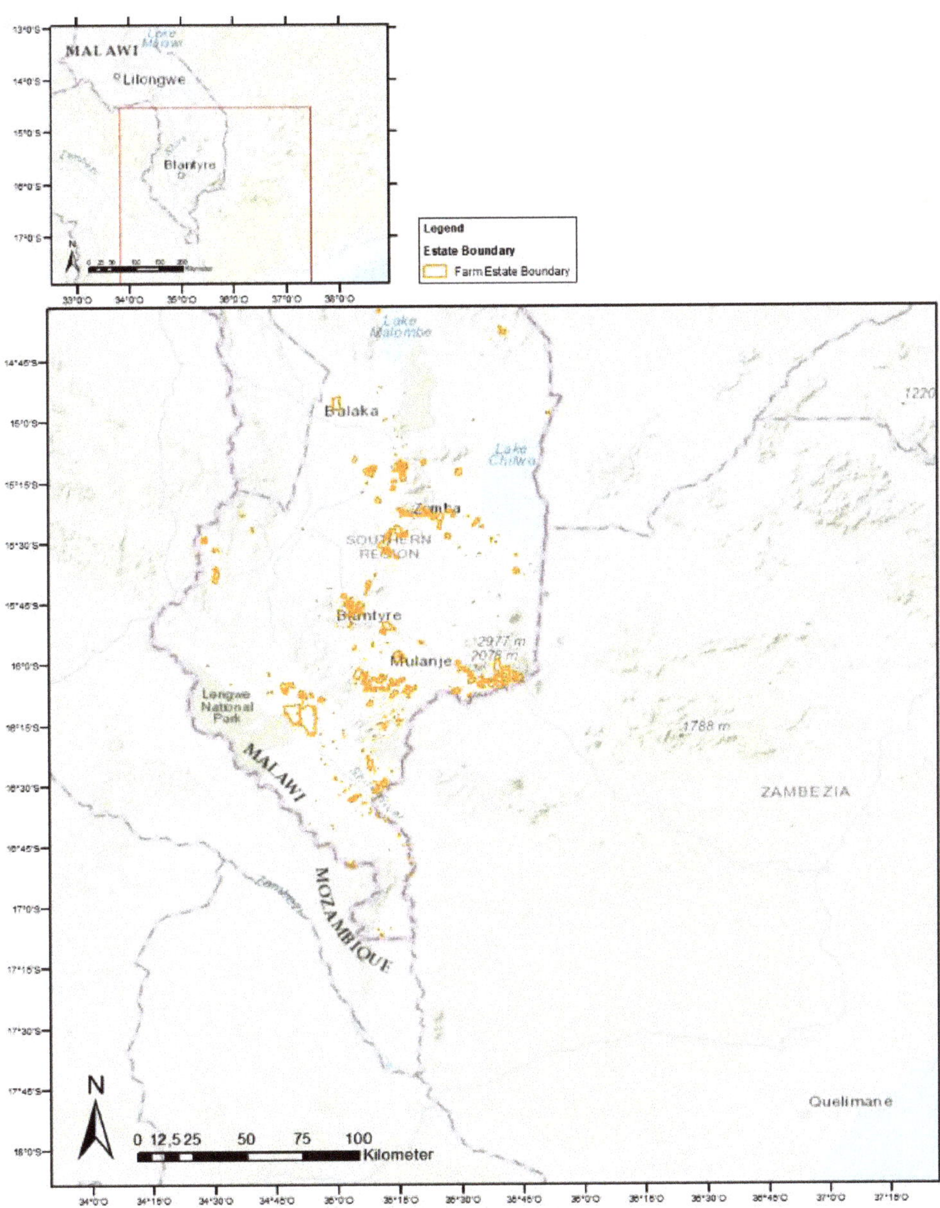

Figure 2. Map of the estate in Southern Malawi.

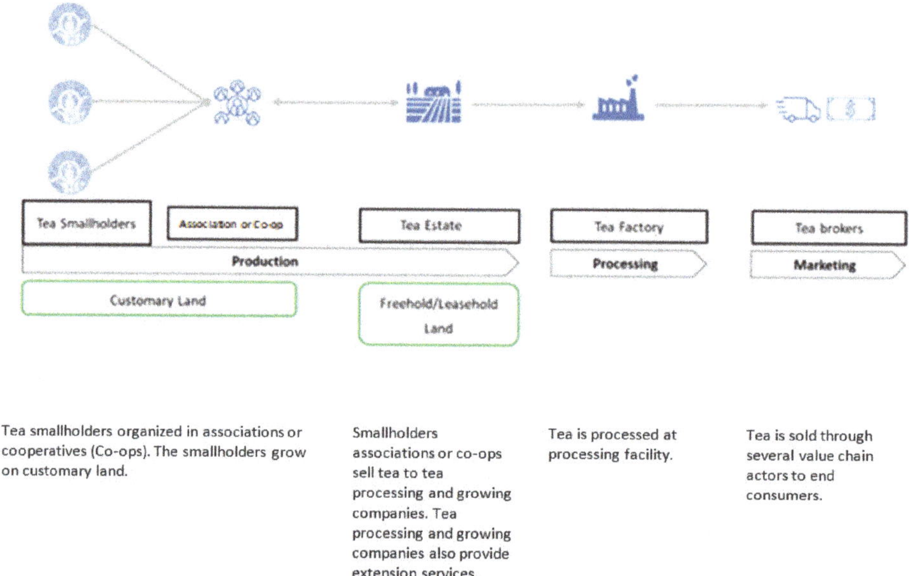

Figure 3. Overview of tea outgrower scheme in Southern Malawi.

Data were collected from four farmer blocks, which represent 16% of the existing farmer blocks in the study area. The residence types are distributed across the four farmer blocks, A, B, C, and D. Farmer Block A has a fairly diverse mix of all residency systems in the study areas and is somewhat spread further away from commercial estate. The patrilocal residence is prevalent in the northeastern Farmer Block B and has close proximity to commercial estate. The matrilocal residence is mostly practiced in the southern Farmer Blocks C and D, with the former being further away from commercial estate.

This study uses a logit and OLS regression to measure the effect of tenure security on investments in tea plants and agroforestry based on a linear model assumption of the independence of variables, homoscedasticity, and normal distribution [24]. There are two separate functions conducted for the analysis, the investment, and productivity functions. These functions are denoted as (Equations (6) and (7)):

$$Y_{tea\ or\ tree} = \beta_0 + \beta_1 X_1 + \beta_2 X_2 + \beta_3 X_3 \ldots + \beta_n X_n + \varepsilon \qquad \rightarrow \text{Logit} \qquad (6)$$

$$Y_{log\ yield\ per\ hectare} = \beta_0 + \beta_1 X_1 + \beta_2 X_2 + \beta_3 X_3 \ldots + \beta_n X_n + \varepsilon \rightarrow \text{OLS} \qquad (7)$$

where $Y_{tea\ or\ tree}$ is the dependent (discrete) variable that illustrates the investment in tea or trees; β_0 is the intercept, while β_1 to β_n are the coefficients of interest of each independent variable X_1 to X_n; $Y_{log\ yield\ per\ hectare}$ is the dependent (continuous) variable that illustrates the (log) yield per hectare of tea, while the other variables are similar to those above with expenditure on tea, i.e., input from the previous year in Table 2 (X6), also expressed in natural log.

Table 2. Overview of variables used for the regression analysis.

Name	Descriptions	Scale	Unit
Dependent Variables			
$Y_{tea\ or\ tree}$ (investment in tea or trees)	1 or 0; invested in tree and/or tea in 2017	Discrete	-
$Y_{yield\ per\ hectare}$ (yield per hectare)	Yield per hectare of farmer i on farm i	Continuous	Kg
Independent Variables			
Tenure security indicator (X1)			
Residence	Residence of farmer	Categorical	-
Marital status	Single, married, or widowed	Categorical	-
Land tenure (X2)			
Land tenure form	Customary or leasehold	Discrete	-
Land acquisition	How farm was acquired (e.g., inheritance)	Categorical	-
Land acquisition time	Years of land acquisition	Continuous	-
Land dispute	Is there dispute about farm	Discrete	-
Household characteristics (X3)			
Household size	Number of adults in household	Continuous	-
Asset	Ownership of items (e.g., radio)	Categorical	-
Housing structure	Type of housing structure	Discrete	-
Transport	Ownership of transport vehicle	Categorical	-
Livestock	Ownership of livestock	Categorical	-
Other income source	Income source other than tea	Categorical	-
Access to finance	Access to loan in the past years	Discrete	-
Number of farms	Number of plots owned	Continuous	-
Demographics (X4)			
Gender	Gender of respondent	Discrete	-
Age	Age of respondent	Continuous	-
Education	Education level of respondent	Categorical	-
Literacy	Literacy of respondent	Discrete	-
Farm Characteristics (X5)			
Farm size	Measured using geoinformation system	Continuous	ha
Land per capita	Farm size per capita of household	Continuous	-
Topography	Topography of the farm	Categorical	-
Land Use (X6)			
Revenue	Revenue from tea last year	Continuous	USD
Expenditure (input)	Expenditure on tea last year	Continuous	USD
Age of tea bushes	Year tea bushes was planted	Continuous	Years
Tress planted	Number of trees planted	Continuous	-
Farmer block (X7)			
Farmer block	Membership of each farmer	Categorical	-

USD: United States dollars.

3. Results

3.1. Descriptive Statistics

The descriptive statistics is presented in Table 3. Overall, 64% of the smallholders surveyed were married, with half of them stating that they were in either a patrilocal or matrilocal residency. An estimated 96% of the surveyed smallholder farmers cultivate their farm under either one or the other forms of customary land tenure arrangement. The average household size in the sample is five members. On average, there are two adults in the family between the working age of 14 and 50 years old. The average age of a farmer is 52 years old. On average, a farmer has two farmlands with an average size of 0.2 hectares. The majority (64%) of the farmlands in this survey are mostly acquired through inheritance. The average age of tea bushes is 28 years, which generated a total of 700 USD in revenue for the farmers under consideration. The average expenditure on the farm by smallholder farmers was 92 USD. Only 15% of the surveyed smallholder reported that they had received a loan in the last years.

Table 3. Descriptive statistics of relevant variables.

Variable	Mean	SD	Min	Max
Dependent Variables				
Investment in tea or trees, $Y_{tea\ or\ tree}$	0.36	0.48	0	1
(Log) Yield per hectare, $Y_{log\ yield\ per\ hectare}$	9.62	0.89	7.49	12.4
Independent Variables				
Residence				
Patrilocal	0.26	0.44	0	1
Matrilocal	0.30	0.46	0	1
Neolocal	0.01	0.09	0	1
Married purchase	0.08	0.26	0	1
Status				
Married	0.63	0.48	0	1
Single	0.07	0.25	0	1
Separated	0.07	0.26	0	1
Widowed	0.22	0.41	0	1
Land tenure				
Land tenure form	0.96	0.18	0	1
Land acquisition			0	1
Gift	0.19	0.4	0	1
Inheritance	0.64	0.48	0	1
Chief	0.03	0.20	0	1
Purchase	0.08	0.29	0	1
Settlement	0.02	0.15	0	1
Share-cropping	0.02	0.13	0	1
Land acquisition time	28	18	0	105
Land dispute	0.01	0.09	0	1
Household characteristics				
Household size	5	2	1	12
Adults	2	1	0	9
Asset				
Phone	0.19	0.39	0	1
Phone and radio	0.46	0.50	0	1
TV	0.02	0.14	0	1
Fridge	0.04	0.19	0	1
Housing structure	0.54	0.50	0	1
Transport				
None	0.22	0.41	0	1
Bicycle	0.59	0.49	0	1
Motorcycle	0.17	0.38	0	1
Car	0.02	0.13	0	1
Livestock				
None	0.35	0.48	0	1
Chicken	0.44	0.50	0	1
Pigs	0.07	0.26	0	1
Others	0.14	0.35	0	1
Other income source				
None	0.25	0.44	0	1
Farming	0.32	0.47	0	1
Labor	0.29	0.46	0	1
Employed	0.07	0.26	0	1
Access to finance	0.15	0.36	0	1
Numbers of farm	2	1	1	5

Table 3. *Cont.*

Variable	Mean	SD	Min	Max
Demographics				
Gender	0.60	0.49	0	1
Age	53	16	22	103
Education				
None	0.29	0.45	0	1
Primary	0.54	0.50	0	1
Secondary	0.15	0.36	0	1
Tertiary	0.02	0.15	0	1
Literacy	0.30	0.46		
Farm characteristics				
Farm size	0.20	015	0	1
Topography				
Flat	0.22	0.42	0	1
Light slope	0.40	0.49	0	1
Steep slope	0.38	0.49	0	1
Land use				
Revenue	700	1343	85	12,366
Tea expenditure (input)	92	89	0	651
Age of tea bushes	28	12	0	67
Tress planted	12	20	1	121
Farmer block				
A	0.23	0.42	0	1
B	0.25	0.44	0	1
C	0.27	0.44	0	1
D	0.26	0.44	0	1

3.2. Regression Results

The results of the regression analysis of Equations (6) and (7) are presented in Tables 4 and 5. Model 1 illustrates the basis model of the analysis, extended by the other listed (relevant) independent variables in subsequent models. The difference in the number of observations between Tables 4 and 5 is due to missing observations for certain independent variables in the respective estimations. In Table 4 we present the marginal effect of the logit regression that is statistically significant, as this can be adequately interpreted. The results suggest that matrilocal residence is ca. 20% more likely to invest in tea or trees compared to patrilocal residency. Conversely, widows are ca. 20% less likely to invest in tea or trees compared to patrilocal residency. Smallholder farmers that acquire land through inheritance were ca. 16% less likely to invest in tea or trees compared to those that receive land as a gift. A female smallholder was 20% less likely to invest in tea or trees compared to their male counterpart. Those with primary education were 15% less likely to invest in tea or trees compared to those without formal education.

The productivity analysis is presented in Table 5. Contrary to the investment analysis above, nonstatistically significant variables were also reported. The results suggest that increase in tea inputs, such as fertilizer, pesticide, results in increases in tea bushes output. Those that inherited land were less productive compared to those that receive farmland as a gift from other sources, e.g., the government. Neolocal farmers were less productive compared to the patrilocal customary system (weak statistical significance). The larger the household size, the more productive the tea-growing smallholder (weak statistical significance). Conversely, the higher the number of working age adults in the household, the less productive the tea outgrowing smallholder. The results also suggest that increasing the farm size of tea-growing smallholders does not necessarily lead to higher productivity. Furthermore, the higher the number of trees on the farms (agroforestry), the lower the productivity. Farmers in Blocks B and D were found to be more productive than farmers in Block A.

Table 4. Significant marginal effects of variables on investment in tea or trees (Logit Equation (6)).

Investment in Tea or Trees $Y_{tea\ or\ tree}$	Model 1	Model 2	Model 3
Matrilocal	0.18 *	0.20 *	0.22 **
	(1.66)	(1.86)	(2.10)
Widowed	−0.23 **	−0.21 *	−0.18 *
	(−2.03)	(−1.89)	(−1.63)
Inheritance	−0.18 *	−0.16 *	−0.16 *
	(−1.89)	(−1.77)	(−1.83)
Gender	−0.23 ***	−0.20 ***	−0.21 ***
	(−2.75)	(−2.45)	(−2.45)
Number of farms		0.11 ***	0.07 *
		(3.16)	(1.75)
Education (primary)			−0.15 *
			(−1.69)
Tea labor expenditure			0.001 *
			(1.88)
Observations	135	135	135

* $p < 0.1$, ** $p < 0.05$, *** $p < 0.01$.

Table 5. Effects of specific variables on yield per hectare (OLS Equation (7)).

$Y_{logyield\ per\ hectare}$	Model 1	Model 2	Model 3
Log Tea Expenditure (Inputs)	0.150 **	0.157 **	−0.0302
	(0.0725)	(0.0789)	(0.0675)
Inheritance	−0.451 ***	−0.419 **	−0.222
	(0.167)	(0.167)	(0.148)
Matrilocal		0.125	0.209
		(0.164)	(0.142)
Neolocal		−0.717	−1.301 **
		(0.514)	(0.209)
Household Size		0.0721 **	0.0661 **
		(0.0341)	(0.0317)
Adult		−0.165 ***	−0.171 ***
		(0.0531)	(0.0505)
Farms Size			−3.27 ***
			(0.32)
Number of trees			−0.01 **
			(0.004)
Farmer Block B			1.083 ***
			(0.180)
Farmer Block D			0.392 ***
			(0.139)
Constant	8.366 ***	8.171 ***	9.647 ***
	(0.776)	(0.806)	(0.677)
Observations	224	224	224
R-squared	0.071	0.124	0.365

* $p < 0.1$, ** $p < 0.05$, *** $p < 0.01$.

Given that certain econometric issues can arise due to omitted variables, the relationship between dependent and independent variable as well as the error term (endogenity) in an OLS regression interpretation of results should be done with caution. This study, however, undertakes some basic tests. The Durbin Watson Test of an OLS regression of 1.83 implies that there is no autocorrelation, i.e., serial correlation between the error terms such that our significant results may be valid. Furthermore, the white test result of 127.98 and a p-value of 0.19 imply there is no major issue with the heteroscedasticity of the error term (robust option). While the use of a two-stage least square estimation (2SLS) would ensure that endogenity is adequately addressed, finding a suitable instrumental variable (IV) in the sample is unlikely.

4. Discussion

This study investigates whether the customary residence of (married) farmers influences agricultural investment in tea and agroforestry under contract farming conditions. Furthermore, it explores which factors, including customary residence systems, influence the productivity of contracted tea farms in Southern Malawi. The results of this study contribute to the literature on the relationship between tenure security, agricultural investment, and productivity among contracted smallholders in Malawi [7,44,46]. While Place and Otsuka [10] and Lunduka [9] found that the probability of investing in agroforestry is higher among patrilocal and neolocal residence households compared to matrilocal residence households, this study found the opposite with regards to investment in tea and/or agroforestry. This could be because the study area of Southern Malawi is dominated by the matrilineal-matrilocal customary system. Matchaya [47] found that female-headed households in matrilineal societies are more land secure than males, which could also incentivize investment in land for its long-term benefit. Lovo [8] found that (male) decision-makers in matrilineal-matrilocal villages were more likely to invest in trees, as this consolidates their tenure security. This study found that widows were less likely to invest in tea or agroforestry. While widows in parts of sub-Saharan Africa (e.g., Zambia) have been found to undertake limited investment in agricultural land that they are unable to inherit, the result for Southern Malawi is somewhat surprising, given the prevalence of the matrilineal-matrilocal system [48]. This could be an indication that women are also disadvantaged under the matrilineal-matrilocal system. Inheritance of land, compared to other forms of land acquisition such as a gift from the village chief, does not seem to drive investment in tea or agroforestry by smallholders in Southern Malawi. Gibson and Gurmu [48] found that land inheritance compared to other forms of land allocation in parts of sub-Saharan Africa (e.g., Ethiopia) was inefficient due to diminishing resources and competition from a large number of siblings. Place and Otsuka [10] attributed fragmentation of land in matrilineal-matrilocal systems in Southern Malawi to a large number of competing heirs. This may likely dissuade investment in tea or agroforestry. The results of this study also suggest that there is gender disparity in investment in tea or agroforestry with women at a disadvantage. Meijer et al. [49] argued that agroforestry and management is a male-dominated sector in Malawi. Chirwa [50] also found a negative relationship between fertilizer investment and/or adoption and the female-headed smallholder agrarian households in Southern Malawi. Smallholders having more than one farm in Malawi were more likely to invest in tea or agroforestry, a finding that is also similar to that of other studies (see [44,51]). Chirwa [50] also found that investment and/or adoption in fertilizer and improved seeds in Malawi was driven by land holdings. This study found that farmers with primary-level education were less likely to undertake agricultural investment in tea and agroforestry in Malawi. This is similar to the findings of Deininger and Jin [52] and Chirwa [53] that well-educated smallholder farmers in Malawi were likely to invest in agricultural technologies compare to less educated smallholders.

This study establishes a basis for the production function by estimating the relationship between inputs (e.g., fertilizer, pesticide) and the output (i.e., yield per hectare). The positive and statistically significant relationship shows that the model predictions are plausible. Lunduka [9] and Place and Otsuka [10] argued that investments in agricultural inputs such as (in)organic fertilizer, hired labor, hybrid seeds, and pesticide in Malawi result in increases in agricultural production efficiency. The results of this study suggest that smallholders in neolocal residence systems were less productive compared to those in the patrilocal system with weak statistical significance. Lunduka [9] found that those households that were tenure insecure in Malawi had lower production efficiency compared to tenure secure households. Considering that the predominant customary residence in Southern Malawi is matrilineal-matrilocal, farms under this system could be expected to have lower productivity relative to others. However, Place and Otsuka [10] found that for maize and tobacco cultivation in Malawi, the diverse tenure systems did not influence productivity. This study found that household size has a positive correlation with tea yield per hectare and may signify more reliance on household labor. This is similar to the results of the study by Chirwa and Kydd [54] for tea outgrowing smallholders in Malawi. This study also found that as the number of working age adults in the household increases, tea yield per hectare declines, which is evidence for the diminishing return to labor. While Benjamin et al. [55] found that a large farm size could lead to favorable financial credit conditions for smallholder farmers in parts of sub-Saharan Africa, this study found that increasing farm size among tea-growing smallholders in Malawi does not result in high productivity. The results also suggest that smallholder farms in Farmer Blocks B and D were more productive compared to those in Farmer Block A. While there is limited information on the differences in the mode of operation between the farmer blocks, a possible explanation is the proximity to estates and a likely spillover effect of agronomic practices. Chirwa [53] and Chirwa and Kydd [54] argued that contractual arrangements between smallholder farmers and commercial estates result in higher productivity for the farmers compared to those that trade with grower-leased factories and state-owned enterprises. Another important outcome of this result is that farmers in the predominately-matrilocal residence system Block D, with close proximity to commercial estates, are more productive, confirming the initial findings of the positive but insignificant relationship between matrilocal residency and productivity. This also implies that farmers in blocks with a predominantly patrilocal residence system can be productive if they are in close proximity to commercial estates.

Some of the variables of the models can be ascertain as ecosystem services (agroforestry) and marketed outputs (yield). According to Benjamin and Sauer [56], the relationship between these variables can be competitive, complementary, or supplementary in nature. The relationship between these variables in parts of sub-Saharan Africa (e.g., Kenya) was found to be more complementary and supplementary rather competitive [56]. Therefore, it is important to investigate such relationships in a Malawian context. In the study area, as the number of trees on farmland increases, the yield per hectare of tea decreases, which may imply a competitive relationship between the crop and ecosystem services once it passes a certain threshold [56]. It is important to mention at this stage that there is a possible existence of selection bias and endogenity in this analysis. Some of the farmers may self-select into specific blocks based on particular attributes such as education, farm size, etc. Thus, future analysis should use certain methods such as propensity score matching (PSM) and/or Heckman selection as well as a two-stage model to address some of these issues.

5. Conclusions

The customary land tenure systems discourse in sub-Saharan Africa is relevant for achieving sustainable agricultural development. Findings from the discourse affect food security and the livelihoods of millions of smallholder farmers. Customary tenure systems across SSA differ and have evolved over time. Their impact on existing farming models, such as contract farming, should also be taken into consideration. Therefore, it is important to conduct country-, region-, and case-specific analyses. This study analyzes the determinants of investment in tea and agroforestry and the factors that influence tea productivity for diverse customary residences under contract farming in Southern Malawi. The results seem to suggest that the predominant matrilocal tenure system in the study area positively influences investment in tea and agroforestry. Other factors such as the proximity of small farms to commercial estates as well as household size also have a similar effect on productivity. This is evidence for the so-called spillover effect where farms close to profitable large-scale commercial estates imitate their operations. Gender bias in agricultural investment is prevalent among the tea outgrowers in Southern Malawi while the trait of having a higher level of education continues to drive investment among smallholders.

There is a need for policy to address gender gaps in Southern Malawi despite the dominance of a matrilineal-matrilocal system, as women may be vulnerable and insecure even in this customary system. Furthermore, emphasis should also be placed on providing less educated tea outgrowing smallholder farmers with the right tools that induces investment in agriculture as a way of safeguarding their livelihood. Thus, the use of extension services and farm management intermediaries could contribute to higher productivity in agricultural (tea) production [54–56]. There is a need to conduct a detailed gender sensitive analysis on how future land reforms that are related to customary systems and contract farming would impact agricultural investment and productivity beyond just the agricultural input in Malawi.

Funding: This research was funded by Jonathan Seipl.

Acknowledgments: Special thanks are given to Jonathan Seipl for the concept and data collection and also Stefania Lovo for taking time to introduce her work and ideas. We would also like to thank Tobias Bendzko from the Technical University of Munich, Germany for supporting the GIS-related work.

Conflicts of Interest: The authors declare no conflict of interest.

References

1. Lowder, S.K.; Skoet, J.; Raney, T. The number, size, and distribution of farms, smallholder farms, and family farms worldwide. *World Dev.* **2016**, *87*, 16–29. [CrossRef]
2. Bromley, D.W. *Environment and Economy: Property Rights and Public Policy*; Basil Blackwell Ltd.: Oxford, UK, 1991.
3. Arnot, C.D.; Luckert, M.K.; Boxall, P.C. What Is Tenure Security? Conceptual Implications for Empirical Analysis. *Land Econ.* **2011**, *87*, 297–311. [CrossRef]
4. Lawry, S.; Samii, C.; Hall, R.; Leopold, A.; Hornby, D.; Mtero, F. The impact of land property rights interventions on investment and agricultural productivity in developing countries: A systematic review. *J. Dev. Eff.* **2016**, *9*, 61–81. [CrossRef]
5. Place, F. Land Tenure and Agricultural Productivity in Africa: A Comparative Analysis of the Economics Literature and Recent Policy Strategies and Reforms. *World Dev.* **2009**, *37*, 1326–1336. [CrossRef]
6. Fenske, J. Land tenure and investment incentives: Evidence from West Africa. *J. Dev. Econ.* **2011**, *95*, 137–156. [CrossRef]
7. Deininger, K.; Xia, F.; Holden, S. Gendered Incidence and Impacts of Tenure Insecurity on Agricultural Performance in Malawi's Customary Tenure System. *J. Dev. Stud.* **2019**, *55*, 597–619. [CrossRef]
8. Lovo, S. Tenure insecurity and investment in soil conservation. Evidence from Malawi. *World Dev.* **2006**, *78*, 219–229. [CrossRef]
9. Lunduka, R. *Land Rental Markets, Investment and Productivity Under Customary Land Tenure Systems in Malawi*; Norwegian University of Life Sciences: As, Norway, 2009.
10. Place, F.; Otsuka, K. Tenure, agricultural investment, and productivity in the customary tenure sector of Malawi. *Econ. Dev. Cult. Chang.* **2001**, *50*, 77–99. [CrossRef]
11. Nankumba, J.S.; Kalua, B.; Kishindo, P. *Contract Farming and Outgrower Schemes in Malawi: The Case Study of Tea and Sugar Smallholder Authorities: Research Report*; Centre for Social Research, University of Malawi: Zomba, Malawi, 1989.
12. Chinigò, D. Rural radicalism and the historical land conflict in the Malawian tea economy. *J. S. Afr. Stud.* **2016**, *42*, 283–297. [CrossRef]
13. FAO. *Analysis of Price Incentives for Tea in Malawi*; Technical notes series; MAFAP; FAO: Rome, Italy, 2015.
14. Kishindo, P.; Mvula, P. Malawi's land problem and potential for rural conflict. *J. Contemp. Afr. Stud.* **2017**, *35*, 370–382. [CrossRef]
15. Holden, S.T.; Otsuka, K. The roles of land tenure reforms and land markets in the context of population growth and land use intensification in Africa. *Food Policy* **2014**, *48*, 88–97. [CrossRef]
16. Matlon, P. Indigenous land use systems and investments in soil fertility in Burkina Faso. In *Searching for Land Tenure Security in Africa*; The World Bank: Washington, DC, USA, 1994; pp. 41–69.
17. Takane, T. Customary land tenure, inheritance rules, and smallholder farmers in Malawi. *J. S. Afr. Stud.* **2008**, *34*, 269–291. [CrossRef]
18. Johnson, O.E.G. Economic analysis, the legal framework and land tenure systems. *J. Law Econ.* **1972**, *15*, 259–276.

19. Feder, G.; Feeny, D. Land tenure and property rights: Theory and implications for development policy. *World Bank Econ. Rev.* **1991**, *5*, 135–153. [CrossRef]

20. Noronha, R. *A Review of the Literature on Land Tenure Systems in Sub-Saharan Africa*; Agriculture and Rural Development Department, World Bank: Washington, DC, USA, 1985.

21. Austin, G. *Labour, Land, and Capital in Ghana: From Slavery to Free Labour in Asante, 1807–1956*; Boydell & Brewer: Suffolk, UK; Woodbridge, UK, 2005; Volume 18.

22. Berry, S.S. *Cocoa, Custom, and Socio-Economic Change in Rural Western Nigeria*; Clarendon Press: Oxford, UK, 1975.

23. Riddell, J.C.; Dickerman, C. *Country Profiles of Land Tenure: Africa 1986*; Land Tenure Center, University of Wisconsin-Madison: Madison, WI, USA, 1986.

24. Bruce, J.W.; Migot-Adholia, S.E. *Searching for Land Tenure Security in Africa*; Kendall/Hunt Publishing Company: Dubuque, IA, USA, 1994.

25. Deininger, K. *Land Policies for Growth and Poverty Reduction*; World Bank Publications: Washington, DC, USA, 2003.

26. Alchian, A.A.; Demsetz, H. The property right paradigm. *J. Econ. Hist.* **1973**, *33*, 16–27. [CrossRef]

27. Miles, M.B.; Huberman, A.M.; Huberman, M.A.; Huberman, M. Qualitative data analysis. In *An Expanded Sourcebook*; Sage: Newcastle upon Tyne, UK, 1994.

28. Ghebru, H.; Khan, H.; Lambrecht, I. *Perceived Land Tenure Security and Rural Transformation: Empirical Evidence from Ghana*; IFPRI Discussion Paper 1545; International Food Policy Research Institute (IFPRI): Washington, DC, USA, 2006.

29. Feder, G. The implications of land registration and titling in Thailand. In Proceedings of the International Association for Applied Econometrics, Buenos Aires, Argentina, 24–31 August 1988; pp. 771–781.

30. Wooldridge, J.M. *Introductory Econometrics: A Modern Approach*; Nelson Education: Toronto, ON, Canada, 2015.

31. Grosh, B. Contract farming in Africa: An application of the new institutional economics. *J. Afr. Econ.* **1994**, *3*, 231–261. [CrossRef]

32. Bruce, J.W. Simple solutions to complex problems: Land formalization as a 'silver bullet'. In *Fair Land Governance: How to Legalise Land Rights for Rural Development*; Leiden University Press: Leiden, Holland, 2012; pp. 31–55.

33. Baumann, P. *Equity and Efficiency in Contract Farming Schemes: The Experience of Agricultural Tree Crops*; Overseas Development Institute: London, UK, 2000; Volume 111.

34. Eaton, C.; Shepherd, A. *Contract Farming: Partnerships for Growth*; Food & Agriculture Organization: Rome, Italy, 2001.

35. Key, N.; Runsten, D. Contract farming, smallholders, and rural development in Latin America: The organization of agroprocessing firms and the scale of outgrower production. *World Dev.* **1999**, *27*, 381–401. [CrossRef]

36. Kishindo, P. Customary land tenure and the new land policy in Malawi. *J. Contemp. Afr. Stud.* **2004**, *22*, 213–225. [CrossRef]

37. World Bank. The World Bank in Malawi. 2020. Available online: https://www.worldbank.org/en/country/malawi/overview (accessed on 15 May 2020).

38. Peters, P.E.; Kambewa, D. Whose security? Deepening social conflict over 'customary' land in the shadow of land tenure reform in Malawi. *J. Mod. Afr. Stud.* **2007**, *45*, 447–472. [CrossRef]

39. World Bank. Population Density (People per sq. km of Land Area). 2007. Available online: https://data.worldbank.org/indicator/EN.POP.DNST (accessed on 15 May 2020).

40. UNDP. Human Development Index and its Components. 2020. Available online: http://hdr.undp.org/en/data (accessed on 15 May 2020).

41. Eggen, Ø. Chiefs and everyday governance: Parallel state organisations in Malawi. *J. S. Afr. Stud.* **2011**, *37*, 313–331. [CrossRef]

42. Chirwa, E.; Kydd, J. Study on Farmer Organisations in Smallholder Tea in Malawi. 2005. Available online: https://agris.fao.org/agris-search/search.do?recordID=GB2012102311 (accessed on 15 May 2020).

43. Peters, P.E. "Our daughters inherit our land, but our sons use their wives' fields": Matrilineal-matrilocal land tenure and the New Land Policy in Malawi. *J. East. Afr. Stud.* **2010**, *4*, 179–199. [CrossRef]

44. Miller, D. Matriliny and social change; how are women of rural Malawi managing. In Proceedings of the CASID, Montreal, QC, Canada, 2 June 1996.

45. Kishindo, P. Emerging Reality in Customary Land Tenure: The Case of Kachenga Village in Balaka District, Southern Malawi. *Afr. Sociol. Rev. Rev. Afr. De Sociol.* **2010**, *14*, 102–111. [CrossRef]

46. Carr, M.K.V. *Advances in Tea Agronomy*; Cambridge University Press: Cambridge, UK, 2018.

47. Matchaya, G. Land ownership security in Malawi. *Afr. J. Agric. Res.* **2009**, *4*, 1–13.

48. Gibson, M.A.; Gurmu, E. Land inheritance establishes sibling competition for marriage and reproduction in rural Ethiopia. *Proc. Natl. Acad. Sci. USA* **2011**, *108*, 2200–2204. [CrossRef] [PubMed]

49. Meijer, S.S.; Sileshi, G.W.; Kundhlande, G.; Catacutan, D.; Nieuwenhuis, M. The role of gender and kinship structure in household decision-making for agriculture and tree planting in Malawi. *J. Gend. Agric. Food Secur. (Agri-Gender)* **2015**, *1*, 54–76.

50. Chirwa, E.W. Adoption of fertilizer and hybrid seeds by smallholder maize farmers in Southern Malawi. *Dev. S. Afr.* **2005**, *22*, 1–12. [CrossRef]

51. Pound, B. *Branching Out: Fairtrade in Malawi: Monitoring the Impact of Fairtrade on Five Certified Organizations*; Fairtrade Africa and the Fairtrade Foundation: London, UK, 2013.

52. Deininger, K.; Jin, S. 'The impact of property rights on households' investment, risk coping, and policy preferences: Evidence from China'. *Econ. Dev. Cult. Chang.* **2003**, *53*, 551–882. [CrossRef]

53. Chirwa, E. Land tenure, farm investments and food production in Malawi. In *Institutions and Pro-Poor Growth (IPPG) Research Programme*; Discussion Paper; University of Manchester: Manchester, UK, 2008; Volume 18.

54. Chirwa, E.W.; Kydd, J. *Farm-Level Productivity in Smallholder tea Farming in Malawi: Do Contractual Arrangements Matter?* Working Paper; University of Malawi, Chancellor College, Department of Economics: Zomba, Malawi, 2006; Volume 3.

55. Benjamin, E.O.; Blum, M.; Punt, M. The impact of extension and ecosystem services on smallholder's credit constraint. *J. Dev. Areas* **2016**, *50*, 333–350. [CrossRef]

56. Benjamin, E.O.; Sauer, J. The cost effectiveness of payments for ecosystem services—Smallholders and agroforestry in Africa. *Land Use Policy* **2018**, *71*, 293–302. [CrossRef]

 *land*

Article

Mapping Environmental Conflicts Using Spatial Text Mining

Jae-hyuck Lee and Do-kyun Kim *

Korea Environment Institute, Sejong 30147, Korea; jaehyuck@kei.re.kr
* Correspondence: dkkim@kei.re.kr; Tel.: +82-44-415-7438

Received: 3 July 2020; Accepted: 20 August 2020; Published: 21 August 2020

Abstract: Mapping the characteristics and extent of environmental conflicts related to land use is important for developing regionally specific policies. However, because it is only possible to verify the frequency of conflicts on a specific predetermined subject, it is difficult to determine the various reasons for conflicts in a region. Therefore, this study mapped the current status of regional environmental conflicts in South Korea using a spatial text mining technique, then proposed relevant management policies. The results were obtained by analyzing environmental conflict data extracted from the online agendas of regional environmental organizations. Air quality-related conflicts in South Korea are concentrated in western municipalities; development-related conflicts are concentrated in the southern region of Jeju Island; and intensive safety-related conflicts occur in metropolitan areas, particularly Ulsan. Thus, the type of conflict is determined by the local environment, in accordance with the definition of environmental conflict, and the distribution is determined by the location of the stakeholder population. This study reveals the issues and locations related to local environmental conflict that require further attention, and proposes more wide-ranging methods for managing the links between conflicts by mapping environmental conflicts on a large scale rather than on an individual basis.

Keywords: South Korea; environmental conflicts; spatial text mining; atmosphere; development; safety

---

1. Introduction

An environmental conflict is a social issue that originates from a difference in views among various stakeholders of environmental resources [1]. Environmental conflict is difficult to resolve, and different demands for limited resources may lead to war between countries and regions [2]. Therefore, many studies have attempted to find an effective solution to environmental conflict, with most research conducted in the form of individual case studies of conflict characteristics and solutions using either qualitative analysis methods, such as interviews and observations, or quantitative analysis methods, such as questionnaires and data from social networking services such as YouTube [3]. However, organizing and summarizing the results can be problematic. For example, content analysis was previously used to systematically organize case studies [1]; however, this attempt was limited as it only examined environmental conflicts that fit a classification framework determined by the researcher.

A potential alternative is to map then organize environmental conflicts. An environmental conflict, by definition, is closely related to land use [4]. The optimal method of land-use analysis involves mapping. As maps enable researchers to explore distinct characteristics and hotspots in a particular space, they represent an alternative to spatial planning-based organization. Carranza et al. [5] used this method to identify areas of fierce environmental conflicts in Chile by organizing and mapping conflicts reported in various newspapers. Their method represented an alternative to solving environmental conflicts by assessing their impact; however, it was not possible to identify the type of problems that intensified and their specific locations. Other studies have mapped major types of environmental conflict. Soytong and Perera [6] mapped environmental conflicts related to air and water quality using the degree of change in the main conflict factors; Kim and Arnhold [7] used the same method for environmental conflicts related to land erosion. Abram et al. [8] attempted to express environmental conflicts by modeling detailed data such as land use, land destruction, carbon content, hydrology, topography, accessibility, and infrastructure. However, these types of analyses can only map macroscopic environmental conflicts for which the required data exist; thus, their ability to capture the differing opinions of local residents is limited. Some studies have analyzed land-use preferences on a micro scale using the public participation GIS method, which reflects the opinion of stakeholders in the field [9]; however, this method does not enable the scope of participation to be widely expanded.

Therefore, this study employs a spatial text mining technique to map and organize the content generated by stakeholders in various environmental conflict case studies. Text mining has been highlighted as a methodology that can replace existing interview and questionnaire techniques, as it can identify the main content of a given text [10]. Furthermore, this developing technique analyzes the characteristics and location of the main content. Häberle et al. [11] confirmed that the spatial distribution of Twitter content can be studied through various machine-learning techniques, and Gulnerman and Karaman [12] presented various techniques for mapping social-media data. Recent advances in these techniques have expanded their applications from comparing the location of McDonald's outlets to obesity incidence by spatializing Twitter data [13], to proposing methods for maximizing joy via an emotional analysis of space in a Disney park [14]. The present study maps environmental conflicts in South Korea, which has experienced various environmental conflicts due to rapid development in recent years, using spatial text mining and, where possible, online environmental conflict data provided by local governments. The study aims to use maps as an alternative analysis method of identifying the spatial context of environmental conflicts.

2. Materials and Methods

2.1. Materials

To analyze environmental conflicts in South Korea, data from the past two years (2017–2018) related to the environmental conflict-focused agendas of 39 active central and regional organizations, including the Korean Federation for Environmental Movement, Green Korea United, and Environmental Justice, were collected and analyzed. The collection was limited to two years of survey data, because the required information was only available for all environmental groups on the Internet for these years. Many environmental organizations had no data prior to 2016, and some failed to update their data in 2019.

These organizations are independent of the government and market, and have been representing critical and alternative voices related to regional environmental conflicts for a long time. Thus, the environmental conflicts studied by these environmental non-governmental organizations (NGOs) clearly reflect the needs of citizens; in other words, they are a major source of data on how environmental conflicts are experienced by the local population. This study collected 330 case studies documented by these organizations, including 155 in 2017 and 175 in 2018.

2.2. Methods

This study used spatial text mining to map environmental conflicts. The technique was designed to verify the content related to the issue of interest via its place in the material generated by the NGOs via keyword factor analysis of each space in the material (Figure 1). The detailed analysis procedure is as follows. First, morphological analysis was performed on the text of the agendas of the different NGOs. Morphological analysis divides text based on parts of speech in order to extract the words used in various types of speech from the text. Second, the main keywords were selected from the text, which were limited to nouns with independent meanings obtained from the morphological analysis. Then, the words in the top 10% according to their frequency were selected as the top keywords [15]. Third, a factor analysis was conducted on the main keywords in each space. After identifying the number of major keywords discussed by region, these values were used to conduct the factor analysis. As the characteristics of each major factor can be verified and the analysis process results in standardization, factor analysis was used as it does not increase the frequency of a keyword simply if there are a large number of discussions related to that keyword [16]. Fourth, the results of the regional factor analysis were mapped using ArcGIS. For each major factor (environmental conflict issue) with a high factor load value, the factor value of the local government was expressed using a color scale. Thus, the spatial distribution of the main environmental conflict issues was identified.

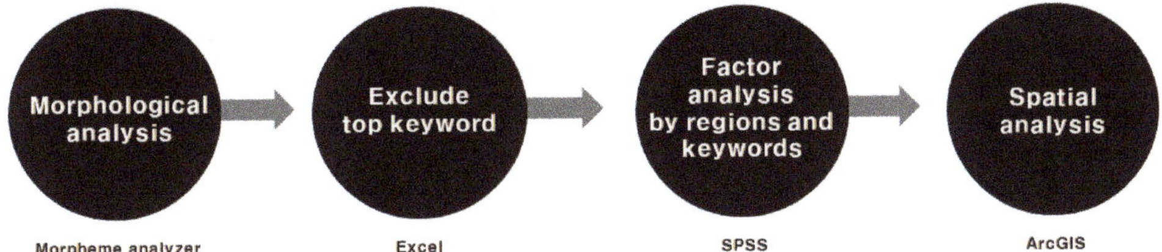

Figure 1. Schematic of the research procedure.

3. Results

First, morphological analysis was performed on the content of all the environmental agendas of the various local governments, and then keywords were extracted. As a result, citizens and residents exhibited contrasting opinions on regional environmental problems through related words such as "project", "region", "environment", "citizen", and "resident" (Table 1). To confirm the regional differences in environmental conflicts, factor analysis was conducted for each region on selected keywords. Varimax rotation, which employs the most common Kaiser normalization technique, was used as the rotation method in factor extraction. As a result of the analysis, three factors with an eigenvalue of one or higher were extracted, with characteristics of "atmosphere", "development", and "safety" according to the related keywords. Distinct regional environmental problems were confirmed based on these three factors (Tables 2 and 3).

Factor 1 predominantly consisted of areas sensitive to the characteristic of "atmosphere." Seoul, Chungbuk, and Gyeonggi were very sensitive to problems related to the atmosphere, and the regions of Gwangju, Jeonnam, Chungnam, and Gangwon showed similar but lower sensitivity than the other three regions. Problems related to factor 1 were more widespread than those related to other factors, which confirms that the problem of fine dust in South Korea was the most serious environmental concern of citizens in the last two to three years. Issues related to the characteristic of "atmosphere" were major environmental concerns for communities in Seoul and Gyeonggi-do, because the inflow of fine dust from China is concentrated in areas with high populations and dense industrial facilities, and in Chungnam and Gangwon because of their high concentration of thermal power plants.

Factor 2 included areas that are sensitive to the characteristic of "development", such as Jeju, Gyeongbuk, and Jeonbuk. The intensity of this characteristic was lower in Gyeongnam and Incheon than in other regions but still high. In particular, the region of Jeju, which is the leading tourist destination in Korea, boasting beautiful natural scenery, has serious conflicts over environmental conservation and development for tourism; for example, construction of the Jeju 2 airport. Furthermore, the regions of Gyeongbuk and Jeonbuk, which are still relatively undeveloped, face strong pressure for development; therefore, the conflict between environmental conservation and development is the central axis of environmental problems in factor 2 areas.

Lastly, Factor 3 areas included those areas sensitive to the characteristic of "safety." Ulsan and Daegu were the most affected areas, followed by Busan and Daejeon. Ulsan, Busan, and Daejeon are adjacent or home to nuclear power plants and related research facilities. Large and small nuclear safety accidents continue to occur in these areas; thus, the stability of the "Hanaro" research reactor in

Daejon, run by the Korea Atomic Energy Research Institute, is a key area of conflict in its community. Therefore, issues of nuclear safety are a major concern for local media and civil society as well as local environmental NGOs in these areas. In addition, asbestos damage and safety concerns have been a major environmental problem in Daegu over the past two years.

Table 1. Frequency (Freq) of major keywords selected for the analysis.

Keyword	Freq	Keyword	Freq	Keyword	Freq	Keyword	Freq	Keyword	Freq
Project	253	Construction	68	Jeju Island	39	Nuclear power plant	33	Restoration	28
Region	235	Result	66	Conservation	39	Education	33	Law	28
Environment	175	Organization	64	Composition	39	Waste	32	Public opinion	28
Citizen	175	Ecology	63	Create	38	Destruction	32	Case	28
Resident	137	Scale	61	Public-private	38	City of Incheon	32	Reinforcement	28
Planning	114	Energy	59	Private	38	State	32	Permission	27
Management	113	unit	58	Participation	37	Protection	32	Nuclear	27
Park	112	Use	55	Nature	37	Public	32	Budget	27
Problem	109	Development	55	Water quality	37	Decision	32	Plan	27
Development	108	Installation	53	Opposition	37	Thermal	31	Open	27
Carry forward	100	Prepare	50	Atmosphere	37	Applicable	31	Korean Federation for Environmental Movement	26
Progress	94	Demand	49	Improvement	37	Review	31	Solution	26
Investigation	94	Emission	49	Confirmation	36	Mud flat	31	Demolition	26
Need	93	Process	48	Administration	36	Open	31	Responsibility	26
Pollution	93	Operation	46	Surrounding	36	Conflict	31	Measure	26
Measure	89	Suspension	45	Accident	36	Bring up	30	Vicinity	26
Government	87	Enforcement	45	Damage	35	After	30	Land	26
Fine dust	85	Announcement	45	Continue	35	Schedule	30	Factory	26
City	82	Situation	43	Implementation	35	Living	30	Detection	26
Construction	82	School	42	Standard	35	Nakdong River	30	Sunlight	25
Power plant	79	Activity	41	Possibility	35	Shin-Kori nuclear power plant	30	Local government	25
Society	77	Ministry of Environment	41	Harm	34	Verification	30	System	25
Policy	76	Expansion	41	Evaluation	34	Chemistry	29	Purification	25
Safety	74	Relation	41	Reduction	34	Consultation	29	Apartment	25
Matter	72	Whole land	40	Asbestos	34	Site	29	Establishment	25
Committee	71	Incheon	40	Dam	34	Health	29	Habitat	25
Facility	69	Approximately	40	River	33	Designation	28	Industry	25
Occurrence	68	Controversy	40	Treatment	33	Concern	28	Extinction	25

Table 2. Results of the factor analysis.

Component	Initial Eigenvalue			Extraction Sums of Squared Load			Rotation Sums of Squared Load		
	Total	% of Variance	Cumulant (%)	Total	% of Variance	Cumulant (%)	Total	% of Variance	Cumulant (%)
1	5.863	36.643	36.643	5.863	36.643	36.643	3.445	21.533	21.533
2	1.417	8.856	45.500	1.417	8.856	45.500	3.007	18.796	40.328
3	1.285	8.032	53.532	1.285	8.032	53.532	2.113	13.204	53.532

Kaiser-Meyer-Olkin = 0.881, Bartlett x^2 = 683.752, df = 120, p = 0.000.

Mapping the environmental conflicts revealed distinct regional characteristics (Figure 2). Regions sensitive to the characteristic of "atmosphere" (Factor 1) exhibited high factor values across the country, although western municipalities relatively close to China were more sensitive. This trend implies that the problem of fine dust is a general problem in Korea; however, urgent measures are required to respond to the needs of western municipalities in particular. Regarding areas sensitive to the characteristic of "development" (Factor 2), the southern regions from Jeju Island had a high factor value. This indicates severe pressure to develop in the southern region, especially Jeju Island. Furthermore, in areas sensitive to the characteristic of "safety" (Factor 3), densely populated rural metropolitan areas in southern regions such as Ulsan, Daegu, Busan, and Daejeon had a high factor value. Moreover, factors with a variance of less than 0.1 and a positive mean were verified to explore the common keywords between them (Table 4). As a result, keywords including "management", "organization", "progress", "ecology", and "committee" were extracted. This confirmed that measures that manage progress via the work of various groups and committees and supplement ecological aspects were commonly applied to Factor 3.

Table 3. Main extracted keywords (with a factor value ≥ 2) and distribution of subject areas by factor value.

Factor	Subject Area	Factor Value	Keyword	Load	Major Conflict
1: Sensitivity to "atmosphere"	Seoul	0.742	Citizen	4.194	The Seoul Federation for Environmental Movement states: "The government's emergency reduction measures in response to high concentrations of fine dust are not practical in Seoul, where high concentrations of fine dust occur on a daily basis due to high standards for issuance. Vehicle regulation and reduced operation are limited to public administrative agencies, and the application standards and follow-up measures in case of violation are unclear. Fine dust in subways and stations is also very serious. Practical and effective measures are required that place citizens' health first. Accordingly, Seoul Federation For Environmental Movement is conducting an intensive citizen action campaign urging additional measures against coal-fired electrical power plants and diesel vehicles, which are the main sources of fine dust."
	Chungbuk	0.674	Region	3.758	
	Gyeonggi	0.673	Fine dust	3.614	
	Gwangju	0.587	Measure	2.442	
	Jeonnam	0.568	Resident	2.295	
	Chungnam	0.567	Power plant	2.230	
	Gangwon	0.562	Park	1.851	
			Project	1.760	
			Occurrence	1.591	
			Policy	1.550	
			Energy	1.404	
			Need	1.237	
			Planning	1.037	
2: Sensitivity to "development"	Jeju	0.723	Project	6.852	The Jeju Federation for Environmental Movement stated: "Jeju OO recreational housing complex development project, which experienced promotion difficulties due to a Supreme Court decision, continues to be promoted by the greed of Jeju Island and OOO. In particular, despite the fact that various trial results acknowledge the fault of Jeju and OOO, the trial that is distressing local residents and landowners is continuing this year with no projected end. Along with criticism that legal costs are being wasted on the trial, which is certain to lose and which is not related to securing publicity or promoting the welfare of residents, there is growing demand to accept the ruling and return land to landowners. This demonstrates how bad development projects, such as OO recreational housing complex development project, can have a huge negative impact on the environment, society, and local economy."
	Gyeongbuk	0.671	Development	2.881	
	Jeonbuk	0.608	Law	2.214	
	Gyeongnam	0.571	Facilitation	2.183	
	Incheon	0.412	Region	2.112	
			Planning	1.768	
			Jeju Island	1.540	
			Facility	1.470	
			Park	1.438	
			Investigation	1.427	
			Dam	1.089	
			City	1.079	
			Resident	1.053	
			Incheon	1.027	
			Pollution	1.025	
			Citizen	4.405	
			Safety	3.001	
3: Sensitivity to "safety"	Ulsan	0.821	Construction	2.707	The Ulsan Federation for Environmental Movement stated: "Regarding the process of public discussion over Shin-Kori unit 5 and 6, in October, the citizen participatory deliberation process was completed, and the public opinion committee's recommendations and the government's final decision were made. The public discussion process was necessary, but insufficient. Each of the following problems had a profound effect on public discussion: pre-leakage of information packages and information stealing; participation of representative pro-nuclear personnel in the expert committee for objective verification; and, as a result, monopolization of the objective verification process, open intervention of government-funded research institutes, which are subordinate agencies of the state, inequality of distribution in Ulsan area, which is the affected area, and inequality for future generations. As Ulsan, which is the affected area, had the smallest citizen participation group (seven people), the anxiety within the area, which called for a safe society, was not properly conveyed to the citizen participatory group. Although the results that came out of the public discussion process are respectable, measures against damage to residents in the vicinity of the nuclear power plant should be presented, along with a declaration to reduce the construction of Shin-Kori unit 5 and 6."
	Daegu	0.670	Unit	2.415	
	Busan	0.569	Kori	2.394	
	Daejeon	0.486	Asbestos	1.921	
			Problem	1.840	
			Damage	1.685	
			Nakdong River	1.555	
			Government	1.538	
			Public opinion	1.493	
			Construction	1.411	
			Nuclear	1.365	
			Law	1.290	
			Project	1.275	
			Project	1.250	
			Demolition	1.102	
			Region	1.065	

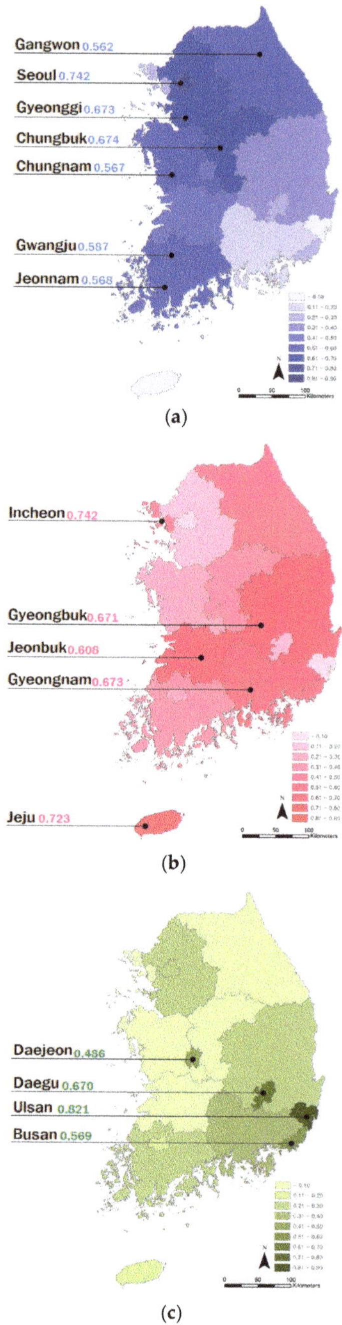

Figure 2. Maps of environmental conflicts for (**a**) areas sensitive to the characteristic of "atmosphere", (**b**) areas sensitive to the characteristic of "development", and (**c**) areas sensitive to the characteristic of "safety".

Table 4. Extraction of commonly recognized keywords with a variance ≤ 0.1.

Keyword	Factor 1	Factor 2	Factor 3	Variance	Mean
Management	0.636	0.703	0.773	0.005	0.704
Group	0.075	0.068	−0.11	0.011	0.011
Progress	0.784	0.389	0.648	0.04	0.607
Ecology	0.253	0.356	0.751	0.069	0.453
Committee	−0.178	0.389	−0.116	0.097	0.032

4. Discussion

This study used spatial text mining to confirm that degrees of sensitivity exist in environmental conflicts and their regional patterns. According to maps of regional environmental conflicts in South Korea, environmental conflict in this country exhibits distinct characteristics whereby local environments and population densities are intertwined. The results imply that environmental conflict is influenced by these elements because of differences in the opinions of various stakeholders regarding the use of environmental resources; this is reflected by the definition of environmental conflict [1]. Moreover, the results reveal the type of conflict that became a particular issue in each region, identifies areas of concern regarding major issues [5], and provides insights into the implementation of environmental management policies on a spatial level [17]. In South Korea, western municipalities require an environmental policy to manage fine dust, southern urban areas require an environmental policy to manage safety issues, and non-urban areas require an environmental policy to manage issues of development. Therefore, in-depth study into regional environmental conflicts and expert participation in management is required. By strengthening the network among regions experiencing the same types of environmental conflict, an environmental management system can be created that would enable regions to respond jointly to a particular environmental problem and difficulties in conflict management; regions could also share cases in which resolutions were found.

Furthermore, commonly recognized keywords indicate that the entire country should focus on solving environmental conflicts. Keywords including "management", "organization", "progress", and "committee" imply that clear participation techniques, such as forming public opinion committees, should be established as methods of mitigating environmental conflicts [18]. Moreover, the keyword "ecology" indicates the need for practical ways to spread and implement ecological values as alternatives to resolving environmental conflicts.

Mapping environmental conflicts in this way is effective for creating a broad-scale alternative to environmental conflict management, because it can verify the relationship between conflicts that may be perceived as different events in each region, thereby enabling a combined response [1]. Furthermore, this technique can predict the types of environmental change that will have significant effects on local populations, as well as the type of environmental conflict that will occur and in which region, by exploring the forms of conflict between the environment and people. This then allows the preemptive preparation of measures to reduce environmental conflicts during planned projects. As such, environmental conflict maps constitute data that diagnose the current status of environmental conflict while providing alternative methods of solving environmental conflict and aiding preparation for the future.

5. Conclusions

This study mapped environmental conflicts in South Korea in order to identify the organic connections between different conflicts instead of treating them as single events, then suggested broad plans for managing such conflicts. First, the major environmental conflicts were verified and their spatial context determined via spatial text mining of key environmental conflicts in South Korea. The results confirmed that environmental conflicts exhibit distinct environmental characteristics and that local environments and population densities are intertwined.

Korea has had numerous civil complaints related to problems with fine dust pollution and development. Accordingly, the Ministry of Environment has responded by implementing various atmospheric and development regulations. However, these common regulatory policies were implemented nationwide, from urban and industrial areas to rural areas, forests, and islands; thus, questions have been raised about the regional effectiveness and hostility toward these regulations. Further, regarding nuclear safety issues, a public hearing was held only in Ulsan area, and no special alternative was prepared for provincial cities where other nuclear power plant facilities are located. This study is significant because it reveals the major environmental conflict issues in South Korea (fine dust pollution, development, nuclear safety) through environmental conflict mapping, and proposes a tool for creating regionally customized land regulation and management plans for each issue.

Moreover, the proposed spatial text mining technique identifies environmental ecological characteristics such as species type, as well as general human and social characteristics such as population, by analyzing existing environmental spatial data. This technique also identifies distinct characteristics of the area based on the needs of local citizens. Furthermore, combining big data (social networking service data such as Twitter, Facebook, YouTube, and Flickr) [19,20] and regional micro data (in-depth interviews and civil complaints) [21,22] would enable the prioritization of land-use policies specialized for each location. When determining such priorities, spatial text mining could provide important data that represents the needs and opinions of local residents.

Since this study analyzed only the environmental agenda at the local government scale in Korea via spatial text mining, it has a limited ability to map and reveal relationships between the environment and human populations on detailed spatial scales. In addition, like the environment, environmental conflicts can change over time; however, this was beyond the scope of this study. Therefore, future environmental conflict mapping research that employs more diverse data, spatial scales, and time periods will produce richer and more detailed measures for managing environmental conflicts.

Author Contributions: Conceptualization, J.-h.L. and D.-k.K.; methodology, J.-h.L.; software, J.-h.L.; formal analysis, J.-h.L.; investigation, D.-k.K.; resources, D.-k.K.; data curation, D.-k.K.; writing—original draft preparation, J.-h.L. and D.-k.K.; writing—review and editing, J.-h.L.; visualization, J.-h.L.; supervision, D.-k.K.; project administration, D.-k.K.; funding acquisition, D.-k.K. All authors have read and agreed to the published version of the manuscript.

Funding: This research was supported by the grant from basic project 'Social model development of environmental pollution damage for recovery of victims' lives' funded by the Korea Environmental Institute (RE2020-17) and 'Urban-based complex plant demonstration study utilizing underground space' program funded by Ministry of Land, Infrastructure and Transport of Korean government (20UGCP-B157945-01).

Conflicts of Interest: The authors declare no conflict of interest.

References

1. Scheidel, A.; Del Bene, D.; Liu, J.; Navas, G.; Mingorría, S.; Demaria, F.; Avila, S.; Roy, B.; Ertör, I.; Temper, L.; et al. Environmental conflicts and defenders: A global overview. *Glob. Environ. Chang.* **2020**, *63*, 102–104. [CrossRef] [PubMed]

2. Le Billon, P. *Wars of Plunder: Conflicts, Profits and the Politics of Resources*; Columbia University Press: New York, NY, USA, 2012.

3. Staniscia, B.; Komatsu, G.; Staniscia, A. Nature park establishment and environmental conflicts in coastal areas: The case of the Costa Teatina National Park in central Italy. *Ocean Coast. Manag.* **2019**, *182*, 104947. [CrossRef]

4. Hanaček, K.; Rodríguez-Labajos, B. Impacts of land-use and management changes on cultural agroecosystem services and environmental conflicts—A global review. *Glob. Environ. Chang.* **2018**, *50*, 41–59. [CrossRef]

5. Carranza, D.M.; Varas-Belemmi, K.; De Veer, D.; Iglesias-Müller, C.; Coral-Santacruz, D.; Méndez, F.A.; Torres-Lagos, E.; Squeo, F.A.; Gaymer, C.F. Socio-environmental conflicts: An underestimated threat to biodiversity conservation in Chile. *Environ. Sci. Policy* **2020**, *110*, 46–59. [CrossRef]

6. Soytong, P.; Perera, R. Use of GIS tools for environmental conflict resolution at Map Ta Phut Industrial Zone in Thailand. *Sustainability* **2014**, *6*, 2435–2458. [CrossRef]

7. Kim, I.; Arnhold, S. Mapping environmental land use conflict potentials and ecosystem services in agricultural watersheds. *Sci. Total Environ.* **2018**, *630*, 827–838. [CrossRef]

8. Abram, N.K.; Meijaard, E.; Wilson, K.A.; Davis, J.T.; Wells, J.A.; Ancrenaz, M.; Budiharta, S.; Durrant, A.; Fakhruzzi, A.; Runting, R.K.; et al. Oil palm–community conflict mapping in Indonesia: A case for better community liaison in planning for development initiatives. *Appl. Geogr.* **2017**, *78*, 33–44. [CrossRef]

9. Brown, G.; Kangas, K.; Juutinen, A.; Tolvanen, A. Identifying environmental and natural resource management conflict potential using participatory mapping. *Soc. Nat. Resour.* **2017**, *30*, 1458–1475. [CrossRef]

10. Lee, J.-H.; Park, H.-J.; Kim, I.; Kwon, H.-S. Analysis of cultural ecosystem services using text mining of residents' opinions. *Ecol. Indic.* **2020**, *115*, 106368. [CrossRef]

11. Häberle, M.; Werner, M.; Zhu, X.X. Geo-spatial text-mining from Twitter–a feature space analysis with a view toward building classification in urban regions. *Eur. J. Remote Sens.* **2019**, *52*, 2–11. [CrossRef]

12. Gulnerman, A.G.; Karaman, H. Spatial reliability assessment of social media mining techniques with regard to disaster domain-based filtering. *ISPRS Int. J. Geo-Inf.* **2020**, *9*, 245. [CrossRef]

13. Ghosh, D.; Guha, R. What are we 'tweeting' about obesity? Mapping tweets with topic modeling and Geographic Information System. *Cartogr. Geogr. Inf. Sci.* **2013**, *40*, 90–102. [CrossRef] [PubMed]

14. Park, S.B.; Kim, J.; Lee, Y.K.; Ok, C.M. Visualizing theme park visitors' emotions using social media analytics and geospatial analytics. *Tour. Manag.* **2020**, *80*, 104–127. [CrossRef]

15. Luhn, H.P. A business intelligence system. *IBM J. Res. Dev.* **1958**, *2*, 314–319. [CrossRef]

16. Taminiau, Y.; Ferguson, J.; Moser, C. Instrumental client relationship development among top-ranking service professionals. *Serv. Ind. J.* **2016**, *36*, 789–808. [CrossRef]

17. Soytong, P.; Perera, R. Spatial analysis of the environmental conflict between state, society and industry at the Map Ta Phut-Rayong conurbation in Thailand. *Environ. Dev. Sustain.* **2017**, *19*, 839–862. [CrossRef]

18. Wittmer, H.; Rauschmayer, F.; Klauer, B. How to select instruments for the resolution of environmental conflicts? *Land Use Policy* **2006**, *23*, 1–9. [CrossRef]

19. Kim, Y.; Kim, C.K.; Lee, D.K.; Lee, H.W.; Andrada, R.I.T. Quantifying nature-based tourism in protected areas in developing countries by using social big data. *Tour. Manag.* **2019**, *72*, 249–256. [CrossRef]

20. Lee, H.; Seo, B.; Koellner, T.; Lautenbach, S. Mapping cultural ecosystem services 2.0–Potential and shortcomings from unlabeled crowd sourced images. *Ecol. Indic.* **2019**, *96*, 505–515. [CrossRef]

21. Lee, C.H. Understanding rural landscape for better resident-led management: Residents' perceptions on rural landscape as everyday landscapes. *Land Use Policy* **2020**, *94*, 104565. [CrossRef]

22. Lee, J.H.; Choi, H. An analysis of public complaints to evaluate ecosystem services. *Land* **2020**, *9*, 62. [CrossRef]

 *land*

Article

Urbanization and Increasing Flood Risk in the Northern Coast of Central Java—Indonesia: An Assessment towards Better Land Use Policy and Flood Management

Wiwandari Handayani [1], **Uchendu Eugene Chigbu** [2,*], **Iwan Rudiarto** [1] **and Intan Hapsari Surya Putri** [1]

[1] Department of Urban and Regional Planning, Diponegoro University, Semarang 50275, Indonesia; wiwandari.handayani@pwk.undip.ac.id (W.H.); iwan.rudiarto@pwk.undip.ac.id (I.R.); intan.hapsari18@pwk.undip.ac.id (I.H.S.P.)

[2] Chair of Land Management, Faculty of Aerospace and Geodesy, Technical University of Munich (TUM), 80333 Munich, Germany

* Correspondence: ue.chigbu@tum.de; Tel.: +49-(0)89-289-22518

Received: 14 August 2020; Accepted: 21 September 2020; Published: 23 September 2020

Abstract: This study explores urbanization and flood events in the northern coast of Central Java with river basin as its unit of analysis. Two types of analysis were applied (i.e., spatial data and non-spatial data analysis) at four river basin areas in Central Java—Indonesia. The spatial analysis is focused on the assessment of LULC change in 2009–2018 based on Landsat Imagery. The non-spatial data (i.e., rural-urban classification and flood events) were overlaid with results of spatial data analyses. Our findings show that urbanization, as indicated by the growth rate of built-up areas, is very significant. Notable exposure to flood has taken place in the urban and potentially urban areas. The emerging discussion indicates that river basins possess dual spatial identity in the urban system (policy- and land-use-related). Proper land use planning and control is an essential instrument to safeguard urban areas (such as the case study area) and the entire island of Java in Indonesia. More attention should be put upon the river basin areas in designing eco-based approach to tackle the urban flood crises. In this case, the role of governance in flood management is crucial.

Keywords: central java; flood; flood management; Indonesia; land policy; land use; land-use change; urbanization

---

1. Introduction

Flood is the most common disaster across the globe [1–4]. Rapid urbanization in low-lying areas leads to higher exposure to various types of floods, in addition to the increase in coastal flooding caused by sea-level rise and rainfall pattern deviation as a result of climate change [5–9]. Urbanization can be clearly indicated by the conversion of land into residential areas based on the premise that the growing urban population requires more land. Land conversion expands both downstream and upstream to accommodate the needs and activities of the growing urban populations. Deng et al. [10] (p. 1341) and Chin [11] (p. 469) assert that urbanization is a significant contributor to changes in the river system and structure as it usually increases flood risk.

Land use policy provides the opportunity to conduct systematic assessment of land and water potential and to identify options to improve flood-prone areas and mitigate flood occurrence. As "a culmination of all activities and decisions concerned with guiding the allocation and use of land in patterns that enable improvements in people's way of living", land use planning policy is a crucial process for mitigating floods [12] (p. 8). On this basis, Hegger et al. [13] propose flood risk prevention as a way to decrease the exposure of people/property using spatial planning policy as a critical approach to Flood Risk Management (FRM). Therefore, flood risk prevention is vital in the flood adaptation cycle. It is related to the capacity to transform and to adapt long-term perspectives in addressing disturbance to achieve sustainable urbanization. The Hyogo Framework for Action 2005–2015 [14] and Sendai Framework for Disaster Risk Reduction 2015–2030 [15] have strengthened the role of land use policy to contribute to disaster risk reduction. Both global commitments prioritize land use allocation through policy instrument to reduce risk factors, considering that the policy will accommodate physical and ecological characteristics in allocating various types of land use.

Many factors affect the occurrence of flooding. However, recent studies in various parts of Asia have shown a significant connection between urbanization (influenced by land-use change) and flooding events [7,13,16–20]. Some of these studies are worth mentioning here. Chen et al. [17] investigated the connection between population growth and land-use changes in relation to natural hazard occurrence in China. They found that the Pearl River Basin is increasingly exposed to floods because of population growth and land conversion. Song et al. [18] assessed the water level dynamics in the Yangtze River Delta and found that precipitation and urbanization caused increased flood risk. Focusing on drainage adaptation, Zhou et al. [19] revealed that land-use changes in Northern China exacerbated the increase in surface runoff due to flooding, which is caused by poor drainage system planning. Zope et al. [20] investigated Land Use-Land Cover (LULC) changes in Oshiwara River Basin in Mumbai—India and revealed that increase in LULC correspondingly led to the increase in flood frequency. In the book Disaster Governance in Urbanizing Asia, Miller and Douglass' [2] argued that that urbanization is a leading factor in the exposure of human settlements to floods and vulnerabilities of various forms. All of the studies mentioned above have influenced this study to infer that controlling urbanization and reducing flood risk cannot be executed separately. The whole urban system at the regional level needs to be considered. After all, most "sites of intense urbanization are prone to natural hazards, such as flood, landslide, drought, and tidal flood" in Indonesia [21] (p. 287).

All of these studies [13,16–21] indicate that flood and urbanization are complex issues. Flood risk is identified based on water system delineation and defined based on gravity-driven river flow pattern following landscape ecology, which then forms a river basin [22]. Accordingly, a river basin is usually characterized by a land area that consists of various types of land use and a number of watersheds that drain from the upstream to downstream area [23]. Water flows without recourse to administrative jurisdictions, and spatial planning (i.e., land use policy) to control urbanization are examined based on the administrative jurisdiction. In Indonesia, it is common that a river basin covers more than one administrative boundary or local government authorities. This means that a river basin may be subject to the management of more than one responsible party. Such a scenario creates a challenge in land use planning and in developing control mechanisms for river management.

This study explores urbanization and flood events in the northern coast of Central Java using the river basin as its unit of analysis. We addressed two main research questions herein: (1) How have urbanization and flood events taken place from the perspective of river basin delineation? (2) To what extent the comprehension of river basin as land and land use could contribute to reducing flood risk through land use policy and better flood management? To answer these questions, this paper is divided into three main sections. Section 2 is a description of the scope and methods used in this study; Section 3 provides an analysis on land-use changes and flood events within the scope of study; and Section 4 discusses issues emerging from the analysis, focusing on the importance of understanding the spatial identity of river basins to contribute to better land use policy and governance mechanisms for flood management.

2. Materials and Methods

2.1. Study Area

Java is the most populous island in Indonesia. Its inhabitants constitute 60% of the total Indonesian population, even though it is less than 7% of the total area in Indonesia [24]. According to the Presidential Decree [25], the Island has around 1200 watersheds and 24 river basins. Some of them are categorized as National Strategic River Basins—meaning that their strategic socioeconomic and environmental functions should be preserved. Our study area is located in the mid-northern part of the Island (see Figure 1), which consists of four river basins. A large part of the area belongs to Central Java Province, which stretches through several local government authorities (or municipalities) that are categorized as either regencies or cities. The existence of arterial and toll roads in the northern corridor is an infrastructural boost that has led to rapid economic development in the area. Accordingly, some emerging threats on the functions of river basins are mostly triggered by uncontrolled population growth. Such growth results in the reduction in non-built-up areas, as forest and agriculture lands are converted to settlement and industrial zones.

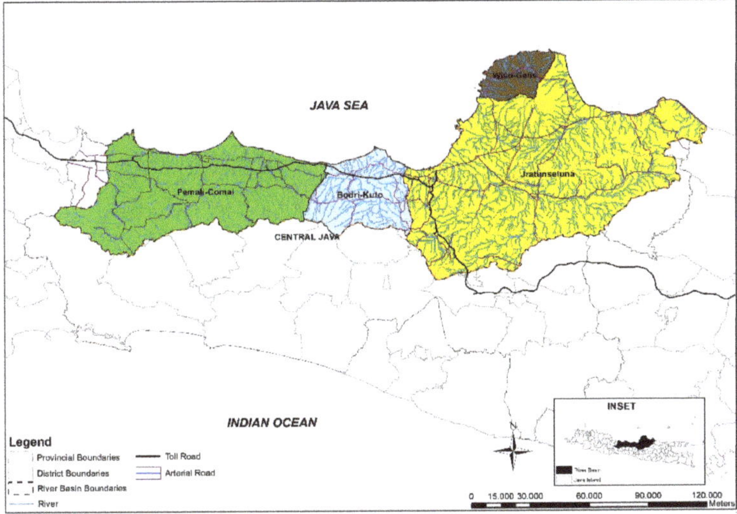

Figure 1. Study Area.

Pemali-Comal, Bodri-Kuto, Wiso-Gelis, and Jratunseluna River Basins cover a total area of 16,403 km^2 that cuts across four cities (Tegal, Pekalongan, Semarang, and Salatiga) and 17 regencies (Brebes, Tegal, Pemalang, Pekalongan, Batang, Kendal, Temanggung, Demak, Jepara, Kudus, Pati, Rembang, Blora, Grobogan, Sragen, Boyolali, and Semarang) (Figure 1). Table 1 highlights the main features of these basins. Jratunseluna is the biggest river basin in the study area. It is a National Strategic River Basin with several vital functions and a significant number of people living in the area. Indeed, proper governance/institutional setting is crucial in managing the basins, considering that the river basin areas are not under local authorities.

Table 1. River Basins in the Northern Coast of Central Java.

River Basin	Area (km^2)	Watershed	Territorial Areas of Jurisdiction	Population
Jratunseluna	9216	69	10 Regencies, 2 Cities (2231 Villages/*Kelurahan*)	8.9 million
Wiso-Gelis	663	27	1 Regency (92 Villages/*Kelurahan*)	1.2 million
Bodri-Kuto	1662	12	3 Regencies (396 Villages/*Kelurahan*)	1.3 million
Pemali-Comal	4860	32	4 Regencies, 2 Cities (961 Villages/*Kelurahan*)	6.9 million

Note: *Kelurahan* refers to a village that is located in a city.

In general, as shown in Figure 2, rainfall in the four river basins fluctuated over nine years period. In most cities and regencies in which all basins, except for Pemali-Comal, are located, the rainfall increased from the previous year and peaked in 2010, followed by a sharp decline the year after. In Jratunseluna, the critical years with the highest frequency of rainfall were 2010, 2013, and 2016. Throughout 2010, 2014, and 2016, the rainfall in Bodri-Kuto continued to increase, peaking at approximately 3600 mm/year. In contrast, the average rainfall in Pemali-Comal River Basin considerably increased in 2012 and 2015 and then remained constant until 2018. However, the rainfall patterns in Wiso-Gelis River Basin, which covers only one regency and was generated from only one climatological station, differ from that of the other river basins. A steady increase was observed from 2009 to 2011 and 2012 to 2015, followed by a decrease in 2016.

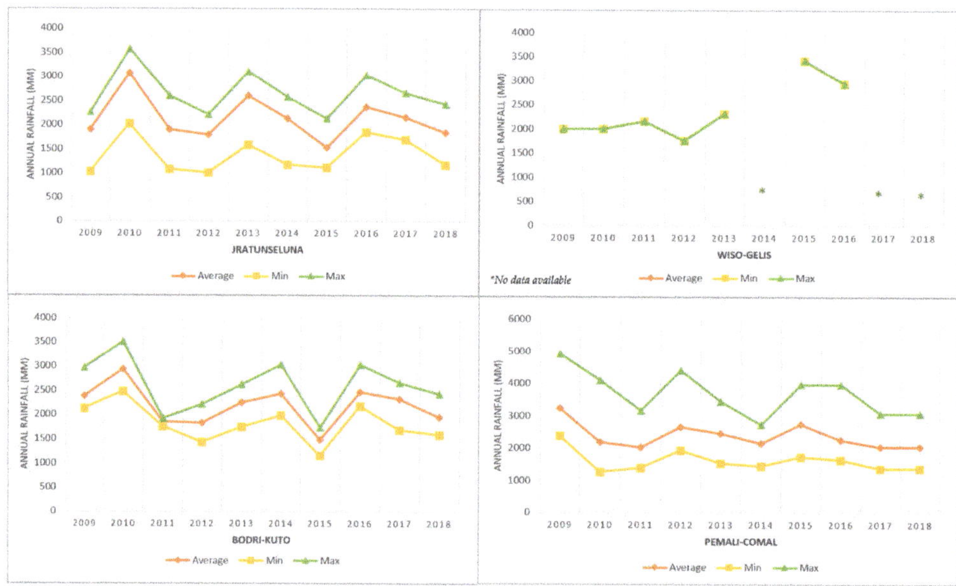

Figure 2. Rainfall in the Study Area in 2009–2018. Source: Meteorological, Climatological, and Geophysical Agency (MCGA) and Central Bureau of Statistics (CBS) 2009–2018. No data available for Wiso Gelis (2014, 2017, 2018). Rainfall data for Wiso-Gelis, Jatunseluna, Bodri-Kuto and Pemali-Comal are collected from 1, 9, 4, and 7 climatology stations, respectively.

2.2. Methods of Data Collection

2.2.1. Spatial Data

Remote sensing data were used to produce Land-Use-Land Cover (LULC) map for 2009 and 2018 30 × 30 m resolution to assess LULC change in Central Java North Coast. In addition, watershed data were used to delineate the river basin area according to Presidential Decree and Ministry Regulation. Table 2 details the spatial data that were processed for the analysis.

Table 2. Spatial Data Collection.

No	Data Type	Year	Data Format	Source
1	Landsat 8 Satellite Image	2009 and 2018	Image	United States Geological Survey (USGS)
2	Watershed delineation	2018	Shapefile	Presidential Decree No. 12/2012 Ministry of Environment and Forestry

2.2.2. Urban and Rural Classification Data

Rural and urban areas are classified based on their administrative jurisdiction, Central Bureau of Statistics (CBS) criteria [26], as well as the direction of built-up area expansion. These resulted in three classifications, namely urban, potentially urban and rural areas. An area is classified as urban when its administrative jurisdiction lies in the city or the capital of a regency. Meanwhile, a potentially urban area refers to any area categorized as rural-urban according to the CBS criteria, in which its rural-urban potential is also considered (see Table 3).

Table 3. Rural and Urban Classifications.

No	Classification	Definition	Delineation
1	Urban area	Consist of *kelurahan* (located in cities) and urban villages (as capital of regency)	Jurisdiction based on government regulation
2	Potentially Urban Area	Consist of villages (*desa*) that are characterized as urban, located in regencies	• CBS scoring [26] based on census data 2010 that is calculated according to selected variables, including population density, percentage of farming households, percentage of households served by electricity, percentage of households served by telephone network, access to main urban facilities, and access to supporting facilities (also explained in [7]) • Neighboring villages of the rural-urban area 2010 that has more than 28.6% built-up area in 2018 (the number is based on the average of built-up in the rural-urban area in 2010 (classification no. 3).
3	Rural Area	Consist of villages (*desa*) that are located in regencies	The rest of the area

To further comprehend the classification explained in Table 3, it is important to note that a village is the lowest administrative jurisdiction in Indonesia. Accordingly, there are three types of villages based on their rural and urban status. The first is *desa*, which are villages located in a regency and characterized as rural. The second is *kelurahan*, which are categorized as urban villages and are located in a city. Third, some villages are characterized as urban according to the CBS criteria, yet they are referred to as *desa* instead of *kelurahan*. Therefore, they are categorized as potentially urban. Another essential difference between *desa* and *kelurahan* is that the local residents elect the head of *desa*, while the head of *kelurahan* is appointed by the mayor or regent, both of which are government employees.

2.2.3. Disaster Data

The primary data source for flood events is the Disaster Management Board (DMB) of Central Java Province. According to DMB, based on the Law concerning Disaster Management [27], flooding is an event or condition where an area or land is submerged due to water volume increase. Flash flood, also known as fluvial flood, involves sudden water discharge in large volume due to river flow obstruction. The DMB flood data are based on a compilation of reports from local (City/Regency) government informing the location (name of villages/*kelurahan*), duration, depth, and damage/loss status. However, not all local governments have reported the events, as it is not an obligatory procedure. Accordingly, this study also investigated data on flood events and fatalities published by mass media websites or other institutions and used them to validate formal data released by the government. The internet-based data were collected using three keywords via Google search engine: flood, name of the district or city concerned, and the year of occurrence.

Data collection on flood events was performed by looking for news articles that contain information on flood location (sub-district and village or *kelurahan*), time of occurrence, height of inundation, the time required for inundation to recede (duration of inundation), and the magnitude of impact or loss due to flooding. Information search regarding flood events in regencies/cities and the specified year were deemed to be completed when the search engine (Google) detected that no more articles related to the keywords were found.

The news reports on flood events from 2009 to 2018 were collected, totaling in 2123 news pieces from approximately 98 sources, including the mass media or institutional website. The total number of flood events was 1925, of which 1609 were reported by one news source (single rapporteur), while the rest were reported by more than one news source (joint rapporteur). Table 4 describes the number of total incidents reported from five sources that had the largest contribution in disaster news. Formal government report only covers around 52% of the total incidents, showing that a significant number of incidents took place yet they were not formally reported to the authorized government.

Table 4. Largest Contribution of Flood Data Sources.

Sources	Total Incidents Reported	Contributions (%)
Formal Government Report		
Disaster Management Board of Central Java Province	1104	52.00
Online Newspapers		
Tribune News	107	5.04
Kompas	86	4.05
Sindo Newa	63	2.97
Detik News	60	2.83
Others Media (85 Media which reported less than 60 incidents)	703	33.11
Total	**2123**	**100**

2.3. Methods of Data Analyses

This study uses two types of analysis: spatial data and non-spatial data analysis (Figure 3). The spatial analysis was focused on assessing LULC change in 2009–2018 based on Landsat Imagery. The non-spatial data (i.e., rural-urban classification and flood events) were overlaid with the results of spatial data analyses.

LULC was classified into five types based on the Indonesian National Standard Regulation [28], namely the built-up, industry, rice fields, forest, and mix plantations (see Table 5). Supervised classification was done on land cover imagery, in which the training sample was determined using Maximum Likelihood Classification in ArcGIS. The accuracy of tentative LULC produced in this step was confirmed through field observations and using the instrument conformity table, totaling in 306 observation points. This was then used to improve LULC interpretation.

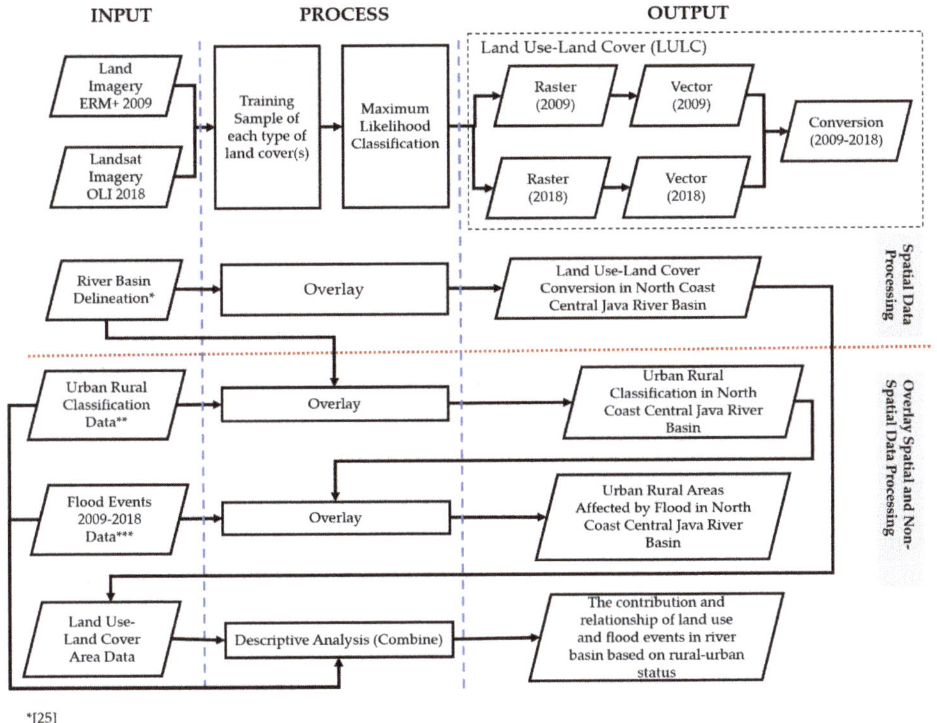

Figure 3. Analytical Method Flowchart.

As illustrated in Figure 3, the result of spatial data analysis was overlaid with two data attributes at the village level; i.e., rural-urban classification and flood events. This step combined urban-rural classification data (as explained in Table 5) and flood events with river basin classification and land-use change in 2009–2018. In the next stage, descriptive analysis was conducted to identify the relationship between land use changes, i.e., from built-up to non-built up areas, with the occurrence of flood over nine years period. In this case, a matrix composed of four elements, including river basin, urban-rural status, land use change and flood events, was generated. The built-up areas consist of settlements and industrial areas, while the non-built up areas include forests, rice fields and mix-plantations.

Table 5. Land Use Classification in Study Area.

Land Use Type	Description
Built-up-Settlement	Land covered by buildings, dominated by grey color, are likely to cluster and/or to be built around the road network.
Built-up-Industry	Land covered by big buildings, dominated by light grey/white color, are likely to cluster and/or to be built around the road network.
Rice field	Land for agricultural with or without slopping terraces, dominated by light green color, mostly characterized as a dike pattern with a smooth texture
Forest	Natural and man-made forests, approximately 75% covered by trees, dominated by dark green color and a rough texture.
Mix Plantation	Different types of vegetation with various density, the color and texture are in between that of the rice fields and forests.

Source: Authors, developed from SNI 7645 [28].

3. Results or Outcomes

3.1. Land Use Change in the Northern Coast of Java 2009–2018

Significant urban expansion has taken place in the Northern Coast of Java. The land conversion rate for each river basin based on its rural-urban classification is listed in Table 6. Each basin has a particular growth rate pattern. Bodri-Kuto River Basin experienced the most critical changes (up to 108%) over nine years compared to the others, which means that massive built-up development occurred in this river basin in terms of settlements and industrial area. It is then followed by Jratunseluna River Basin, in which the built-up area has expanded from 1222 to 1581 km^2 since 2009 to 2018. Most of the expansion took place in the urban area at approximately 137%. Meanwhile, the development of Pemali-Comal River Basin mostly occurred in the potentially urban area (43.31%), specifically in Tegal and Pekalongan Regency. The growth of built-up area in Wiso-Gelis was significantly higher in the rural area (36.13%) than in the potentially urban (17.54%) and urban (10.08%) areas. This scenario is indicative that substantial urbanization within the study area [7] has led to a significant land conversion that expanded to rural areas surrounding the urban centers. On the governance side, administrative autonomy, which devolves the authority over land use allocation to local government, has led to uncontrollable land conversion due to a lack of coordination among local governments.

The land conversion status for each river basin varies. Bodri-Kuto River Basin experienced the highest rate of land conversion to built-up areas over nine years, followed by Jratunseluna, Wiso-Gelis and Pemali-Comal, respectively. The highest increase in built-up area is in the urban area of Jratunseluna River Basin. In contrary, in Pemali-Comal and Bodri-Kuto River Basin, there was a significant increase in built-up area in the potentially urban area, specifically in Tegal and Kendal Regency.

As illustrated in Figure 4, the increase in built-up area was not only concentrated in urban areas. Among the three river basins, there was a significant development in potentially urban and rural areas during the 2009–2018 period. Thus, growth in potentially urban areas is also influenced by nearby urban activities. Accordingly, urban expansion is extended to areas surrounding the city centers even though there are more vacant lands available for use. Toll road development and industrial zone establishment have very much influenced the growth and direction of land conversion. To illustrate this, the land allocation for industrial lands in Bodri-Kuto increased significantly, from less than 10 km^2 to more than 60 km^2. This is then followed by the expansion of residential and commercial activities in surrounding areas to accommodate the needs of industrial employees.

Table 6. Land Conversion in the Selected River Basins 2009–2018.

River Basin	Area (km^2)		Change (%)	Average Annual Growth Rate (%)
	2009	2018		
	Built-Up	Built-Up		
Jratunseluna	1222.58	1581.26	29.34	3.26
Urban	89.19	211.93	137.62	15.29
Potentially Urban	330.88	467.69	41.35	4.59
Rural	802.51	901.64	12.35	1.37
Wiso-Gelis	70.72	89.81	26.99	3.00
Urban	2.38	2.62	10.08	1.12
Potentially Urban	31.42	36.93	17.54	1.95
Rural	36.92	50.26	36.13	4.01
Bodri-Kuto	117.17	244.56	108.72	12.08
Urban	12.34	17.4	41.00	4.56
Potentially Urban	41.51	88.39	112.94	12.55
Rural	63.32	138.77	119.16	13.24
Pemali-Comal	556.14	670.71	20.60	2.29
Urban	63.27	69.7	10.16	1.13
Potentially Urban	204.47	293.02	43.31	4.81
Rural	288.4	307.99	6.79	0.75

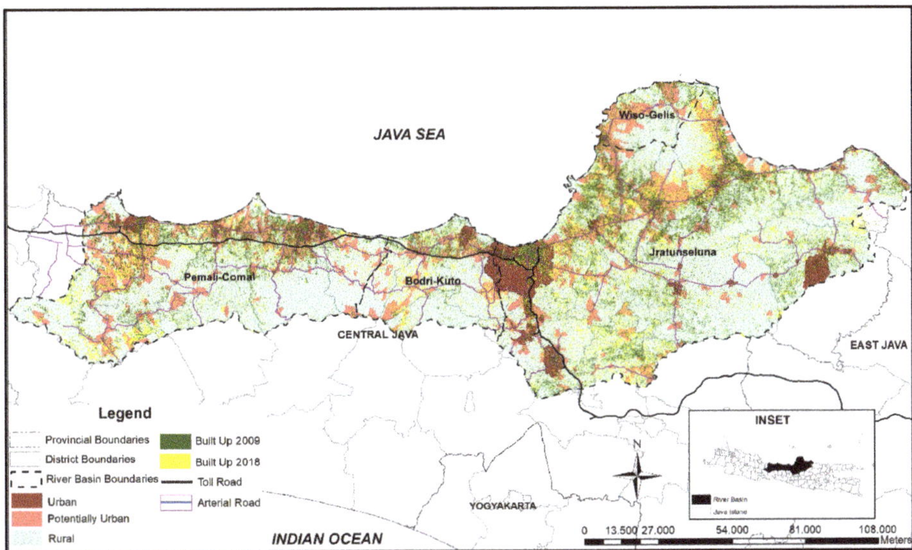

Figure 4. Land Conversion, 2009–2018.

Figure 5 further illustrates land-use changes in several types of land allocation. Rice fields, forests and mix plantations are dominant throughout the river basins. However, there was a considerable loss of mix plantation land in most of the river basins, except in Jratunseluna. Bodri-Kuto experienced the highest loss of mix plantation areas (that is up to 221 km^2 between 2009 and 2018), followed by Pemali-Comal (13.98%) and Wiso-Gelis (3.28%). In all river basins, the loss of mix plantation areas occurred in potentially urban areas, for example in Tegal, Pekalongan and Demak Regency. Despite the significant reduction in land use for mix plantations, Figure 4 depicts that there was a slight increase in the rural forest area in Bodri-Kuto which went up to approximately 53 km^2 over nine years. Growth of forest area within some river basins in Central Java is in line with the enacted regulations of the Governor of Central Java [29] and the Minister of Environment and Forestry Regulation [30]. The regulation provides evidence that a policy should serve as a strategic instrument in controlling land allocation and improving river performance.

The increase in built-up area, especially in certain regencies or rural areas, occurred because land has been converted into rural-urban potential areas, and at some point, urban areas. Within this scenario, in the near future Java will become an urban island, on which built-up areas will expand downstream to upstream, overall creating problems in the environment. Based on the built-up ratio in 2018 (which categorizes the area as a rural-urban potential area), more than 600,000 inhabitants are spread throughout approximately 250 villages. This indicates that more rural areas have been urbanized due to the increasing population and expanding built-up area. For example, Pati Regency has the highest number of villages belonging to the potentially urban area (129 villages), followed by Kudus Regency, where 113 villages are potentially categorized as urban areas. Both regencies are located at the downstream of Jratunseluna River Basin.

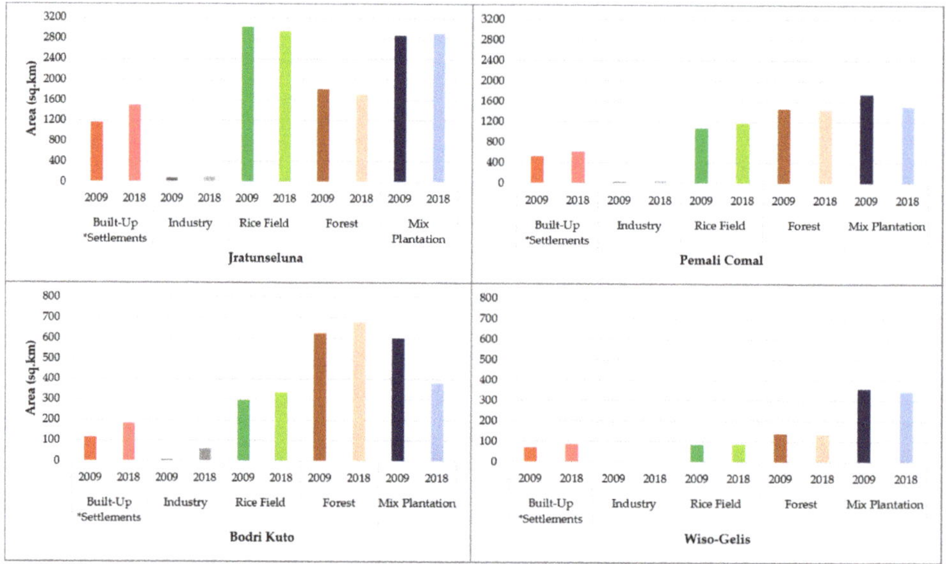

Figure 5. Land Use Change in Selected River Basins, 2009–2018.

3.2. Flood Events in the River Basins

The frequency of flood in the four river basins fluctuates. There are limited data for flood events in the initial years (2009–2011) because the local disaster management board has yet been established and the mass online media have not been widely used in reporting disaster news in details. Accordingly, in 2009–2011, a small number of floods occurred in all areas, including in the urban, rural-urban and rural areas (Figure 6). The number of floods in the rural and urban areas has increased in 2012 and 2013 since the disaster data report has been updated. In addition, in 2014, flood occurrence sharply increased, especially for the rural-urban areas. In the rural areas alone, there were more than 200 incidents of flooding reported. Two of the most prominent river basins in the study area, Jratunseluna and Pemali-Comal River Basin, contribute to a high number of flood events. Specifically, up to 2014, a majority of flood events in the Jratunseluna River Basin happened in namely Pati, Kudus and Jepara Regency. Meanwhile, in the Pemali-Comal River Basin, the Pekalongan Regency contributes to a massive number of flood incidents.

It is notable that the peak of flood events on the North Coast of Central Java occurred in 2014, followed by a dramatic drop in 2015 and a steady increase afterwards up to 2018. On average, the height of flood in the study area is 20–40 cm. The flood duration varies from less than one hour to more than 24 h. In more detail, 56 out of 925 flood events analyzed in this study reached a height of 1.5 m or more and are categorized as severe flooding. This occurred mainly in Pati Regency, Rembang Regency and Semarang City, which are part of the Jratunseluna River Basin, and in several cities or regencies within the Pemali-Comal River Basin, including Pekalongan City and Pemalang Regency. In particular, the worst flood reached up to 3.5 m in Pemalang Regency in 2018. In addition, 259 flood events were up to 1-m high, most frequently in Kudus, Pekalongan and Jepara Regency.

Figure 6 presents the number of floods in the urban, potentially urban and rural areas surrounding river basins over the past nine years. It is evident that flood mostly took place in the rural areas rather than in the urban and rural-urban areas. The total number of flood events in urban, rural-urban, and rural areas in 2009–2018 was 485, 642 and 798 incidents, respectively. Flooding is very much influenced by rainfall intensity. The expansion of urbanization promotes flooding due to the increase in total impervious areas, leading to excessive rainfall. In addition, disaster risk reduction initiatives are also of importance.

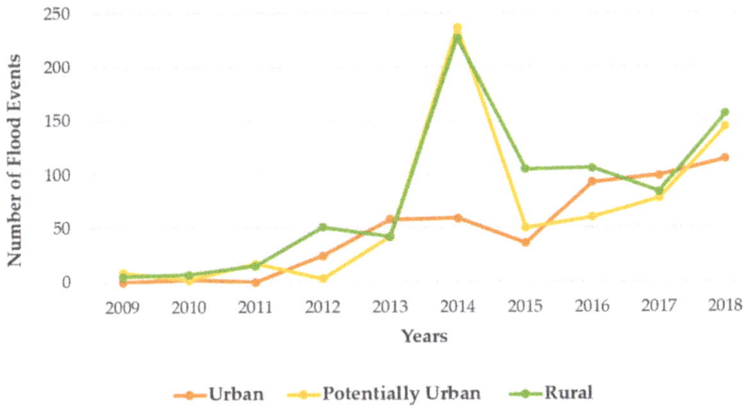

Figure 6. Flood Events in Urban, Potentially Urban and Rural Areas in the North Coast of Central Java 2009–2018. Flood events data in 2009–2011 are highly depend on online news due to limited data from DMB.

In the study area, flood events occurred during the rainy season. It was identified that 70% of flood events happened in January to March, during which the highest frequency of rainfall was reported, especially in 2014. For example, Pekalongan Regency contributed to the most significant flood events in 2014 (around 45%). According to the Meteorological, Climatological, and Geophysical Agency [31] the highest rainfall in Pekalongan Regency was recorded in January–February, at 991 mm and 1117 mm per month, respectively. In contrast, the rate of rainfall in the same month of the previous and following years was lower, at approximately 500–800 mm per month [32,33]. As revealed by other studies on rainfall patterns, since 2003, Java Island had a shorter term and higher intensity of rainfall [34]. Siswanto and Supari [35] revealed that extreme rainfall in Java tends to be irregular, in which such event is spatially distributed across the island and the positive and negative trends are proportional.

Figure 7 further illustrates flood events in each river basin. Jratunseluna, the biggest river basin, experienced the highest number of floods compared to other river basins. The flood events were concentrated in specific flood-prone areas, namely the Pati Regency, Kudus Regency and Semarang City, all of which represent the rural, potentially urban, and urban characteristics within Jratunseluna River Basin. In total, there were 1057 flood events in Jratunseluna River Basin, accounting for up to 48% of total flooding in the rural area over nine years. The rural area of Pati Regency contributed to the highest frequency of flood events, amounting to 219 out of 509 events spread out through 88 rural villages. In addition, up to 31% of flood events in Jratunseluna happened in the potentially urban area, with Kudus Regency experiencing the highest flood frequency with 102 flood events spread throughout 28 villages. In the urban areas, the highest frequency of flood events was recorded in Semarang City, which contributed up to 80% of total urban flood events spread throughout 26 *kelurahan* in 2009–2018. Flood events in the urban area continued to increase considerably over nine years, in contrast to the fluctuating flood in rural and potentially urban areas. The worst flood event in Jratunseluna River Basin took place in Grobogan Regency in 2013, which inundated approximately 5000 houses due to broken embankments. Demak Regency was also hit by severe flooding (1–2 m in height) in 2017, forcing 1450 households to abandon their homes. For the case of Jratunseluna, floods mostly hit the rural area compared to the urban and potentially urban areas.

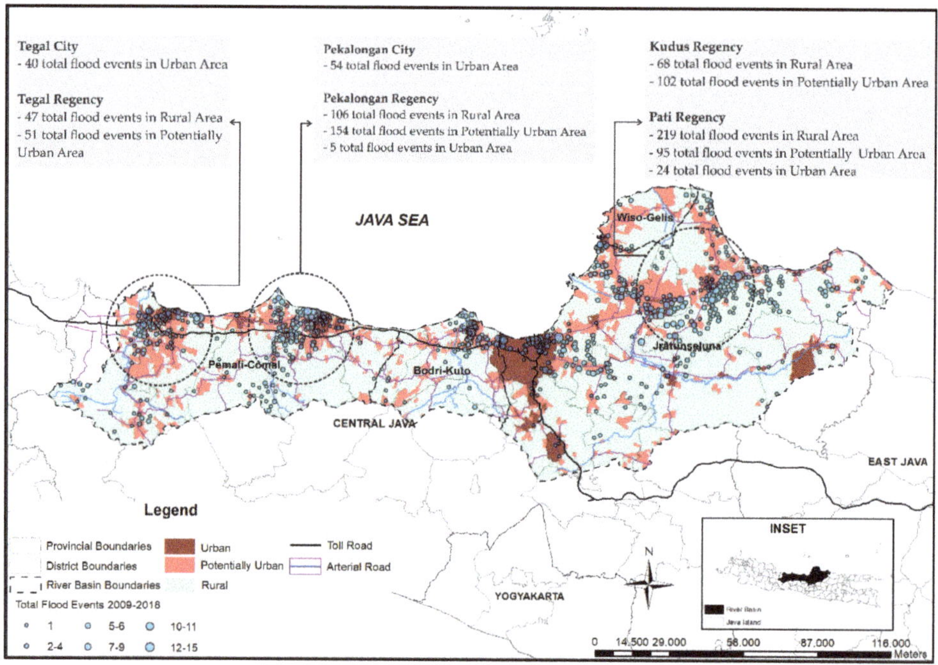

Figure 7. Urbanization and Flood Events in the Four Selected River Basins.

Similarly, the flood events in Pemali-Comal River Basin peaked in 2014, with up to 236 incidents. In total, there were 671 incidents throughout 2009–2018 in the region, spreading through 290 villages dominantly categorized as potentially urban areas. In detail, the number of flood events in potentially urban area within Pemali-Comal River Basin was up to 269 flood events, which contribute to approximately 40% of total incidents. Significant built-up land expansion (39%) in the potentially urban areas within Pemali-Comal River Basin was also observed, followed by a rise in flood events in those areas. Pekalongan had the most significant number of flood events compared to other regencies, with 154 incidents spread throughout 46 potentially urban villages. In respect to flood events in rural areas, Pekalongan Regency also contributed to the most significant number (43%) of total incidents. In Pemali-Comal and the Jratunseluna River Basin, flood incidents in the urban areas were less frequent than in the rural and potentially urban areas. Interestingly, flooding was more frequent in Pekalongan City compared to in other cities/regencies, which contributed up to 30% of total incidents in the urban area. Additionally, Tegal City and Brebes Regency also contributed to a high number of floods at approximately 25% and 24% of overall urban floods, respectively. This indicates that most villages on the coast of Pekalongan and Tegal Greater Area are prone to flood. More than 10 floods events occurred over nine years in the most flood prone villages within both river basins. Severe flooding in Pemali-Comal River Basin took place in Pemalang Regency in 2018. Due to river runoff following heavy rainfall, a flood as high as 3.5 m inundated thousands of houses in several villages.

With their smaller size compared to Jratunseluna and Pemali-Comal, there were fewer flood events in Bodri-Kuto and Wiso-Gelis River Basins. In Bodri-Kuto, there was a considerable fluctuation of flood events from 2009 to 2017, which peaked in 2018 with 53 flood events. Approximately 90% of flood events in this river basin occurred in Kendal Regency, while the rest took place in Semarang City. The urban area of Bodri-Kuto River Basin has the highest contribution of flood events at 60% of total flood events spread throughout 16 villages. It mainly occurred in Kendal Regency, where 13 flood events were reported at the village level. As noted by the Disaster Management Agency [36], flooding in Kendal urban area was caused by river runoff and low drainage capacity for water conveyance. Meanwhile, there was a slight increase in flood events in potentially urban and rural villages in Bodri-Kuto, amounting to 36 and 37 flood events during the nine-year period, respectively. The worst flood in Bodri-Kuto River Basin took place in Kendal Regency in early 2014, during which almost ten districts were affected by a 1.5-m flood.

Similarly, Wiso-Gelis as the smallest river basin experienced the worst flood around 1–1.5 m in height in 2014. The flood submerged 990 houses in Jepara Regency. In total, 17 flood events were recorded from 2009–2018. This shows that flood events mainly occurred in four villages within the rural and urban areas, in which with the number of flood events in the rural area was slightly higher than in the urban area. Accordingly, there is an indication that floods occur only occasionally in the potentially urban areas of the river basin. Since 2016, flooding in the Wiso-Gelis River Basin has been trending negatively, as shown by the decreasing number of flood events in the urban, potentially urban and rural areas of this river basin.

3.3. Land Use Change and Flood Phenomenon in River Basins

The population of Java has significantly grown from four million (at the beginning of the 19th century) to 40 million (in the early 20th century), to more than 150 million inhabitants in 2018 [37,38]. Moreover, the population within our study has increased from 17.1 million in 2009 to 18.3 million lives in 2018. Population growth led to significant land conversion and deforestation, which creates an impact on water cycle and rainfall pattern. Longer dry seasons lead to significant water supply problems, as the area keeps developing and experiencing rapid population growth. A previous study [39] showed that during the dry season (June and July), the rainfall patterns in most parts of Indonesia, including Central Java, tend to deviate from its normal conditions.

The overall contributions of land use and flood events over nine years (between 2009 and 2018) within four river basins are shown in Table 7. Approximately 80% of the river basin areas belong to the non-built-up area, which consists of rice fields, forests and mixed plantation areas. However, at the same time, the overall built-up area also increased significantly in all river basins, while the non-built up area decreased. Bodri-Kuto River Basin showed the highest loss in non-built up areas. In 2018, the non-built up area contributed to 85% of the total area within this river basin, while in 2009 the percentage was higher. Nonetheless, there was an upward trend of non-built up area in urban part of Jratunseluna in 2018, which was sharply expanded up to 30%. In contrast, the non-built up area in all other river basins showed a downward trend. The rise of the non-built up area in urban part of the Jratunseluna River Basin was caused by the transformation of settlement areas into mixed plantations or wetlands. For example, in the shoreline of Semarang City and Demak Regency, the increase in non-built up area occurred due to erosion in the area [40]. Accordingly, coastal erosion and inundation have caused a substantial loss of coastal land surrounding Demak Regency. Water as Leverage for Resilient Cities Asia Program Report [41] explained that Semarang's dynamic shoreline has been shifting faster over the last decade due to the changing climate and land subsidence, eroding mangrove areas, fishponds, villages and city assets. Moreover, the area of Demak Regency has experienced the most significant coastal erosion and loss of mangroves and aquaculture.

242

Table 7. Contribution of Land Use and Flood Events in the Study Areas.

| River Basin | Land Use Contribution (%) | | | | Flood Events Contribution (%) | | | | | | | | | | Total Flood Events Contribution 2009–2018 (%) |
| | 2009 | | 2018 | | | | | | | | | | | | |
	Built-Up	Non-Built Up	Built-Up	Non-Built Up	2009	2010	2011	2012	2013	2014	2015	2016	2017	2018	
Jratunseluna															
Urban	1.0	3.7	2.3	4.8	0.0	16.7	20.4	20.4	35.4	7.7	16.7	30.2	27.0	23.0	20.3
Potentially Urban	3.7	14.6	5.1	12.7	42.9	0.0	6.1	6.1	31.3	46.9	23.3	18.8	31.2	31.0	31.6
Rural	9.0	68.0	9.9	65.1	57.1	83.3	73.5	73.5	33.3	45.4	60.0	51.0	41.8	46.0	48.2
Total	13.7	86.3	17.3	82.7											
	100		100		100	100	100	100	100	100	100	100	100	100	100
Wiso-Gelis															
Urban	0.4	0.3	0.4	0.2	0.0	0.0	0.0	0.0	0.0	40.0	100	50.0	0.0	0.0	35.3
Potentially Urban	4.8	31.1	5.6	28.9	0.0	0.0	0.0	0.0	0.0	20.0	0.0	33.3	0.0	0.0	17.7
Rural	5.6	57.8	7.6	57.3	0.0	0.0	0.0	0.0	100	40.0	0.0	16.7	0.0	100	47.1
Total	10.8	89.2	13.6	86.4											
	100		100		0.0	0.0	0.0	0.0	100	100	100	100	100	100	100
Bodri-Kuto															
Urban	0.8	3.9	1.1	3.6	0.0	100	0.0	87.5	42.3	54.5	100	57.7	85.7	50.9	59.4
Potentially Urban	2.5	16.9	5.4	14.1	100	0.0	0.0	0.0	26.9	22.7	0.0	11.5	9.5	24.5	20.0
Rural	3.9	72.0	8.5	67.4	0.0	0.0	0.0	12.5	30.8	22.7	0.0	30.8	4.8	24.5	20.6
Total	7.1	92.9	14.9	85.1											
	100		100		100	100	0.0	100	100	100	100	100	100	100	100
Pemali-Comal															
Urban	1.3	2.3	1.5	2.2	0.0	0.0	0.0	0.0	68.4	10.6	9.8	38.5	43.4	26.1	23.6
Potentially Urban	4.2	19.8	6.1	17.8	0.0	50.0	100	0.0	26.3	46.2	36.1	34.6	32.3	43.9	40.1
Rural	6.0	66.5	6.4	66.1	100	50.0	0.0	100	5.3	43.2	54.1	26.9	24.2	29.9	36.4
Total	11.5	88.5	14.0	86.0											
	100		100		100	100	100	100	100	100	100	100	100	100	100

A significant exposure to flood has taken place in the urban and potentially urban areas following the increase in built-up areas (Table 7). Flood events are more frequent in the urban and potentially urban areas, most of which are located nearby the coastal line. This is also in line with a previous study by Rudiarto et al. [7,42], where 40% of flooding events were found within the range of 10 km from the coastline, while 80% of tidal flooding was distributed mostly in the areas of less than 5 km from the coastline. Flooding is a result of various factors, and urban flooding is not only mostly caused by water overflowing from the river (fluvial flooding) but also by land conversion in combination with weak drainage systems (pluvial flooding). As flooding is more common in the urban areas, which means that densification has a significant influence on the increasing event of pluvial flooding. Densification typically occurs due to the conversion of agricultural land into settlement and industrial land, leaving a lot of the areas vulnerable to flooding [43]. It is likely that the number of rainy days significantly decreases with higher rain intensity. This very much influences the surface water runoff and put more pressure on the river and drainage systems. Robust and adaptive drainage arrangements, therefore, is of importance in this circumstance.

Aside from pluvial flooding (which happens mostly in urban fabric area), fluvial flooding (which occurs because of the overbank of the water from the river) also has a significant influence in the rural-urban and rural areas. In this regard, it is caused by land-use change that transforms forests in the rural area into a built-up area. Some rural areas have been hit by intensive flooding, especially in the Jratunseluna and Wiso-Gelis River Basin, which contributed to more than 40% of the total flood events over nine years (Table 7). However, flooding in this region is not only connected to deforestation, but also to in situ urbanization. Handayani [44] found that industrialization at the rural level is happening in Central Java. This type of industrialization may potentially lead to the increase in flood risk in areas that are not necessarily located in the big urban center.

4. Discussion and Issues Emerging from the Study

The results that emerged from this study have some implications to the urban development and policies. This study reflects that land use dynamics would depend on policy decisions and implementations for improvement. Sufficient comprehension on policies at the river basin level is an essential prerequisite in flood risk management. For the sake of solution-oriented discussions, two issues emerged from this study. First, the need to create a better understanding on urbanization and flooding phenomena, to raise more solution-oriented awareness through land use policy. This is crucial, especially in countries like Indonesia. For that reason, the authors have put an emphasis on Indonesia through case studies. Second, there is a need to identify the role of governance in flood management, particularly in curbing urban flooding. Both issues are discussed further in the following sections.

4.1. River Basins Have Dual Spatial Identity that Embraces Policy and Land Use Issues in the Urban System

Knowledge concerning river basins dynamics in the urban system is still vague. It is, in most cases, simply viewed as a landscape or an appendage of water bodies [45]. Though it is often used as a point of departure in discussing several issues related to urbanization, it is unduly taken for granted in terms of its functions in the urban system. The analytical aspect of this study offers a renewed systematic way of looking at river basins as a sub-ecosystem embedded within the urban system, and as a concept in the urban discourse. As can be deduced from the case presented in this study (at least in the context of Java), river basins have a dual spatial identity in the urban system. It is both a natural land object, as well as a form of land use.

A river basin is a part of the land because it is a section of the "earth surface with all physical, chemical and biological features" [46] (p. xix). It can be viewed as a land object because it is uniquely embedded to (as well as a natural embodiment of) the physical urban system, and yet are distinguishable in legal (invisible) ecosystems recognized in policies, laws and statutes. In fact, within the land administration system, the river basin can be categorized as a cadastral object and as a unique legal entity, which can be both fiat (i.e., invisible) and bona fide (i.e., visible). It has a boundary and can be surveyed and measured in physical, ecological, socioeconomic, and cultural terms. It can also be viewed as a "property", because it is the embodiment of several "set of rights and a set of duties or obligations" (including interests and privileges) that subsist in the urban land, which the urban people expect to leverage or enjoy [47] (p. 2). Hence, it has various forms of values attached to it—including ecological, economic, political, cultural, social, touristic, aesthetic, and other urban functional values. As a result, a river basin should be viewed as a portion of land meant to be administered, managed, and controlled to ensure that it fulfils its function within the urban system. In this regard, flooding is a negative consequence of the relationship between a river basin and its urban surroundings, which makes the area unavailable to urban people.

River basins also constitute an essential type of land use in the urban system. The perspective of conceiving the river basins for land use is best illustrated by answering the question: why do urban people want to live around a river basin? In the context of Java, the river basin is a sub-system that embodies vegetation and waterways required for food, energy, water, biodiversity, and shelter, among many others. It serves a cooling effect in the urban heat island concerns [48]. Hence, river basins constitute land use because they are part of the decisions people make regarding land or natural resources available to them within permissible natural and administrative restrictions. Land use is, therefore, a purposeful intervention made by humans concerning what and how to exploit, explore, protect or conserve aspects of the land system [49,50]. Urban river basins are, therefore, subject to land use adoptable by urban people according to permissible natural and legal (or administrative) characteristics, leading to transformations in the way they live in the urban system.

How does the above idea relate to tackling urban flooding? The dual spatial identity of river basins (both as a land object and land use in the urban system) offers an opportunity to mitigate flooding, mostly in coastal areas. However, it will also pose a threat if it is not managed well. In the case of Java, it can be argued that built-up area expansion to the upstream area of the river basins lead to significant negative consequences. Not only it threatens the food and water supply sustainability (referring to river basin as land), but it also generates issues in infrastructure provision (such as to manage flood) and ownership due to rapid settlement growth in areas that play a strategic role in the river system (conflict of interest regarding land use).

Chen et al. [17] argue that, based on the experience in China, the sprawling built-up land increases the difficulty and costs to deploy and manage hazard-resistant infrastructures, construction-wise. Accordingly, the compact city concept is perceived as the most sustainable urban form to limit the uncontrolled effects of infrastructure provision caused by the need to contain urban growth. Global urban sprawl usually leads to the increase in emission load due to the increased use of transportation. However, what is usually not written much about is that sprawl development causes problems that limit water conveyance and supply. Rudiarto et al. [7] stated that the urbanization of north Central Java has been very significant since the 1990s. This is followed by the increase in climate disasters, as shown by incessant floods. Handayani and Rudiarto [51] have further examined this phenomenon in Semarang Metropolitan, the biggest urban center in the area. In this regard, Douglass [2] argued that it creates an urban disaster in Asia, a situation where agglomerations affect urban areas. Thus, there lies an urgent call to focus on urban growth management in an integrated framework following an ecosystem-based (or eco-based) regional approach. An eco-based approach would involve conceiving the river basins as a unique ecosystem and employing a wide range of ecosystem management activities to reduce the vulnerability of urban people and urban environment due to flooding. In this regard, the approach would tackle urban challenges that arise from the location of river basins. Hence, whereas flooding is a critical problem linked to the river basins, it can be mitigated as part of broader ecological system management.

Focusing on the use of urban infrastructure provision to check flood events, Zhou et al. [19] have shown through their study that the drainage system is vital in reducing the risk of urban flood. Based on several cases in major cities in Northern China, they [19] revealed that the frequency of flooding is caused by the lack of or failure in the urban drainage system. A similar case in the UK [52] revealed that drainage is essential in reducing flood risk, since flooding is very much influenced by urban densification and changing rainfall patterns. Accordingly, a proper drainage system is very critical to accommodate water conveyance during intensive rainfall. Even though there is still much debate on this matter in Indonesia, just as in the UK, there is evidence of a change in rain patterns in Java due to rapid urban growth and deforestation [37,53]. The number of rainy days is likely to decrease significantly but with a higher intensity of rainfall. This very much influences the surface water runoff and puts more pressure on the river and drainage systems. Such situation requires robust and adaptive drainage arrangements.

4.2. There Are Several Opportunities to Broaden the Role of Governance in Flood Management

Any serious effort to tackle urban flooding induced or influenced by the river basins demands the problematization of river basins, that is, viewing them as a problem that requires a solution. This is important in urban policymaking or urban reform efforts that are targeted for urban flood management. Historically, river basin development "has been used to structure water resource management" [54] (p. 839). Evidence from cases presented in this study shows that the management of river basins, if geared towards solving the flood problems, would have a mitigative effect in controlling the situation.

Understanding the opportunities for flood management through governance should be a critical aspect of urban development. The governance of river basins in specific, or water resources in general, would allow urban administrators to explore various technical and socio-political strategies to mitigate flood at various levels (basin, local, regional and national). Consequently, a governance approach capable of addressing both general urban issues and flood challenges is imperative. Governance-related urban policy instruments can serve as an essential factor in ensuring proper flood intervention to manage urbanization and flood prevention. In this regard, Friend et al. [55] argued that there is always a gap between policy planning and implementation, while there is a need for communication and negotiation among actors.

In the context of Java, such interactions are even more critical in respect to flood prevention, as there are many authorities with different roles and functions that manage the river basins (Figure 8). Both vertical and horizontal coordination are needed to ensure integrated policies. Vertical coordination is essential because the National Government (i.e., the Ministry of Public Works through the River Management Centre) is responsible in managing the rivers from upstream to downstream, while the drainage systems that cross through two different regencies are under the responsibility of the Provincial Government. Institutions at different levels of authority need to work intensively with the local governments (cities and regencies) in regard to the river basin management. This involves spatial planning policies that include various infrastructure provisions under the local government authority. Accordingly, horizontal coordination is also crucial, mostly because urban expansion due to rapid urbanization takes place beyond the administrative jurisdiction. Indeed, integration and collaborations would enable more sustainable urbanization.

In principle, the governance arrangement reflects subsidiarity. Each level deals with a specific role and decision making is made at both the top and the lowest level. However, in practice (and focusing on the river basins), decision making in water management is not made at the lowest level, where water is used. The national and provincial authorities are the ones who carry out the roles of river management and drainage systems, respectively. This, therefore, leads to the need for better interagency collaborations to allow for effective communication and the co-designing of strategies for action.

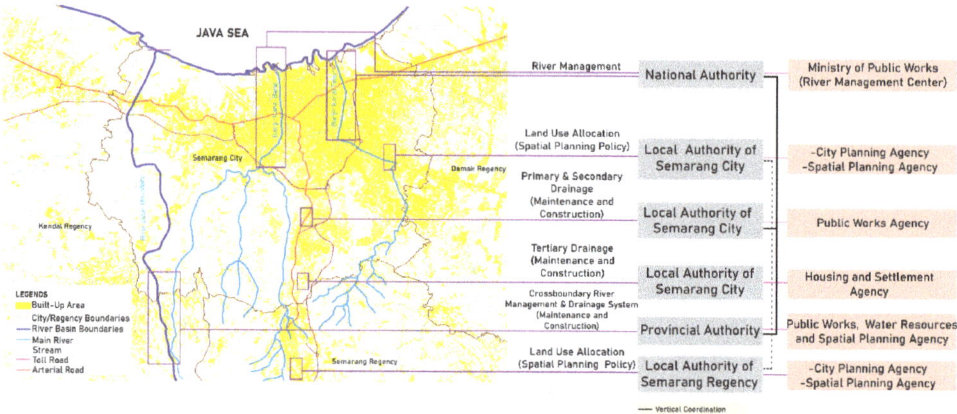

Figure 8. Roles and Responsibilities Related to River Management and Land Use Allocation.

It is essential to introduce governance in flood management. However, this is a relatively new concept that needs to be discussed further [2,56]. The term becomes vital in current situation because the issue of flooding cannot be solved by one sole organization. After all, it is a multifaceted challenge that affects housing, farming and forestry as well as transport, among many others. It requires an effective decision-making process that involves various sectors and authorities. It also prompts integrated approaches that are certainly not limited to infrastructural work [57,58]. For this reason, Hegger et al. [13] categorized five types of strategies in flood risk management: flood risk prevention, flood defense, flood risk mitigation, flood preparation and flood recovery. According to Hegger's strategy typology, land-use change should be controlled through proper spatial planning and this may serve as a policy instrument for flood risk prevention. However, it is interesting to note that based on Hegger et al.'s [13] and Raikes et al.'s [59] investigation in selected countries across the globe, there is still a lack of integration among the different types of strategies in place. Policies related to water supply, flood management and spatial planning are also fragmented. Despite that, both scholars [13,60] also argue that fragmentation is inevitable, because there are many strategies involved and each country has their own policy direction on flood management, with varying strengths and weaknesses.

Raikes et al. [59] reveal that most government policies are more focused on infrastructural work rather than on comprehensive flood prevention (through spatial planning policy). Handayani et al. [16] hold a similar position based on their two case scenarios, also done in Indonesia. Accordingly, Pardoe et al. [58] argued that infrastructural work will not be sufficient to accommodate the balance among land, people, and water interactions. Instead, Pardoe et al. [7] proposed a holistic policy instrument, which is vital to ensure the availability of a sustainable space for people and water. Many countries around the world have their own country-specific strategy for managing flood situations. The Netherlands with the concept of "Room for the River" [60] (p. 369) and the UK with "Making Space for Water" [61] (p. 534) approach demonstrate a case on how flood could be managed through suitable land use allocation.

There is, indeed, a responsibility of the government to provide public infrastructures to mitigate flood events. Infrastructure provision requires not only technical capacity and funding but also proper coordination among different government institutions. A considerable amount of investment allocated for significant infrastructural work for flood prevention will only act as a short-term and reactive solution rather than a long-term one. On the other hand, there is an increasing role of developers, since most of the land is owned privately. They are dominant players in developing industrial and housing estates, which are regarded as major land conversion within the study area. Accordingly, collaborations with private sectors (landowners) is an excellent opportunity to further manage river basins, mostly to mitigate flood, which is unavoidable in the situation where urbanization (i.e., land conversion) has spread through the whole area.

5. Conclusions

Limited attention has been paid to the potential effects of river basins on urbanization-associated floods. In general terms, it is well known that "most cities are historically developed near rivers or oceans to ensure the supply of water" [62] (p. 1). It is therefore not surprising that Indonesia—a country surrounded by waters from many rivers—has cities that are located around waters. Therefore, this study confirms Zhang et al.'s [63] (p. 384) thesis that the process of urbanization "exacerbates flood responses" in low-lying areas. Using this case study, we identified possible urban land use components of the global urban flooding crisis. This also implies that rapid urbanization, in addition to the lack of land-use planning (or inappropriate implementation of the plan), have increased the amount of land exposed to floods [62] (p. 1). One key issue deduced from this study is that there is a relationship between flooding and urbanization. However, such a relationship may not always be straightforward. It can vary from country to country depending on their respective planning and development strategies, human behavior or response to flood and urbanization scenarios; and most importantly, the role of governance in the management of floods. From the context of land use and management, this study has shown that the river basin is a linkage factor or object in the flood–urbanization relationship.

The study highlights the importance of investigating the role of river basins in impacting flood events in highly urbanized areas. Disaster risk reduction through proper land use planning and controlling is an essential instrument for safeguarding urban areas (such as the case study area, and the entire island of Java in Indonesia). This provides an opportunity to sustain coastal or island settlements and prevent them from being converted into urban islands, which may face complex environmental issues, including extreme precipitation or water-related disasters, and hydrometeorology-associated events. However, without proper management measures, technical measures alone are not sufficient to improve this situation. In this regard, the role of governance in flood management is crucial. This aspect is a missing link in the urban environmental risk management strategy in Indonesia. The country's decentralization policy, which has been in operation since 1999, has led to a cumbersome coordination process for land use allocation instead of a solution-oriented one. Due to the decentralization policy, the local government is lacking in authority in this context. Instead, each local or municipality government is focused on the economic development, which is highly dependent on massive land conversion with no recourse to geographical delineations of the river basins. Such action will continue to bring dire environmental consequences, especially given that no appropriate actions have been taken to alleviate them. Hence, the effect of urban land-use on extreme precipitation and flooding should be studied more explicitly. The study presented in this paper is an urgent call to comprehend urbanization beyond a mere administrative-based process. Urban environmental risks generated from urbanization can be mitigated by understanding their land use components. We hope that this study will motivate other scholars from the Global South to investigate the role of river basins in other urban areas in search of solutions for sustainable environmental risk governance in urban areas.

Author Contributions: Conceptualization, W.H. and U.E.C.; methodology, W.H. and I.R.; Resources, W.H., I.R., U.E.C., and I.H.S.P.; Formal analysis, W.H., I.R., and I.H.S.P.; Investigation, W.H., I.R., and I.H.S.P.; writing—original draft preparation, W.H., I.R., and I.H.S.P.; writing—review and editing, W.H. and U.E.C.; Validation, W.H., I.R., U.E.C., and I.H.S.P.; visualization, I.H.S.P. All authors have read and agreed to the published version of the manuscript.

Funding: This research was funded by Diponegoro University and Ministry of Research, Technology, and Higher Education Indonesia, and the APC was funded by Technical University of Munich.

Acknowledgments: Acknowledgment is given to the Director General of Higher Education, Ministry of Research, Technology, and Higher Education Indonesia, Diponegoro University, and Technical University of Munich for supporting this research. This publication was supported by the German Research Foundation (DFG) and the Technical University of Munich (TUM) in the framework of the Open Access Publishing Program.

Conflicts of Interest: The authors declare no conflict of interest.

References

1. Deng, X.; Xu, Y. Degrading Flood Regulation Function of River Systems in the Urbanisation Process. *Sci. Total Environ.* **2018**, *622*, 1379–1390. [CrossRef] [PubMed]
2. Miller, M.A.; Douglass, M. (Eds.) *Disaster Governance in Urbanising Asia*; Springer: Singapore, 2016; pp. 1–12.
3. Sarauskiene, D.; Kriauciuniene, J.; Reihan, A.; Klavins, M. Flood Pattern Changes in the Rivers of the Baltic Countries. *J. Environ. Eng. Landsc. Manag.* **2015**, *23*, 28–38. [CrossRef]
4. Adhikari, P.; Hong, Y.; Douglas, K.R.; Kirschbaum, D.B.; Gourley, J.; Adler, R.; Brakenridge, G.R. A digitized global flood inventory (1998–2008): Compilation and preliminary results. *Nat. Hazards* **2010**, *55*, 405–422. [CrossRef]
5. Bertilsson, L.; Wiklund, K.; de Moura Tebaldi, I.; Rezende, O.M.; Veról, A.P.; Miguez, M.G. Urban flood resilience—A multi-criteria index to integrate flood resilience into urban planning. *J. Hydrol.* **2019**, *573*, 970–982. [CrossRef]
6. Buchori, I.; Pramitasari, A.; Sugiri, A.; Maryono, M.; Basuki, Y.; Sejati, A.W. Adaptation to coastal flooding and inundation: Mitigations and migration pattern in Semarang City, Indonesia. *Ocean Coast. Manag.* **2018**, *163*, 445–455. [CrossRef]
7. Rudiarto, I.; Handayani, W.; Setyono, J.S. A Regional Perspective on Urbanisation and Climate-Related Disasters in the Northern Coastal Region of Central Java, Indonesia. *Land* **2018**, *7*, 34. [CrossRef]
8. McGranahan, G.; Balk, D.; Anderson, B. The rising tide: Assessing the risks of climate change and human settlements in low elevation coastal zones. *Environ. Urban.* **2007**, *19*, 17–37. [CrossRef]
9. Neumann, B.; Vafeidis, A.T.; Zimmermann, J.; Nicholls, R.J. Future Coastal Population Growth and Exposure to Sea-Level Rise and Coastal Flooding—A Global Assessment. *PLoS ONE* **2015**, *10*, e0118571. [CrossRef]
10. Deng, X.; Xu, Y.; Han, L.; Song, S.; Yang, L.; Li, G.; Wang, Y. Impacts of Urbanisation on River Systems in the Taihu Region, China. *Water* **2015**, *7*, 1340–1358. [CrossRef]
11. Chin, A. Urban Transformation of River Landscapes in A Global Context. *Geomorphology* **2006**, *79*, 460–487. [CrossRef]
12. Chigbu, U.E.; Kalashyan, V. Land-use planning and public administration in Bavaria, Germany: Towards a public administration approach to land-use planning. *Geomat. Land Manag. Landsc.* **2015**, *4*, 7–17. [CrossRef]
13. Hegger, D.L.; Driessen, P.P.; Wiering, M.; Van Rijswick, H.F.; Kundzewicz, Z.W.; Matczak, P.; Crabbé, A.; Raadgever, G.T.; Bakker, M.H.N.; Priest, S.J.; et al. Toward more flood resilience: Is a diversification of flood risk management strategies the way forward? *Ecol. Soc.* **2016**, *21*, 52. [CrossRef]
14. ISDR. Hyogo Framework for Action 2005–2015: Building the Resilience of Nations and Communities to Disasters. In Proceedings of the World Conference on Disaster Reduction, Kobe, Japan, 18–22 January 2005.
15. UNISDR. *Sendai Framework for Disaster Risk Reduction 2015–2030: United Nations International Strategy for Disaster Reduction*; UNISDR: Geneva, Switzerland, 2015.
16. Handayani, W.; Fisher, M.R.; Rudiarto, I.; Setyono, J.S.; Foley, D. Operationalizing resilience: A content analysis of flood disaster planning in two coastal cities in Central Java, Indonesia. *Int. J. Disaster Risk Reduct.* **2019**, *35*, 101073. [CrossRef]
17. Chen, Y.; Xie, W.; Xu, X. Changes of Population, Built-up Land, and Cropland Exposure to Natural Hazards in China from 1995 to 2015. *Int. J. Disaster Risk Sci.* **2019**, *10*, 557–572. [CrossRef]
18. Song, S.; Xu, Y.P.; Wu, Z.F.; Deng, X.J.; Wang, Q. The relative impact of urbanisation and precipitation on long-term water level variations in the Yangtze River Delta. *Sci. Total Environ.* **2019**, *648*, 460–471. [CrossRef]
19. Zhou, Q.; Leng, G.; Su, J.; Ren, Y. Comparison of urbanization and climate change impacts on urban flood volumes: Importance of urban planning and drainage adaptation. *Sci. Total Environ.* **2019**, *658*, 24–33. [CrossRef]
20. Zope, P.E.; Eldho, T.I.; Jothiprakash, V. Impacts of land use–land cover changes and urbanisation on flooding: A case study of Oshiwara River Basin in Mumbai, India. *Catena* **2016**, *145*, 142–154. [CrossRef]
21. Handayani, W.; Rudiarto, I.; Setyono, J.S.; Chigbu, U.E.; Sukmawati, A.M. Vulnerability assessment: A comparison of three different city sizes in the coastal area of Central Java, Indonesia. *Adv. Clim. Chang. Res.* **2017**, *8*, 286–296. [CrossRef]
22. Parkes, M.W.; Morrison, K.E.; Bunch, M.J.; Hallström, L.K.; Neudoerffer, R.C.; Venema, H.D.; Waltner-Toews, D. Towards Integrated Governance for Water, Health and Social-Ecological Systems: The Watershed Governance Prism. *Glob. Environ. Chang.* **2010**, *20*, 693–704. [CrossRef]

23. Gregersen, H.M.; Ffolliott, P.F.; Brooks, K.N. (Eds.) *Integrated Watershed Management: Connecting People to Their Land and Water*; CABI: Wallingford, UK, 2007.

24. Handayani, W.; Waskitaningsih, N. *Kependudukan dalam Perencanaan Wilayah dan Kota*; Teknosain: Yogyakarta, Indonesia, 2019.

25. Indonesia Government. *Presidential Decree No. 12/2012: The Determination of River Areas*; Indonesia Government: Jakarta, Indonesia, 2012.

26. Central Bureau of Statistics (CBS). *Head of Central Bureau of Statistics Regulation No. 37/2010: Urban Rural Classification in Indonesia*; CBS: Jakarta, Indonesia, 2010.

27. Indonesia Government. *Law No. 24/2007: Disaster Management*; Indonesia Government: Jakarta, Indonesia, 2007.

28. National Standardization Agency of Indonesia. *Indonesian National Standard Regulation Number 7645: 2010 Land Cover Classification*; National Standardization Agency of Indonesia: Jakarta, Indonesia, 2010.

29. Central Java Government. *Governor of Central Java Regulation No. 46/2012: The Provincial Level of Forestry Plan 2011–2030*; Central Java Government: Central Java, Indonesia, 2012.

30. Indonesian Ministry of Environment and Forestry. *Minister of Environment and Forestry Regulation No. P.41/MENLHK/SETJEN/KUM.1/7/2019: The National Level of Forestry Plan 2011–2030*; Indonesian Ministry of Environment and Forestry: Jakarta, Indonesia, 2019.

31. MCGA (Meteorological, Climatological, and Geophysical Agency). Data Online—Data Iklim: Pusat Database BMKG, Semarang-Indonesia, 2014. Available online: http://dataonline.bmkg.go.id/data_iklim (accessed on 15 April 2019).

32. MCGA (Meteorological, Climatological, and Geophysical Agency). Data Online—Data Iklim: Pusat Database BMKG, Semarang-Indonesia, 2015. Available online: http://dataonline.bmkg.go.id/data_iklim (accessed on 15 April 2019).

33. MCGA (Meteorological, Climatological, and Geophysical Agency). Data Online—Data Iklim: Pusat Database BMKG, Semarang-Indonesia, 2016. Available online: http://dataonline.bmkg.go.id/data_iklim (accessed on 15 April 2019).

34. MCGA (Meteorological, Climatological, and Geophysical Agency). Data Online—Data Iklim: Pusat Database BMKG, Semarang-Indonesia, 2017. Available online: http://dataonline.bmkg.go.id/data_iklim (accessed on 15 April 2019).

35. Siswanto, S.; Supari, S. Rainfall Changes Over Java Island, Indonesia. *J. Environ. Earth Sci.* **2015**, *5*, 1–10.

36. Kendal Disaster Management Agency. Banjir Limpas Sungai Kendal. 2019. Available online: http://bpbd.kendalkab.go.id/berita/id/20190127001/banjir_limpas_sungai_kendal (accessed on 15 April 2019).

37. Pawitan, H. *Perubahan Penggunaan Lahan dan Pengaruhnya Terhadap Hidrologi Daerah Aliran Sungai (Effect of Land Use Change to the Hydrology in Watersheds)*; Laboratorium Hidrometeorologi FMIPA IPB: Bogor, Indonesia, 2004.

38. Central Bureau of Statistics (CBS). *Statistik Indonesia 2018 (Indonesian Statistic 2018)*; Central Bureau of Statistics: Jakarta, Indonesia, 2018.

39. Avia, L.Q. Change in Rainfall Per-Decades over Java Island, Indonesia. *IOP Conf. Ser.: Earth Environ. Sci.* **2019**, *374*, 012037. [CrossRef]

40. Dewi, R.S.; Bijker, W. Dynamics of Shoreline Changes in the Coastal Region of Sayung, Indonesia. *Egypt. J. Remote Sens. Space Sci.* **2020**, *23*, 181–193. [CrossRef]

41. Water as Leverage for Resilient Cities Asia Program. *One Resilient Semarang: Volume II Concept Design Proposals Final Report*; Water as Leverage Consortium Project Report: Semarang, Indonesia, 2019.

42. Rudiarto, I.; Pamungkas, D. Spatial Exposure and Livelihood Vulnerability to Climate-Related Disasters in the North Coast of Tegal City, Indonesia. *Int. Rev. Spat. Plan. Sustain. Dev.* **2020**, *8*, 34–53. [CrossRef]

43. Rudiarto, I.; Handayani, W.; Wijaya, H.B.; Insani, T.D. Land Resource Availability and Climate Change Disasters in the Rural Coastal of Central Java–Indonesia. *IOP Conf. Ser. Earth Environ. Sci.* **2018**, *202*, 012029. [CrossRef]

44. Handayani, W. Rural-urban transition in Central Java: Population and economic structural changes based on cluster analysis. *Land* **2013**, *2*, 419–436. [CrossRef]

45. Das, S.; Teron, R.; Duary, B.; Bhattacharya, S.S.; Kim, K.H. Assessing C–N balance and soil rejuvenation capacity of vermicompost application in a degraded landscape: A study in an alluvial river basin with Cajanus cajan. *Environ. Res.* **2019**, *177*, 108591. [CrossRef] [PubMed]

46. Fleischhauer, E.; Eger, H. Can Sustainable Land Use be Achieved? An Introductory View on Scientific and Political Views. In *Towards Sustainable Land Use: Furthering Cooperation between People and Institutions*; Blume, H.P., Eger, H., Fleischhauer, E., Hebel, A., Reij, C., Steiner, K.G., Eds.; Catena Verlag GMBh: Reiskirchen, Germany, 1998; Volume 1, pp. xix–xxxii.

47. Gwaleba, M.J.; Chigbu, U.E. Participation in Property Formation: Insights from Land-Use Planning in an Informal Urban Settlement in Tanzania. *Land Use Policy* **2020**, *92*, 104482. [CrossRef]

48. Yao, R.; Wang, L.; Gui, X.; Zheng, Y.; Zhang, H.; Huang, X. Urbanisation Effects on Vegetation and Surface Urban Heat Islands in China's Yangtze River Basin. *Remote Sens.* **2017**, *9*, 540. [CrossRef]

49. Zhanlu, Z.; Zhang, Z. *Land Use Planning*; China Renmin University Press: Beijing, China, 2006.

50. Wang, W.; Han, T. *Land Use Planning*; China Agriculture Press: Beijing, China, 2013.

51. Handayani, W.; Rudiarto, I. Dynamics of urban growth in Semarang Metropolitan-Central Java: An examination based on built-up area and population change. *J. Geogr. Geol.* **2014**, *6*, 80. [CrossRef]

52. Miller, J.D.; Hutchins, M. The Impacts of Urbanisation and Climate Change on Urban Flooding and Urban Water Quality: A Review of the Evidence Concerning the United Kingdom. *J. Hydrol. Reg. Stud.* **2017**, *12*, 345–362. [CrossRef]

53. Haryani, G.S. Ecohydrology in Indonesia: Emerging Challenges and Its Future Pathways. *Ecohydrol. Hydrobiol.* **2016**, *16*, 112–116. [CrossRef]

54. Fatch, J.J.; Manzungu, E.; Mabiza, C. Problematising and Conceptualising Local Participation in Transboundary Water Resources Management: The Case of Limpopo River Basin in Zimbabwe. *Phys. Chem. Earth Parts A/B/C* **2010**, *35*, 838–847. [CrossRef]

55. Friend, R.; Jarvie, J.; Reed, S.O.; Sutarto, R.; Thinphanga, P.; Toan, V.C. Mainstreaming urban climate resilience into policy and planning; reflections from Asia. *Urban Clim.* **2014**, *7*, 6–19. [CrossRef]

56. Tierney, K. Disaster Governance: Social, Political, and Economic Dimensions. *Annu. Rev. Environ. Resour.* **2012**, *37*, 341–363. [CrossRef]

57. Pardoe, J.; Penning-Rowsell, E.; Tunstall, S. Floodplain Conflicts: Regulation and Negotiation. *Nat. Hazards Earth Syst. Sci.* **2011**, *11*, 2889–2902. [CrossRef]

58. Heintz, M.D.; Hagemeier-Klose, M.; Wagner, K. Towards A Risk Governance Culture in Flood Policy—Findings from the Implementation of the "Floods Directive" in Germany. *Water* **2012**, *4*, 135–156. [CrossRef]

59. Raikes, J.; Smith, T.F.; Jacobson, C.; Baldwin, C. Pre-disaster Planning and Preparedness for Floods and Droughts: A Systematic Review. *Int. J. Disaster Risk Reduct.* **2019**, *38*, 101207. [CrossRef]

60. Rijke, J.; Van Herk, S.; Zevenbergen, C.; Ashley, R. Room for The River: Delivering Integrated River Basin Management in The Netherlands. *Int. J. River Basin Manag.* **2012**, *10*, 369–382. [CrossRef]

61. Jones, P.; Macdonald, N. Making Space for Unruly Water: Sustainable Drainage Systems and the Disciplining of Surface Runoff. *Geoforum* **2007**, *38*, 534–544. [CrossRef]

62. Bae, S.; Chang, H. Urbanisation and Floods in the Seoul Metropolitan Area of South Korea: What Old Maps Tell Us. *Int. J. Disaster Risk Reduct.* **2019**, *37*, 101186. [CrossRef]

63. Zhang, W.; Villarini, G.; Vecchi, G.A.; Smith, J.A. Urbanisation Exacerbated the Rainfall and Flooding Caused by Hurricane Harvey in Houston. *Nature* **2018**, *563*, 384–388. [CrossRef] [PubMed]

 *land*

Article

Gender Inequality and Symbolic Violence in Women's Access to Family Land in the Southern Highlands of Tanzania

Justin Lusasi [1,*] **and Dismas Mwaseba** [2]

1   Department of Policy Planning and Management, College of Social Sciences and Humanities, Sokoine University of Agriculture, Morogoro Box 3035, Tanzania
2   Department of Agricultural Extension and Community Development, College of Agriculture, Sokoine University of Agriculture, Morogoro Box 3002, Tanzania; dismasmwaseba@gmail.com
*   Correspondence: julusasi@gmail.com

Received: 14 October 2020; Accepted: 5 November 2020; Published: 22 November 2020

Abstract: We set out to unveil gender inequality with respect to women's access to family land following the surge in tree-planting in selected villages in the Southern Highlands of Tanzania. Specifically, the study describes land-transaction procedures at the household level and shows how the lack of women's involvement in such land transactions affect their access to and control over family lands. Gender inequality is portrayed in a variety of social and economic activities, with women being deprived of access to, control over, and ownership of land. Although the current land laws address gender inequalities pertaining to women's access to, ownership of, and control over land, the impact of such reforms has been minimal. Drawing on Bourdieu's concept of symbolic violence, we reveal how women suffer symbolic violence through traditional practices of land management and administration. Societies in the studied villages are strongly patriarchal, with men being dominant and women subordinate. In such a patriarchal system, women's empowerment is urgent. Women require knowledge and awareness of the laws and regulations that affirm their rights not only to family lands, but also to participation in decision-making processes regarding family assets. We recommend non-oppressive approaches to natural-resource management. As such, we call for existing authorities at the village and district levels, Non Governmental Organisations (NGOs) and legal bodies to promote gender equality in land-management practices. We also advocate dialectical communication between women and men in order to reveal and heal practices of symbolic violence, and enhance gender equality in respect of access to land and its control and ownership in villages in the Southern Highlands of Tanzania. Effective implementation of existing land laws and regulations that address gender inequality and associated violence is unavoidable.

Keywords: gender; inequality; access; land; symbolic violence; Southern Highlands; Tanzania

---

1. Introduction

Gender mainstreaming constitutes a fulcrum for development planning and an entry point to the successful management of natural resources in developing countries [1]. This has been advocated by the United Nations General Assembly through its fifth Sustainable Development Goal (SDG), which targets achieving gender equality and empowering all women and girls by 2030 [2]. Gender equality, achievement of which still requires global efforts from small units in rural areas to highly developed communities, is a pillar for attaining a peaceful, prosperous and sustainable world.

Notwithstanding geographical disparities involved, most societies exhibit gender inequalities in access to and control over and the management of land and landed resources, including forests, with women being the victims in this situation [3,4]. In most areas of rural Tanzania, land ownership is guided by traditions, customs, and taboos that legitimize some groups of people as landowners, while excluding others [5]. Although some societies use criteria, such as age, marital status, and wealth to identify those who are entitled to access land [6], others use gender differences and traditional taboos as criteria determining access to, use of, and control over this natural resource [1]. In the book *Gender, Environment, and Development*, Heleen van den Hombergh describes the importance of gender concepts in environmental debates. Although Hombergh admits that men and women use natural resources differently and at different rates [7], other scholars argue in addition that men and women do not have equal rights to resources, nor can they draw equal benefits from them [8].

Gurung and colleagues, studying a society of the Hindu Kush in the Himalayas, described women as having access to and being involved with low-value non-timber products, including collecting and fetching firewood, fodder, and medicinal herbs, which serves immediate social demands, while men undertake the more laborious and heavier tasks of felling trees and cutting branches [8]. Similar observations were made by Follo and colleagues [4] in their analysis of gender in relation to forest ownership in Europe. They described private forests as forming a male system that is experiencing some changing dynamics, as women emerge as new forest owners. As Aelst and Holvoet observed [9], unlike men, women are restricted from owning important resources, which forms an obstacle to the development of sustainable rural livelihoods. Despite such marginalization, through gender, women's triple roles of reproduction, production, and community services, can be seen as having more interaction with nature, than is the case with men, giving them broader knowledge of land productivity, possible natural tree species, and the deleterious impacts of the mismanagement of natural resources, including land [10]. Moreover, women endure violations of conscious choices in their day-to-day practices because the violations are misrecognized and, hence, normalized. This is because of symbolic violence: *"in order to be socially recognized, symbolic violence must get itself misrecognized"* [11].

For rural development to be realized, synergistic efforts involving both men and women to use the existing natural resources of land, water, biodiversity, and the like, effectively are unavoidable. Unfortunately, women do not enjoy full involvement in the management of natural resources because of traditions and taboos that oppose this [1]. When it comes to positions of power, most African women are subordinate to men, which, in the long run, influences social relations, and, hence, the management of natural resources, community development, and family livelihoods [12].

Several studies [3,6,13,14] have addressed women's access to land in Tanzania and Africa. Building on these studies, and drawing in particular on Pierre Bourdieu's symbolic violence, an insidious and invisible form of male domination over women, this study analyses the gendered access, control and management of family land in the Southern Highlands of Tanzania following a recent surge in tree-planting that requires the acquisition of more land. More specifically, the study describes land transaction processes at the household level and the manner in which the lack of women's involvement in such transactions affects their access to and control over family land. The paper is organized as follows. The background section assesses the availability of village land for tree-planting and the rush for timber. This is followed by a presentation of the conceptual framework, followed by a description of the methods of data collection and analysis. The findings and a discussion are presented next, followed by the conclusion at the end of the paper.

2. Background

The surge in tree-planting activities in the Southern Highlands of Tanzania, just mentioned, is largely a result of the dwindling supply of timber from the Sao Hill state plantation in the 2000s [15], which has motivated both private companies and individuals to fill the shortage. Early sales from private woodlots generated enormous earnings, making the timber trade a lucrative business. This has in turn stimulated the participation of different categories of domestic investor, ranging from village residents to urban dwellers, to rush to the villages and acquire village land on which to plant trees [16]. The multitude of tree-growers that have now surfaced has fuelled formal and informal transactions of village lands between indigenous smallholders and tree-growers, practices that have transformed potential cropland into woodlots [17,18]. While urban-based investors see tree-planting as a lucrative business, smallholders have found that their land has suddenly become a profitable commodity, leading them to become involved in land transactions to make money. However, as these transactions are conducted and/or supervised by men, they benefit them more than they benefit women. Although socially recognized, the complementary domination of men over women that can be observed amounts to symbolic violence [11].

Tanzania's land reforms, such as the Land Act and the Village Land Act of 1999, have the aim of eradicating gender inequalities in land ownership. However, these reforms have yet to be fully realized [6]. Although they have decentralized the land administration system to lower-level state organs such as village councils (VCs) to ensure, among other things, equality in land transfers and ownership between men and women, these reforms have still not reached rural communities. Where they have done so, significant barriers have been encountered that limit their implementation. Since land is an indicator of economic well-being, the traditions and taboos that restrict women from managing land cement their marginalization and discrimination in the management of natural resources.

At the beginning of the African land rush, Sam Moyo and Yeros asserted that the scramble for agricultural land and reforms to land tenure being promoted by neoliberal policies have had significant negative impacts on livelihoods and welfare, especially among women, who in most cases take the lead in the social reproduction of households [19]. Similarly, a scholarly work on land reforms in Africa by Rasmus Pedersen showed this scramble for land going hand in hand with the individualization of land parcels [6], a practice that further strengthens men's as opposed to women's control over land. The on-going tree-planting activities in the Southern Highlands of Tanzania have economic, religious, cultural, social, and environmental impacts regarding the loss of medicinal herbs, natural food stuffs, fuelwood, fodder, grazing lands and ritual sites, as well as the degradation of catchment areas, with the eventual impacts being skewed towards women [8,20].

Traditional practice in some African societies dictates that women cannot inherit land or landed properties but must access land through their male relatives, such as their fathers and husbands [1,3,6]. Although they form the largest proportion of adults in Tanzania, and although agriculture is their main form of employment, women do not have rights to immovable resources, such as land [5]. Thus, their marginalization is traditionally created and maintained by societies that recognize men as the dominant group and women as subordinate [21].

The assumption that access to land is becoming less gendered, that is, that it now relies less on a woman's relations with her male relatives, is still subject to qualification [6]. In most of Sub-Saharan Africa, access to land is a masculine right, which still causes women to access land through marriage or other male relatives. Although this situation affects women differently, their effects are no less profound [22]. In the wave of land reforms in different African states in the 1990s, Tanzania enacted two pieces of legislation: the Land Act and the Village Land Act, both of 1999. While recognizing existing customary rights, these laws decentralized land administration system to village councils, and prohibited discriminatory practices preventing access to land by women [23]. Specifically, Section 3 (2) of the Village Land Act stipulates: *"the right of every adult woman to acquire, hold, use, deal with and transmit land by or obtain land through the operations of a will, shall be to the same extent and subject to the same restrictions as the right of every adult man."* To safeguard women's rights to land, the acts provide for women's participation in organs of land administration. Thus, the Village Council must include more than two women, the Village Adjudication Committee more than three [24]. The law insists that any custom that denies land rights to women, children or individuals with disabilities shall be void and inoperative [23].

Despite such interventions, land access remains skewed more to men than to women, although practice has improved somewhat [6]. Even though more men register for land ownership than do women, land rights to both genders are firmly recognized. In fact the current reforms, though far from being fully implemented, are fostering the individualization of land parcels. Coupled with the increased pace of tree planting, which increases the pressure on land, land transactions have disrupted customary practices that ensured access to land for divorced women and widows. Thus in many traditional societies, women still suffer from discrimination [14].

3. Conceptual and Methodological Approaches

3.1. Conceptual Framework

The current study is inspired by rampant transactions of village lands in the Southern Highlands of Tanzania. Specifically, this study was conducted in Mufindi, Makete, Wanging'ombe, Kilolo, and Njombe Districts (Figure 1), where land transactions in support of tree-planting have become very frequent. Land is being transformed into a commodity, one that is readily available for sale to domestic investors, who use it to grow exotic trees of pine and eucalyptus. As observed earlier, most land-sellers are men, who undertake these transactions even when women (wives or mothers) oppose it, a further demonstration of the power imbalance between men and women, with women ranking low in both power and status [21,22]. The rush for timber and, hence, for tree-planting, is not so widely observed in other parts of Tanzania as in the Southern Highlands, especially in the Njombe and Iringa regions, where planting timber trees was a state affair, until supplies of timber from state estates, such as Sao Hill, dwindled to their lowest levels [15].

There is a long history of struggles against female discrimination, as is reflected in the use of different terms [25]. Several concepts, such as psychoanalytical feminism, liberal feminism, radical feminism, radical cultural feminism, and liberal political theory, advocate the equality of all, in spite of differences of gender [26,27]. However, societies in different parts of the world have embraced practices that perpetuate the power imbalance between men and women [11]. Pierre Bourdieu describes the dominant practices that deprive subordinates of agency and voice as symbolic violence, a non-physical form of violence that is manifested in the power asymmetry between social groups, which is 'exerted for the most part through the purely symbolic channels of communication and cognition, recognition or even feeling' [27]. Symbolic violence exists in either recognized or misrecognized forms, may or may not be challenged, and may be changed or unchanged [28]. Thus, misrecognition is the highest level of symbolic violence, occurring in most natural settings and social practices, and demonstrating the highest level of a lack of consciousness regarding such practices [29]. James David unpacked the meaning of misrecognition in Bourdieu's view, seeing it as a result of day-to-day practices that give meanings, with the result that things, situations, and processes are ultimately not recognised for what they really are because of previous cognition within the habitus of those confronting it [30]. Since a thing, situation, or act is attributed to another realm of meaning, interests, inequalities, domination, and other effects continue as natural and covered. Nancy Fraser called being misrecognised as being "denied the status of full partner in social interaction and prevented from participating as a peer in social life ... ." [31]. Misrecognition engenders subjects "unworthy of respect or esteem". When accepted as normal social practices, actions of disrespect and a lack of esteem interfere with equal participation among social groups, inequalities that amount to symbolic violence. Thus, misrecognition is a form of knowledge derived from social practices that act as capital to a person, granting power and prestige to the possessor. For instance, the androcentric practices of natural resource management among global societies are taken for granted (doxic) practical values that receive little questioning from either men as dominant or women as dominated [28]. Similarly, the traditional arrangements of most rural societies in the global south grant powers to men as the owners of immovable natural resources and, hence, as owners, decision-makers, and the rightful owners of land [29,32]. Concrete and symbolic violence are both active mechanisms of social life tied to the order of domination and destruction [33]. Though not visible, practices of symbolic violence are equally destructive, as they create a dichotomy of freedom and constraint through misrecognition, condescension, consent and the complicity of one group in subordinating another [11,27,34]. It is through misrecognition that practices of symbolic violence are perceived differently from what they really are, hence being accepted as a natural way of life in the society concerned [35]. In practice, women's silence regarding men's dominant practices not only generates unequal power relations, it also makes acceptance more difficult and, hence, leads to the reproduction of subordinates' own subjection [36]. Societies have accepted and internalized such social control and complementary domination over women on the part of both men and women. Symbolic violence exceeds the covert inequality and domination of women perpetrated by men, as it extends to commodity market systems where economic conditions exclude, marginalize, disenfranchise and promote gender inequalities [33]. Thus, disparities in the parental care given to daughters versus sons, restrictions in the form of do's and don'ts, and submission in marriage, groom symbolic violence, which is nonetheless perceived by societies as natural, given and unchangeable [11].

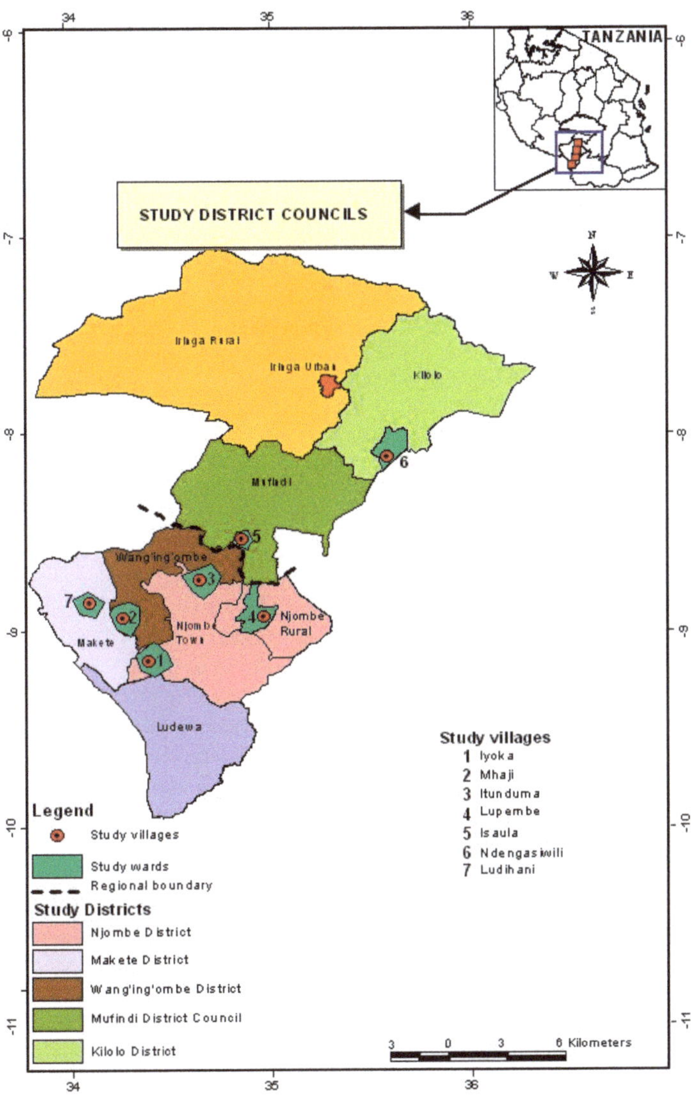

Figure 1. Map of Tanzania showing the study districts and their respective villages. (Source: Timber Rush Project).

Violence in its different physical and symbolic forms exists in different areas of human activity, from laws to gender relations and racial discrimination [33]. Durey identifies at least three areas where social practices manifest symbolic violence and hence misrecognition, thus, undermining women's potential, namely unequal pay between men and women for the same type and amount of work, restrictions on women having further education and professional careers, and full-time employed women bearing primary responsibilities in the household, including for children care [35]. Although collective voices demanding women's empowerment and approaches, such as Women in Development (WID) and Gender and Development (GAD), try to foster women's recognition and empowerment in different countries, they are far from being successful when measured globally. However, practices of symbolic violence can be challenged and sometimes changed once they are recognized [27]. Social competition is one strategy that had been adopted by courageous but dissatisfied members of subordinate groups when they call for change by challenging the dominant group. This has led to changes in the social identity, status, material disposal and symbolic inequalities of some of the reported cases [3,6,21,29].

3.2. Methods

The study adopted a qualitative approach involving data collection through in-depth interviews and focus-group discussions with women specifically selected from households involved in land transactions. Before the interviews and focus-group discussions started, the respondents were informed about the aims of the study, and their consent to participate in it and to have their voices recorded was sought verbally. After agreeing to participate, they were assured that their identities would be kept anonymous and their responses confidential. In addition, they were told they had the right to withdraw from participation at any time if they felt it appropriate to do so. The collected data were transcribed verbatim and analysed by means of directed content analysis [37]. The study involved ten female respondents from seven villages (Figure 1) with the following distribution: Isaula (3), Mhaji (2), and one from Lupembe, Iyoka, Itunduma, Ndengisivili, and Ludihani villages. Since the study aimed at understanding the nature of land transactions and women's attitudes to them, each respondent was involved in a semi-structured interview that focused on issues concerning land transactions. More specifically, the interview inquired after their involvement in land transactions, their attitudes to them and the impacts of such transactions on their livelihoods. Although the sample for the study was small, limiting its generalizability to issues of symbolic violence in the Southern Highlands, it highlights and, thus, enables one to understand the women's experiences in relation to symbolic violence driven by land transactions in the study areas.

Moreover, the analyses in this study are situational in nature, with interviews being conducted with specific targets, namely women, with respect to their experiences regarding access to land as impacted by rampant land transactions for tree-planting. The caveat made by Follo [4], namely that when data are collected from a family or the head of a household as a unit of analysis, one obtains more from men than from women, proved applicable in this case. As a result, husbands and wives were interviewed separately, thus giving women adequate space to explain things in detail. Thus, husbands were interviewed about their motives for selling land, while women were interviewed regarding their involvement in land transactions. Thapar-Bjorkert and colleagues have described interviews as a source of knowledge that, though created by individual perspectives, makes social structures and collective processes through individual narratives that are never ordinarily direct become direct and be constructed discursively [11]. The interview tool consisted of a checklist of guiding questions on land ownership, transactions and violent practices associated with land and the acquisition of landed resources. More specifically, wives were interviewed on their involvement in land transactions and on the manner in which these have had some influence on their welfare. On average, it took about an hour to complete an interview with one respondent. Four focus group discussions (FGDs) with eight to nine women in a group were also conducted in Isaula, Mhaji, Itunduma and Ndengisivili villages. A convenient sampling method was employed in choosing the women for the

FGDs. Women of different ages, status and education were recruited. It took about an hour and a half to complete an FGD. Generally, data-collection processes through interviews and FGDs were open, critical and included dialogical interactions [28]. During interviews with men, we sought their opinions on women's participation in land transactions. Most men insisted that women "know nothing" or that women cannot say anything important about these matters because they are not supposed to be involved. This is an example of the social practice of Othering and of the testimonial epistemic violence of silencing [36], where men tend to disparage the knowledge and intellect of women, thus disabling them from speaking. To overcome the challenges involved here, we avoided having men in FGDs. This promoted the epistemic agency of the women, who eventually felt comfortable and able to tell everything [36].

4. Results and Discussion

4.1. Patriarchal Systems and Gender Inequality in Accessibilty to Land

The phrase symbolic violence refers to a situation in which powerful actors continue to enjoy unchallenged privileges in accessing resources and power through which they dominate social interactions [11]. Different forms of domination and violence against women, both physical (e.g., beating) and psychological, also called symbolic violence, are widely reported from across the globe [10,11,28,29,33,35], including from communities in the Southern Highlands of Tanzania. This study uncovered social practices that maintain the status of men as a dominant social group whose plans and decisions regarding resource use and misuse are seldom challenged by women. In fact, marriage, which brings individuals together as couples, is implicated in many social settings, including raising children and making a family and a household, but not in the management of economic assets such as land. In the study villages, the patriarchal system of human relations, which subjugates women in matters related to resource ownership, still prevails. When it comes to land and landed resources, men in all cohorts are symbolically entitled to ownership through inheritance or allocation by the elders of the clan or tribe, while women are disadvantaged. When asked the reasons for unequal land distribution practices, a woman at Mhaji Village stated men's views of land ownership among girls and women by saying:

> "A girl is not given land because upon marriage she will go away and lose a family identity and perpetuate her husband's identity. Therefore, she will get land where she gets married. Boys will remain in the family, hence are given land to establish settlements and for crops to feed their families". (Interview at Mhaji village on 9 November, 2019).

Thus, although women are involved in the management of family lands, they are not entitled to own the resource, but only to use it for the purposes of improving the household's well-being. A similar account was given by women in an FGD in Isaula village, who complained with one voice against the traditional social practices that limit women and daughters from owning immovable resources such as land and trees, on the basis that, as women do not contribute to the development of the clan, they should not participate in its inheritance.

Although rural families sustain the doxa regarding domination and subordination[1], some married women problematize these social practices by demanding equal treatment and consideration, and hence a share of the benefits from economic assets like land. In one of the interviews, an obviously aggrieved woman shared the bitter words she had received from her husband when she demanded a share of the gains from the land he had sold to tree planters,

> "You want the share of money, but when we got married, did you come with the soil from your father?" (Interview at Isaula village, 17 April 2018).

---

[1] A society's taken for granted, unquestioned truths (after Bourdieu).

This response, apart from psychologically abusing the woman, negates her agency and voice by treating her as a nothing or nobody. This misrecognition of her potential position and her silence legitimizes the power of men over women, thus perpetuating symbolic violence, which is harder to recover from than physical violence [11,35]. Such attempts by women to challenge these long-standing dominant practices are in vain. A woman cannot make any further demands because no one will support her claim against her husband, thus making her demand unreasonable. This kind of male practice portrays male dominance in society as a natural phenomenon, hence legitimizing women's subordination [35] and treating them as unworthy of respect and esteem [36].

4.2. Land Transactions for the Surge in Tree Planting Have Perpetuated the Disadvantaging of Women in Respect of Land Ownership

According to the relevant literature, land is an asset that can raise women's economic well-being in the form of money or income earned from the lease or sale of land [3,32,38] and, thus, reduce their poverty [39]. Some daughters and women in the study villages have realized the opportunities presented by land and, therefore, claim their share of land from their fathers. Their efforts to show their agency and voices are nonetheless in vain because of the misogynistic behaviour of their fathers. It is striking that mothers are not asked to intervene in matters of control over resources because they too are victims of similar situations. An interview with a young married woman revealed that, after being refused by her husband because she is not a member of his clan, she and her young sister, who has married elsewhere, resorted to claiming ownership of land from their biological father. As could be expected, they were disappointed to hear their father denying them access to land because they were married. Instead, he told them that their husbands should handle all issues regarding land. After weighing up the negative statements made by her husband and father, the older sister blamed her father by saying:

> "Why is my father treating me like this? I am his daughter like my brothers. I deserve a share of land, even if it's just a little plot." (Interview in Isaula on 17 April 2018).

Although cases of women's marginalization persist, some of them acknowledge that some changes have been made in society and that some families treat their children equally. Asked about her experience of land acquisition at Isaula village, one woman (61 years old) said:

> "In the past, it was a system that land is given to boys only. Girls were told that they would get land from the families to which they get married. Even for crops, there was a strong monopoly that a woman had no rights to sell her farm produce, and this practice still exists in some families here. We are thankful to people from Participatory Ecological Land Use Management (PELUM)[2] for helping a lot to change this." (Interview on 17 April 2018).

Women are subjected to social pressure that restricts them in realizing their potential, including the expression of their rights and interests. According to Idris, factors that limit women's economic potential include time poverty, ignorance, reproductive pressure, a lack of assets, a lack of financial services, a lack of male support and labour, and cultural norms [5]. Due to symbolic violence, which restricts women in furthering their education, rural women remain potentially less educated and are hence subjected to social pressure that deters them from claiming their rights and limits their awareness of what to do when they are denied their rights [40]. Although Tanzania's land laws [41] provide for access to land resources by marginalized groups, including women, the factors mentioned in this paper render the provisions unproductive.

---

2   PELUM is an acronym for Participatory Ecological Land Use Management. It is a USAID-funded NGO that established the Isaula Village Land Use Plan, and also surveyed land parcels for individuals. They thus sensitized married women being co-owners of land plots with their husbands and women to being the sole owners of land parcels.

Some women accept the restrictions imposed by men simply in order to preserve harmony in the family, which contributes to their subjugation. In one case, a divorced woman started her life in the house she had built on land given her by her father. Since marrying another man, who moved into her house, she has complained of the domination she experiences in her own house. Although she is proud of having her own house, she still faces similar challenges in her marriage. She recognizes the domination she is experiencing, but her respect for her husband has led into her debt and dependence and exposed her to ingratitude [11]. When asked about her experience of her marriage, she had this to say:

"Sometimes he overreacts to some issues as if am staying at his house. He intervenes in some of my plans with trees that he found there after marrying me. I do not like that. (Interview at Mhaji, 9 November 2019).

When asked why she does not stand firm and insist on her plans, she said,

"A man needs a woman who is submissive, otherwise he will be stressed. So I just cool down and listen just to make him feel honoured."

Although she is not in agreement with her husband's practices, she decided to accept his mistreatment of her so as not to prolong the disagreements, thus accepting the man's wrong-doing but falling complicit [11] in her own subjugation. Thus, submission to symbolic violence turns women into the involuntary servants of the norm of male domination. However, this finding on woman's experience of gender relations contrasts with the findings reported by Panda and Agarwal, who considered land ownership to be a factor in reducing tolerance to violence, when they say, 'Women owning immovable properties—land and a house—are found to face a significantly lower risk of marital violence than property-less women' [29]. These properties are considered a security guarantee that can support a woman in escaping from a violent environment. Land and a house can provide a minimum of basic needs, shelter, food and clothing. Ownership of such properties provides freedom from social deprivation, poverty and poor economic opportunities, which, according to Amartya Sen, is by itself development [42]. This observation was substantiated by a woman with four children who was divorced after staying in her marriage for nine years. During her first marriage, she did not own a piece of land or a house and instead had to depend on her husband, who mistreated her in several ways, including insults, slaps, beating, threats of abandonment and eventually a divorce. Lack of assets such as land and a house compelled her to seek refuge with a man who had showed interest in her, and she hence ended up as a second wife to him, which in her opinion, was not a good thing to do, but she had been driven to it out of desperation. She said,

"If you don't own a house, you will be driven away like a dog, and because of life's hardships you will find yourself establishing relationship with married men, which is bad." (Interview at Mhaji, 9 November 2019)

Now that she owns these properties, she evaluates her past and current situation as follows,

"Some men are not liberated by education. They severely oppress their wives, and the wives cannot do anything. If you fight for your rights, they create situations that will eventually harm you so that you don't benefit from compensation and properties." (Interview at Mhaji, 9 November 2019)

Recently, it has been widely advocated that women should own assets, such as land, and should exercise control over them. Since the ownership of assets such as land and a house provides a woman with security and tends to moderate men's violent practices [29], married women misrecognize the domination associated with it. However, having a plot to farm does not guarantee freedom to make use of the products of that land, nor control over expenditure of the income generated from the harvests as long as the harvests are taken to a house built by a man or built on his land. Some men use asset ownership by a woman as a control mechanism. Men pretend to surrender control of some plots of land by transferring ownership to their wives, who eventually feel loved and secure by having control of production. As a result, women put a great deal of effort into tilling, planting, weeding, and the like to improve yields and, thus, increase harvests. However, normally such ownership and control fades at the beginning of the harvest season, when men resume their dominance and exercise control over the farm's production. In one case, a woman from Mhaji village who is married as a second wife was made aware of misrecognition of her rights when she was given 1.5 acres of land as hers to produce for the benefit of her household. She put more effort into her farm, hoping to earn a substantial income at harvest, but this did not materialize, as her husband controlled the produce and the income generated from the harvests. Narrating her experience, she said:

> "This house is like a government asset. When you are given a farm, you will go and cultivate it, but the harvests belong to the government. You cannot plan for utilization of the harvests because though he told you the farm is under your control, he takes the harvests. When you sell some products, he takes the money. If you ask why this is, he asks you back if that is your property. He says it is in his house." (Interview 9 November 2019).

The amount of coercive power exerted by the husband establishes consent to and complicity with power relations by both the dominator and the subordinate, which ultimately legitimizes and internalizes symbolic violence. Inequality and violent practices are spread across the villages that experience rampant tree planting activities. Such irregularities are manifested as the misrecognition of women's potential, rights, and responsibilities, their consent to socially defined and accepted power relations between men and women, and their complicity with normalized symbolic power.

Consent and complicity on the part of women regarding practices that amount to symbolic violence have perpetuated decision-making practices that are skewed in the interests of men. Men's desire for financial resources has fuelled transactions of land and landed resources, leaving most families with sparse productive lands. Women are threatened and/or maltreated when they voice concerns over resource control. This long-standing patriarchal and misogynist system has established a skewed relationship between men and women. Apart from being deprived of the rights to resource ownership and control, women are regarded as homemakers, men as breadwinners [10]. With symbolic power, the man decides what to do with the land, and since women are not involved in decision-making, they are equally uninformed of the decisions that have been made, including which plots of land have been sold to domestic investors, leading to a loss of family lands. In this regard, one 53-year-old woman reported one particular bad experience with male domination. In her case, her husband sold one of her favourite parcels of cropland without her consent or awareness. When asked how it happened, she said:

> "It was the rainy season, so I went to prepare the farm for growing maize. Arriving at the farm, I was astonished that our farm was invaded by tree growers, which made me come home quickly to ask him. He then told me that he sold it too." (Interview at Isaula, April 2018).

When asked what her reaction was, she simply said,

> "I felt very bad but could not do anything. When a man decides, it is over. You cannot disagree."

Silence is one of the options women take to show their complicity with a potentially coercive situation. Disapproving a man's actions, which would be more appropriate, implies misbehaving. Misrecognizing their potential, women comply with the domination that is exerted over them to avoid more violence from their oppressive partners [11]. This is a practice whereby symbolic violence is recognized but not challenged, thus legitimizing its proliferation and influence over the subordinate [27]. If left unchallenged, a dominant person will continue the same practices as if they are legitimate.

Tree planting has led to the conversion of much cropland into woodlots. Such conversions are done by families either to diversify their incomes for their households by undertaking tree businesses, or to sell cropland to domestic investors for tree-planting. When asked about tree-planting in her household, one woman from Ludihani village confirmed that they have several woodlots established by her family. However, since her husband does not participate in crop-farming, he has expanded tree-planting to family croplands. This is worrying her, as she sees their croplands dwindling, which will have adverse effects in the near future. Talking of her husband's practices, she said:

> "He has planted trees on some of my maize farms. He does not know the risks we are going to face when we run out of food. He does not stay at home [but] rather spends most of his extra time at the pub for bamboo juice." (Interview 17 April 2018)

The dwindling of croplands is evidence of the land hunger that families will be facing in the near future. Women are worried about the impending hunger and poverty that might hit them when land and food are eventually gone. Men's planning, apart from being discriminatory, exacerbates poverty and hunger, thus undermining the global efforts stipulated in SDGs 1 and 2 to achieve zero poverty and zero by 2030 [2].

Drawing on Bourdieu's scholarly work *Distinction* of 2002, Angela Durey argues that the characteristics and practices of dominant groups can be changed or moderated through the critical practices of conscious subordinate groups, especially when such subordinates actively engage in activities that can influence change [35]. Similarly, Ojha stressed the status of misrecognition as dynamic, allowing it to be recognized, challenged and changed. It is thus for subordinate groups that are agents of change to recognize their potential and opportunities to instigate the desired changes in their interests [28].

5. Concluding Remarks

In this paper, we set out to uncover gender inequality with respect to women's access to family lands, in the context of the tree-planting surge in selected villages in the Southern Highlands of Tanzania. Drawing on Bourdieu's concept of symbolic violence, we observe the subjugation of women's voice and agency in claiming resources, hence, their limited involvement in land transactions. This is despite reforms to the system of land administration that have been introduced in Tanzania [23,41], especially in the study area, where land ownership, control, and utilization are still dominated by the patriarchal system, which discriminates against women. Women become victims of symbolic violence through traditional practices of inheritance, which limit their ability to have control over land. This happens because of their misrecognition of the violence inflicted on them. However, even in cases where symbolic violence is recognized, the affected women decide not to challenge it for fear of causing matrimonial disturbances. Moreover, although Colaguori [33] denies that violence is a destructive practice per se when he says, *"Violence is therefore not only an active mechanism of social life, it establishes the political ontology of social life"*, he still acknowledges, as we do, that violence greatly affects women in terms of their access to both land and its associated benefits.

Effective implementation of the existing land laws and regulation is unavoidable. However, success will depend on overcoming the symbolic violence that disadvantages women. This calls for local governments at the village and district levels, NGOs such as PELUM, legislatures and policy-makers to promote gender equality in land management practices, as well as empowering women against dominant and discriminatory traditions and customs. This entails, among other things, capacity-building, regarding knowledge of the law among village land administrators—that is, the Village Council—and best practice of the same among villagers. Legal frameworks, laws, and regulations can affirm their right not only to family lands, but also to participation in decision-making process regarding family resources.

Finally, although the conflict between men and women over resource ownership and control is nothing new, it provides room for debate and negotiation for a better balance of power. In a study of epistemic violence on Othering, Bunch [36] recommended that communities actively develop non-oppressive practices in respect of gender and social positions among others, as a way of stopping violence. Moreover, contrary to radical feminist perspectives that focus on women alone in dealing with society's gender challenges, we prefer to promote dialectical communication between women and men. This will both reveal and heal practices of symbolic violence and enhance gender equality in the management of land and other resources in the selected villages in the Southern Highlands of Tanzania.

Author Contributions: J.L. was involved in conceptualizing the study, writing the draft introduction, methodology and data collection, analysis, writing of findings and discussion. D.M. is involved as a supervisor, secured funding for the project, data curation, proofreading, discussion of the findings and conclusion. All authors have read and agreed to the published version of the manuscript.

Funding: This research was part of the Timber rush project that is funded by DANIDA with grant number 15-P02-TAN.

Conflicts of Interest: The authors declare no conflict of interest.

References

1. Fonjong, L.N. Gender Roles and practices in natural resource management in the North West Province of Cameroon. *Local Environ.* **2008**, *13*, 461–475. [CrossRef]
2. UNDP. *Sustainable Development Goals (SDGs): Empowered Lives, Resilient Nations*; UNDP: New York, NY, USA, 2015.
3. Wineman, A.; Liverpool-Tasie, L.S. Land markets and the distribution of land in northwestern Tanzania. *Land Use Policy* **2017**, *69*, 550–563. [CrossRef]
4. Follo, G.; Lidestav, G.; Ludvig, A.; Vilkriste, L.; Hujala, T.; Karppinen, H.; Didolot, F.; Mizaraite, D. Gender in European forest ownership and management: Reflections on women as "New forest owners". *Scand. J. For. Res.* **2016**, *32*, 174–184. [CrossRef]
5. Idris, I. Mapping Women's Economic Inclusion in Tanzania. KD4 Helpdesk Report 332; 2018. Available online: https://assets.publishing.service.gov.uk/media/5b432d9e40f0b678bc5d01c1/Barriers_to_womens_economic_inclusion_in_Tanzania.pdf (accessed on 2 December 2019).
6. Pedersen, R.H. A Less Gendered Access to Land? The Impact of Tanzania's New Wave of Land Reform. *Dev. Policy Rev.* **2015**, *33*, 415–432. [CrossRef]
7. Homberg, H.v.d. *Gender, Environment and Development: A Guide to the Literature*; University of Amsterdam: Amsterdam, The Netherlands, 1993.
8. Gurung, B.; Thapa, M.T.; Gurung, C. *Briefs/Guidelines on Gender and Natural Resources Management*; Organisation Development Centre ODC: Kathmandu, Nepal, 2000; Volume 6.
9. Van Aelst, K.; Holvoet, N. Intersections of Gender and Marital Status in Accessing Climate Change Adaptation: Evidence from Rural Tanzania. *World Dev.* **2016**, *79*, 40–50. [CrossRef]
10. Alliyu, N. Patriarchy, Women's Triple Roles and Development in Southwest Nigeria. *Int. J. Arts Humanit. (IJAH) Bahir Dar-Ethiop.* **2016**, *5*, 94–110. [CrossRef]
11. Thapar-Björkert, S.; Samelius, L.; Sanghera, G.S. exploring symbolic violence in the everyday: Misrecognition, condescension, consent and complicity. *Fem. Rev.* **2016**, *112*, 144–162. [CrossRef]
12. Kirk, G. Women Resist Ecological Destruction. In *A Diplomacy of the Oppressed: New Directions in International Feminism*; Zed Books; Ashworth, G., Ed.; Lehtinen: London, UK, 1995.
13. Alemu, G.T. Women's Land Use Right Policy and Household Food Security in Ethiopia: Review. *Int. J. Afr. Asian Stud.* **2015**, *12*, 56–66.
14. Fox, L.; Wiggins, S.; Ludi, E.; Mdee, A. *The Lives of Rural Women and Girls What Does an Inclusive Agricultural Transformation That Empowers Women Look Like?* Overseas Development Institute: London, UK, 2018.
15. Ngaga, Y.M. A Platform for Stakeholders in African Forestry. *For. Plant. Woodlots Tanzan.* **2011**, *1*, 80.

16. Lusasi, J.; Friis-Hansen, E.; Pedersen, R.H. A typology of domestic private land-based investors in Africa: Evidence from Tanzania's timber rush. *Geoforum* **2020**. [CrossRef]

17. Olwig, M.; Noe, C.; Kangalawe, R.; Luoga, E. Inverting the moral economy: The case of land acquisitions for forest plantations in Tanzania. *Third World Q.* **2015**, *36*, 2316–2336. [CrossRef]

18. PFP. Forest Plantation Mapping of the Southern Highlands. Iringa, Tanzania. 2017. Available online: http://www.privateforestry.or.tz/uploads/Forest_Plantation_Mapping_SH_Final_Report_3.pdf (accessed on 27 October 2019).

19. Moyo, S.; Yeros, P. Rethinking the Theory of Primitive Accumulation: Imperialism and the New Scramble for Land and Natural Resources. Social Science. In Proceedings of the IIPPE Conference, Istanbul, Turkey, 20–22 May 2011.

20. Locher, M. 'How come others are selling our land?' Customary land rights and the complex process of land acquisition in Tanzania. *J. East. Afr. Stud.* **2016**, *10*, 393–412. [CrossRef]

21. Bourdieu, P.; Wacquant, L. *An Invitation to Reflexive Sociology*; Polity Press: Oxford, UK, 1992.

22. Aveling, E.-L.; Cornish, F.; Oldmeadow, J. Diversity in sex workers' strategies for the protection of social identity: Content, context and contradiction. In *Symbolic Transformation: The Mind in Movement through Culture and Society*; 2013; pp. 302–322. Available online: http://eprints.lse.ac.uk/47797/ (accessed on 17 May 2019).

23. URT. *The Village Land Act*; URT: Dar es Salaam, Tanzania, 1999.

24. Wily, L.A. *Community-Based Land Tenure Management: Questions and Answers about Tanzania's New Village Land Act, 1999*; IIED: London, UK, 2003; Volume 120.

25. Calás, M.B.; Smircich, L.; Bourne, K.A.; Bilimoria, D.; Piderit, S. Knowing Lisa? Feminist Analyses of 'Gender and Entrepreneurship'. In *Handbook on Women in Business and Management*; Edward Elgar Publishing: Cheltenham, UK, 2013; pp. 78–105.

26. Eckert, P.; McConnell-Ginet, S. *Chapter 1: An Introduction to Gender*; Eckert, P., McConnell-Ginet, S., Eds.; Cambridge University Press: Cambridge, UK; New York, NY, USA, 2001.

27. Bourdieu, P.; Nice, R. Masculine Domination. *Contemp. Sociol. A J. Rev.* **2002**, *31*, 407. [CrossRef]

28. Ojha, H.R.; Cameron, J.; Kumar, C. Deliberation or symbolic violence? The governance of community forestry in Nepal. *For. Policy Econ.* **2009**, *11*, 365–374. [CrossRef]

29. Panda, P.; Agarwal, B. Marital violence, human development and women's property status in India. *World Dev.* **2005**, *33*, 823–850. [CrossRef]

30. James, D. How Bourdieu bites back: Recognising misrecognition in education and educational research. *Camb. J. Educ.* **2015**, *45*, 97–112. [CrossRef]

31. Fraser, N. From Redistribution to Recognition? Dilemmas of Justice in a 'Post-Socialist' Age. *New Left Rev.* **1995**, *212*, 68–94.

32. Snyder, K.A.; Sulle, E.; Massay, D.A.; Petro, A.; Qamara, P.; Brockington, D. "Modern" farming and the transformation of livelihoods in rural Tanzania. *Agric. Hum. Values* **2019**, *37*, 33–46. [CrossRef]

33. Colaguori, C. Symbolic Violence and the Violation of Human Rights: Continuing the Sociological Critique of Domination. *Int. J. Criminol. Sociol. Theory* **2010**, *3*, 388–400.

34. Okali, Christine. Gender Analysis: Engaging with Rural Development and Agricultural Policy Processes. In *Future Agricultures (No. 026)*; Okali, Christine: 2012; Available online: www.future-agriculture.org (accessed on 19 April 2018).

35. Durey, A. Rural medical marriages: Understanding symbolic violence in the social practice of gender. *Women's Stud. Int. Forum* **2008**, *31*, 73–86. [CrossRef]

36. Bunch, A.J. Epistemic Violence in the Process of Othering: Real-World Applications and Moving Forward. *Sch. Undergrad. Res. J. Clark* **2015**, *1*. Available online: http://commons.clarku.edu/surj/vol1/iss1/2 (accessed on 4 June 2020).

37. Shannon, S.E. Three Approaches to Qualitative Content Analysis. *Qual. Health Res.* **2005**, *15*, 1277–1288. [CrossRef]

38. Meinzen-Dick, R.S.; Quisumbing, A.; Doss, C.; Theis, S. Women's land rights as a pathway to poverty reduction: Framework and review of available evidence. *Agric. Syst.* **2019**, *172*, 72–82. [CrossRef]

39. Mi, C.; Park, C.M.Y.; White, B. Gender and generation in Southeast Asian agro-commodity booms. *J. Peasant. Stud.* **2017**, *44*, 1105–1112. [CrossRef]

40. UNDP; URT. *Tanzania Human Development Report 2014*; Economic Transformation for Human Development; UNDP: Dar es Salaam, Tanzania; URT: Dar es Salaam, Tanzania, 2014.
41. URT. *The Village Land Regulations*; URT: Dar es Salaam, Tanzania, 2001.
42. Sen, A.K. *Development as Freedom*; Oxford University Press: New York, NY, USA, 1999.

Erratum

Erratum: Lee, J.-H.; Kim, D.-K. Mapping Environmental Conflicts Using Spatial Text Mining. *Land* 2020, 9, 287

Jae-hyuck Lee and Do-Kyun Kim *

Korea Environment Institute, Sejong 30147, Korea; jaehyuck@kei.re.kr
* Correspondence: dkkim@kei.re.kr; Tel.: +82-44-415-7438

Received: 2 November 2020; Accepted: 4 November 2020; Published: 5 November 2020

The authors would like to correct the following section of this paper [1]:

Before: **Funding:** This research was funded by the Korea Environmental Institute (RE2020-17).

Corrected version: **Funding:** This research was supported by the grant from the basic project 'Social model development of environmental pollution damage for recovery of victims' lives' funded by the Korea Environmental Institute (RE2020-17) and the 'Urban-based complex plant demonstration study utilizing underground space' program funded by the Ministry of Land, Infrastructure and Transport of the Korean government (20UGCP-B157945-01).

The authors would like to apologize for any inconvenience caused to the readers by these changes. The changes do not affect any scientific result of the paper.

Reference

1. Lee, J.-H.; Kim, D.-K. Mapping Environmental Conflicts Using Spatial Text Mining. *Land* **2020**, *9*, 287. [CrossRef]

Publisher's Note: MDPI stays neutral with regard to jurisdictional claims in published maps and institutional affiliations.

MDPI

St. Alban-Anlage 66

4052 Basel

Switzerland

Tel. +41 61 683 77 34

Fax +41 61 302 89 18

www.mdpi.com

Land Editorial Office

E-mail: land@mdpi.com

www.mdpi.com/journal/land